Sustainable Design from Vision to Action

This book offers a comprehensive guide to the sustainable design of products, services, or related systems. It goes beyond concept explanations to provide you with practical instructions on how to apply the tools and methods to your own designs. The world is finally waking up to the necessity of sustainability, especially in the design and engineering of all the physical products that surround us every day.

Sustainable Design from Vision to Action not only provides a thorough understanding of the high-level values and goals of sustainable design, but also gives readers actionable step-by-step guides for how to implement them on the ground, in daily practice. This includes quick reference tables and other resources for practical design, with lists of resources for greater depth. The activities can be used by designers and engineers, in classrooms, or in industry. This innovative textbook focuses primarily on physical product development, but also applies to

services, systems, and digital products. It includes a thorough range of quantitative and qualitative methods across the whole product life cycle, including material choice, energy use, systems thinking, design for recycling, user behavior change, business models, equity and inclusion, and more. The book aims to change your design practice to help build a world that is healthy, abundant, beautiful, and fulfilling for all species, for all time.

This highly illustrated text will provide an excellent introduction to sustainable design in practice for industrial design and mechanical engineering students. It will also be useful for professional designers, engineers, and managers in industry.

Jeremy Faludi is Assistant Professor of Design for Sustainability at the faculty of industrial design engineering, TU Delft, the Netherlands.

"Many books provide suggestions on how to address today's ecological issues. These are essential, but they leave out the details necessary to transform the visions into action. The title of this badly needed text tells it all: this book has practical tools to make the visions real."

Don Norman, *author of* Design for a Better World: Meaningful, Sustainable, Humanity Centered

"This captivating book breaks down complex sustainability topics into actionable ways of designing, using straightforward language and many illustrations. It's the sustainable design partner that you will want to go back to again and again."

Cindy Pereira Cooper, *Engineering for One Planet co-founder and initiative leader, The Lemelson Foundation*

"If we're to redesign our world to be more just, equitable and sustainable, we'll need a roadmap to show the way. There may be no better one than this to leverage the full suite of tools, from biomimicry to behavior change to business strategy, needed for the journey."

Joel Makower, *co-founder and chairman, Trellis Group and GreenBiz*

"I've followed Jeremy's sustainable design thought leadership for more than 15 years and recommend this book to anyone looking for guidance on how to put theory into practice."

KoAnn Vikoren Skrzyniarz, *founder/chairwoman of SustainableBrands*

"I have reformulated [my course] to follow your book chapter by chapter and with the suggested activities. I am pleased to report that it is going extremely well. Personally, I love the materials, and the students appear to be very engaged. I dare say 'riveted.'"

Benoit Cushman-Roisin, *Professor of Engineering Sciences at Dartmouth College*

Sustainable Design from Vision to Action

Edited by Jeremy Faludi

Routledge
Taylor & Francis Group

LONDON AND NEW YORK

Designed cover image: © Jeremy Faludi

First published 2026
by Routledge
4 Park Square, Milton Park, Abingdon, Oxon OX14 4RN

and by Routledge
605 Third Avenue, New York, NY 10158

Routledge is an imprint of the Taylor & Francis Group, an informa business

British Library Cataloguing-in-Publication Data
A catalogue record for this book is available from the British Library

ISBN: 978-1-032-82475-8 (hbk)
ISBN: 978-1-032-82474-1 (pbk)
ISBN: 978-1-003-50467-2 (ebk)

DOI: 10.4324/9781003504672

Typeset in Helvetica LT Std
by codeMantra

This book is dedicated to all those making the world happier, healthier, and more beautiful. As the old poem says, we are the music makers, and we are the dreamers of dreams.

Contents

Part I Vision

Part II Action

Contributors

Conny Bakker is Professor of Design Methodology for Sustainability and Circular Economy at the faculty of industrial design engineering, TU Delft, the Netherlands.

Ruud Balkenende is Professor of Circular Product Design at the faculty of industrial design engineering, TU Delft, the Netherlands.

Jeremy Faludi is Assistant Professor of Design for Sustainability at the faculty of industrial design engineering, TU Delft, the Netherlands.

Cynthia Lawson Jaramillo is Professor of Integrated Design at The New School's Parsons School of Design, New York, USA.

Michele Kahane is Professor of Practice in Management and Social Innovation at The New School's Parsons School of Design, New York, USA.

Ruth Mugge is Professor in Design for Sustainable Consumer Behaviour at the faculty of industrial design engineering, TU Delft, the Netherlands.

Evren Uzer is Associate Professor of Urban Planning at The New School's Parsons School of Design, New York, USA.

Acknowledgments

Thanks to those who helped edit and create this book, especially Adam Menter, Justé Moutuzaité, and Jessica Papa. Thanks to the many reviewers and advisors, including Benoit Cushman-Roisin of Dartmouth College; Jelle Joustra, Sonja van Dam, and Stefan Persaud of TU Delft; Beth Ferguson of UC Davis; Cindy Anderson of Alula Consulting; Wendy Jedlicka of Minneapolis College of Art and Design; Lynda Grose of California College of the Arts; Jeff Zeman, Theresa Millard, Matthew Neiman, and Emily Vanderheyden of True North consultants, Cindy Cooper of the Lemelson Foundation; Ben Allen of Tonic Books; and Don Norman of UC San Diego.

Thanks also to the friends and family who supported us during the writing, and for the many faculty and sustainable design pros who kept asking for it to be written. We hope it sparks your creativity and boosts your action. Best wishes as we all build a better world together!

PART I

Vision
Goals and Metrics

You never change things by fighting the existing reality. To change something, build a new model that makes the existing model obsolete. – Buckminster Fuller

Look around you—how much of what you see was designed and engineered? Unless you're camping, almost all of it. Engineers and designers have enormous responsibility and enormous opportunity to redesign the world. Together, we can change it from a world of ever-increasing pollution, climate crisis, resource conflict, and human exploitation, to a world of increasing health, nature restoration, abundance, and justice. These are design choices, business choices, and policy choices. There are many reasons to do this, from survival to empathy to finances to beauty. This is sustainable design.

The aim of this book is for you to learn how to form bold visions of sustainability with goals relevant to your product or system, to use metrics to define those goals concretely and specifically, then achieve your goals by taking action with helpful design tools and methods. The book's purpose is not to introduce new theories of sustainability—for decades, we have had good enough theories. What we lack most is action bringing those theories into reality at scale. We hope that everyone reading this book will not merely read it, but will apply its lessons to your own products, services, and systems.

Reaching bold goals is difficult. While you shouldn't be satisfied with mediocrity, you should also not let the perfect be the enemy of the good. Aim high, and when you fail, accept it and plan to fail better next time, because failures are a necessary part of a good design process that you can learn to appreciate.

DOI: 10.4324/9781003504672-1

You can apply this book's lessons to many fields, from power engineering for improving efficiency, to influencing user behavior for reducing consumption, to system design for product recovery, and more. Whatever your specialty and motivations, we hope you will use these visions, methods, and metrics to make your work part of the solution. To change from unconsciously designing a future of loss, malady, and inequity to consciously designing a future of bounty, flourishing, and joy. Here in Part I of the book, we help you form your vision, with the goals and metrics that make your vision concrete.

Vision

There are many visions of sustainability; the chapters here summarize the UN Sustainable Development Goals, Planetary Boundaries (also with Doughnut Economics), Ecodesign, Cradle to Cradle, and the Circular Economy. Each of these goes from broad outlines (the vision) to concrete design specifications (the goals). These goals can be brainstorming tools, like some of the design strategies in Part II, or they can be deliverables you need to meet. To turn them into deliverables, you attach metrics to the goals.

Metrics

Your sustainability metrics are your working definition of sustainability—everything else is rhetoric. You can assess qualitatively or quantitatively or both: with numbers, yes/no checkboxes, or whatever you need. Evidence-based metrics help you avoid greenwashing and help you spend your time, money, and creative effort on what matters most. Assessments can benchmark current design performance, set goals for future designs,

measure whether or not those goals were met by a certain design, and choose between design options that get closer to your goals.

Here, the main metric described is Life Cycle Assessment (LCA), though you can also use all the visions listed above as assessments, too. There are even circularity calculators, Sustainable Development Goal checklists, and a Cradle to Cradle certification eco-label, though they're outside the scope of this book. Goals and assessments are best made at the whole-system level, so they should include some kind of systems thinking. Here, the Whole System Mapping method is suggested to integrate metrics into ideation, so rather than sustainability requirements removing ideas after ideation, they become the source of new idea generation.

Integrating into Your Design Process

Sustainable design tools and methods don't replace standard design methods, they work alongside them. In a standard "double diamond" design process, sustainability should appear multiple times: in the initial Discovery stage, especially in researching which problems to solve; in the Define stage, where goals (design specifications) are set; in the Develop stage, where new solution ideas are generated; and in the Deliver stage, when new solutions are evaluated (Figure 0.1).

As Figure 0.1 shows, different sustainability tools and methods are useful in different stages for different activities: Those defining an overall vision by providing environmental and social literacy are used almost entirely for initial context research (e.g., planetary boundaries). Those that define a vision

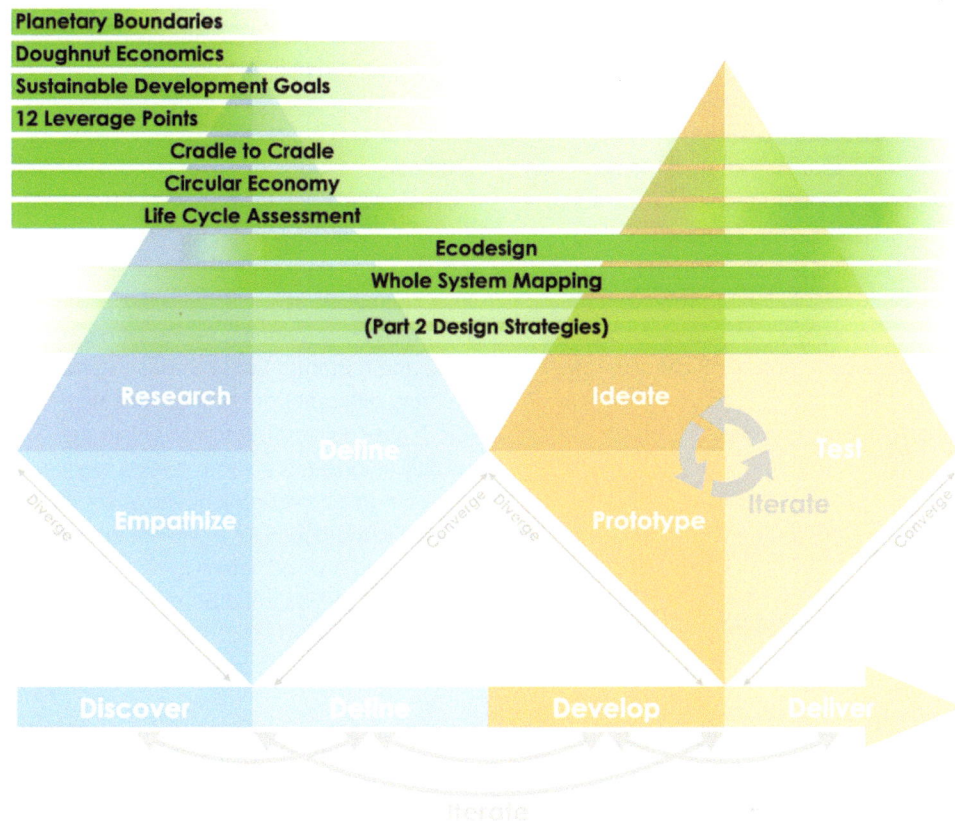

Figure 0.1 Sustainability tools and methods overlaid on the product development "double diamond"

but also provide metrics are also useful in defining concrete design specifications, and assessing whether those metrics were met at the end of the design process (e.g., life cycle assessment). Some even help define goals, ideate solutions, and assess the solutions (Ecodesign and Whole System Mapping). The sustainable design strategies in Part II of this book mostly relate to ideation and assessment, though some also help define goals. All the tools in this book can be used in multiple activities, with greater or lesser usefulness; that's why their lines are blurred, to show there aren't strict limits. However, sustainable design tools and

methods generally aren't used to empathize or prototype because traditional design tools already do both of those very well.

Not every design project needs every tool or method. For example, designing the next version of an existing physical product might have almost no research phase, with the vision already set years ago by company executives and a life cycle assessment already done; your design team might only use Whole System Mapping to define goals, ideate, and assess them. The ideation could be supplemented by Ecodesign or methods from Part II. For a different example, if you're starting a new company to solve one of the world's major sustainability problems, you might begin with a Sustainable Development Goal, decide your own metrics for it, and use nothing but standard design methods to achieve it. Companies like Lucid Motors do this, but many find that other sustainable design methods help, like Fairphone using material choice and design for repair strategies discussed in Part II. You need to choose the best tools for the job at hand, and this book provides a toolbox to choose from.

Innovating Without Products

Some of these tools and methods are useful even if you are not developing a product, but trying to reduce overconsumption by reducing the number of products developed. Governments, nonprofits, and communities can especially use planetary boundaries, the UN SDGs, Circular Economy, and Doughnut Economics to define their overall vision for regulations, community programs, and more. They can also use LCA or other metrics to find top priorities and set targets for what to improve, as well as measure success. And they can ideate solutions with Ecodesign and Whole System Mapping, along with other tools and methods. While the economics are very different from product development, and large coalitions of stakeholders are usually involved, the same design tools can drive progress through innovative solutions. Design is a process for envisioning and manifesting a world where sustainability is not a compromise but a beautiful, just, and healthy way of life.

SUSTAINABLE
DEVELOPMENT G

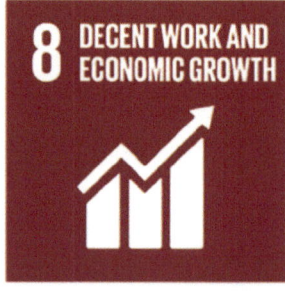

Source: UNDP (2015)
(https://www.un.org/
sustainabledevelopment)

CHAPTER 1

Sustainable Development Goals

Conny Bakker, Jeremy Faludi, and Ruud Balkenende

Goals

- Identify environmental, economic, and social concerns in design, and the meaning of "sustainability"

- Use the UN Sustainable Development Goals to identify the environmental, economic, and social concerns in a product or service

- Articulate the professional responsibilities of designers regarding environmental, economic, and social concerns

DOI: 10.4324/9781003504672-2

Why It Matters

Visions of sustainability are critical, for guiding design targets, setting priorities, and measuring success (or how far you are from success). Some visions of sustainability are qualitative, others are quantitative. All of them involve many factors that are interconnected and interdependent. When you face challenging tradeoffs later in the design process, these visions will help guide your critical thinking and judgement.

Summary

- Sustainable development is usually defined as "development that meets the needs of the present without compromising the needs of future generations to meet their own needs." This should include not just humans, but generations of all life.

- Sustainability comprises environmental ("planet"), social ("people"), and economic ("prosperity") values, also referred to as the "triple bottom line."

- The United Nations developed 17 goals for sustainability (UNDP, 2015). These goals create a comprehensive picture of what global sustainability could look like, including all three aspects of environment, society, and economics.

1.1 Defining Sustainability

People have been trying to define sustainability since the 1972 report of the Club of Rome, *The Limits to Growth* (Meadows et al., 1972), which said that, without substantial changes in resource consumption, "the most probable result will be a rather sudden and uncontrollable decline in both population and industrial capacity." Needless to say, the report was heavily criticized. Did its prediction of doom not happen because it was wrong, or because people heeded its warnings and redesigned material society to avoid doom? Subsequent research into the report's premises continues to confirm that society is currently not on a pathway to sustainability, but progress is also being made. Sustainability is not a black and white question between doom and paradise, but a complex multidimensional question where each dimension measures a degree of better or worse. Over the past few decades, sustainability has gone from a niche activity, a "nice to have" item, to a mainstream must-have item. Governments and companies publish sustainability visions, targets, and regulations.

The commonly used definition of sustainable development is from the 1987 report, *Our Common Future*: "Sustainable development is development that meets the needs of the present without compromising the needs of future generations to meet their own needs" (World Commission on Environment and Development, 1987). This human-centric definition has been expanded to better include ecosystems in later definitions, as in Cradle to Cradle, "how do we love all the children of all species — not just our own — for all time?" (McDonough & Braungart, 2002). Either way, time is key.

Earth is a finite planet, so anything that draws down finite resources or adds waste faster than it can be metabolized by the Earth's ecosystems is not sustainable, because we cannot keep doing it forever.

The book, *Natural Capitalism* (Hawken et al., 1999), phrases sustainability using the metaphor of financial capital: If you have a large sum of money in the bank that earns interest at a certain rate, and you never spend more per year than the interest earned per year, you can keep doing that forever (presuming stable interest rates). But if you spend more, you spend your capital, and eventually are left with nothing. Industry and society are currently spending nature's capital far faster than it is earned. Earth Overshoot Day, the day when humanity has exhausted nature's budget for the year, was 25 July in 2024 (overshootday.org). For the remainder of the year, humanity's demand for ecological resources and services exceeded what Earth can regenerate.

1.2 The Triple Bottom Line

A more detailed definition of sustainability is the "Triple Bottom Line." It says that organizations should measure sustainability by their environmental, social, and economic impacts (or, People, Planet, and Prosperity). The idea was originally proposed by John Elkington in 1994 (elaborated in his 1999 book). He wanted organizations to focus not only on the economic value that they add, but also on the environmental and social value that they add (or destroy).

Solutions that benefit the planet as well as the economy are considered "viable." They may, however, negatively affect social structures.

Benefiting both people and the planet is "bearable": it always requires external financial support, but it improves lives and sustains the planet. Benefiting both people and prosperity is "equitable": it helps rebalance society. Only if a product or service benefits all three bottom lines, i.e., becomes bearable, viable, and equitable, is it considered truly sustainable. The Triple Bottom Line is often drawn as three overlapping circles to illustrate the intersections (Figure 1.1).

In reality, these three factors are not a Venn diagram but concentric circles (see Chapter 2 on Planetary Boundaries). However, as we design things, we often design for only one or two of the factors (bearable or equitable or even merely environmental or social), so the Venn diagram gives a useful language for where our work fits and indicates what could be improved. In pursuing ideal sustainability, we should not let the perfect be the enemy of the good—it's impossible for one product to fix the whole world, it's still good to develop products and services with weaker sustainability benefits. But we should push for stronger sustainability whenever we can.

Figure 1.1 The Triple Bottom Line

1.3 Sustainable Development Goals

In 2015, the United Nations created a series of comprehensive sustainable development goals for organizations in the public and private domain (UN General Assembly, 2015). Today, the UN Sustainable Development Goals (UN SDGs) are considered the most encompassing vision of what global sustainability means. Their 17 goals for 2030, shown in Figure 1.2, combine a high-level perspective with dozens of specific "targets" and "indicators" that designers, engineers, supply chain managers, and others can use to guide design (UNDP, 2015).

For a first glance understanding, the SDGs can be summarized as follows:

1. **No poverty**: End poverty in all forms and dimensions by 2030.

2. **Zero hunger**: End all forms of extreme hunger and malnutrition by 2030.

3. **Good health and well-being**: Ensure healthy lives and promote well-being for all at all ages.

4. **Quality education**: Ensure inclusive and equitable quality education and lifelong learning opportunities for all. Education is a key to escaping poverty.

5. **Gender equality**: Achieve gender equality and empower all women and girls.

6. **Clean water and sanitation**: Ensure access to clean water and sanitation for all.

7. **Affordable and clean energy**: Ensure access to affordable, reliable, sustainable and clean energy.

8. **Decent work and economic growth**: Promote inclusive and sustainable

Figure 1.2 UN Sustainable Development Goals

Source: https://www.un.org/sustainabledevelopment.

economic growth, employment, and decent work for all.

9. **Industry, innovation, and infrastructure**: Build resilient infrastructure, promote sustainable industrialization, and foster innovation.

10. **Reduced inequalities**: Reduce inequalities in income as well as those based on age, sex, disability, race, ethnicity, origin, religion, or economic or other status within a country.

11. **Sustainable cities and communities**: Make cities inclusive, safe, resilient, and sustainable.

12. **Responsible consumption and production**: Ensure sustainable consumption and production patterns.

13. **Climate action**: Take urgent action to combat climate change and its impacts.

14. **Life below water**: Conserve and sustainably use the oceans, seas, and marine resources.

15. **Life on land**: Sustainably manage forests, combat desertification, halt and reverse land degradation, halt biodiversity loss.

16. **Peace, justice and strong institutions**: Promote just, peaceful, and inclusive societies.

17. **Partnerships for the goals**: Revitalize the global partnership for sustainable development.

(https://www.un.org/sustainabledevelopment/)

1.4 SDGS as Design Goals

Presenting the SDGs as a set of goals to be met (and thus, problems to be solved),

means designers have their work cut out for them. If you want to use the SDGs to guide your design practice, you may struggle to make sense of the many different goals. To make the SDGs more concrete and check their relevance to your work, look less at the top-level goals and more at the targets and indicators. Each SDG has a list of them. Finding the ones relevant to your product or service helps you formulate design goals and metrics. For example, SDG 12 (Responsible consumption and production) has 11 targets, such as "substantially reduce waste generation through prevention, reduction, recycling and reuse" with its indicators "National recycling rate, tons of material recycled"; and the target "Encourage companies, especially large and transnational companies, to adopt sustainable practices and to integrate sustainability information into their reporting cycle" with its indicator "Number of companies publishing sustainability reports."

To use the SDGs as design goals, follow these three steps: (1) define priorities, (2) set your goals, and (3) integrate into your work plan. The SDG compass (GRI et al., 2015) was developed to help business (and designers) do this. A summary of it follows.

1.4.1 Defining Priorities

Assess the impacts of your business or product on the SDGs, both positive and negative, both current and potential. You may ask yourself the following questions:

- **Overall**: Across the life of your product (each life cycle stage listed below), where do the most important environmental and/ or social impacts happen? And what are those impacts? Does the company have a strategy to support SDGs or to minimize negative impacts on them?

- **Raw materials and production**: What materials do you use that cause a high environmental and/or social impact to mine, grow, or refine? What about your production processes?

- **Distribution and retail**: Which processes and materials (e.g., transport, packaging, energy for cooling) have a high impact?

- **Use and reuse**: Does your product or service use much electricity, gas, water, or other resources during its operating life? Why? How long does your product last, and why do people stop using it?

- **Recovery and disposal**: At the end of your product's life, what happens to it? How much of it is recovered? What materials and processes cause high impacts during recovery or disposal?

In the following chapters, we will explain in more detail how you can assess sustainability impacts and then improve them.

1.4.2 Setting Your Own Goals

If your product or service causes negative impacts in certain parts of the life cycle, how can you improve these, or even turn them into positive impacts? If you have positive impacts, how can you multiply them? All negative impacts should become part of the corporate sustainability agenda, with evidence-based goals, to prevent cherry-picking and potential accusations of greenwash.

1.4.3 Integrating

The final step is putting the plan into action. This means that sustainability must be integrated into the core business and design process. This is your responsibility as a professional designer or engineer, especially when you manage a team. How will you do it? Some goals you can accomplish yourself though better design and engineering. Some will be more systemic, requiring you to seek new partnerships and collaborate, for instance, across the value chain, or with governments and civil society organizations. Increasingly companies use the SDGs as guiding principles—just search for any big brand in combination with "SDG." The SDGs are used to identify new (green) growth opportunities and/or to lower companies' risk profiles (GRI et al., 2015). Figure 1.3 shows an example of positive and negative SDG impacts, with suggestions for improvement, for a particular product (a digital tablet).

Integrating into your design process requires more than just goals, it requires actions and metrics for success. The rest of this book helps start you with basic activities, mindsets, and metrics to pursue these goals. For example, Chapters 10–13 on material choice and Chapters 15 and 16 on material recovery support SDGs 9, 12, and others. Chapter 25 on Equity and Inclusion supports SDGs 5, 10, and more. Chapter 22 on the Business Model: Presidio Booster supports SDGs 8, 9, 12, and others.

1.5 Criticism

Working with so many goals and targets can be cumbersome, and there is a risk that this leads to prioritizing some goals at the cost of others. For instance, the sheer number of socio-economic goals (14 in total) can result in these being perceived as more important than the planetary goals (three or four in

Figure 1.3 Example of how to assess an electronic tablet against the SDGs, with improvement options

total, #13, 14, and 15, perhaps also #6). Randers et al. (2019) calculated the impact of the efforts to achieve the 14 socio-economic goals and concluded that this would cause problems for the three environmental SDGs, leading to a more equitable but less viable and bearable world, using Triple Bottom Line terms. They say: "Conventional efforts to achieve the 14 socio-economic goals will raise pressure on planetary boundaries, moving the world away from the three environmental SDGs. Extraordinary efforts will be needed to achieve all SDGs within planetary boundaries."

A second point of critique is that the SDGs perpetuate the idea that economic growth and technological innovation are needed for sustainability, for instance, through SDGs 8 and 9. It can be questioned if the same model of growth, competitiveness, and profit-making that has brought us continued expansion at the cost of environmental well-being, and often also social well-being, can also be used to solve these problems. Instead, critics argue, we need alternative economic models, for instance, models stressing de-growth or green growth.

Resources and References

Resources for Further Study

- United Nations: the 17 Sustainable Development Goals, https://sdgs.un.org/goals (Click on a goal, then click "targets and indicators" for details.)

- The Engineering for One Planet Framework, https://engineeringforoneplanet.org

References

Elkington, J. (1999). *Cannibals with forks: Triple bottom line of 21st century business*. John Wiley & Sons Ltd.

GRI, UN Global Compact, & World Business Council for Sustainable Development (WBCSD). (2015). SDG Compass. The guide for business action on the SDGs. Available at: https://unglobalcompact.org/library/3101

Hawken, P., Lovins, A., & Lovins, H. L. (1999). *Natural capitalism*. Little, Brown and Company.

Meadows, D. H., Meadows, D. L., Randers, J., & Behrens, W. W., III. (1972). *The limits to growth: A report for the Club of Rome's project on the predicament of mankind* (1st ed.). Universe Books.

McDonough, W., & Braungart, M. (2002). *Cradle to cradle: Remaking the way we make things*. Macmillan.

Randers, J., Rockström, J., Stoknes, P., Goluke, U., Collste, D., Cornell, S. E., & Donges, J. (2019). Achieving the 17 Sustainable Development Goals within 9 planetary boundaries. *Global Sustainability* 2, e24, 1–11.

UNDP. (2015). Sustainable Development Goals. Available at: https://www.undp.org/sustainable-development-goals

UN General Assembly. (2015, October). *Transforming our world: The 2030 agenda for sustainable development* (A/RES/70/1).

World Commission on Environment and Development (1987). *Our common future*. Oxford University Press.

How to Apply #1: Use the Sustainable Development Goals to Improve Your Design
Time Estimate: 1–2.5 Hours

Evaluate your product or service through the lens of the UN's Sustainable Development Goals.

If you're not working on your own project/product, choose a consumer product you are interested in and would like to research. If you don't have an idea of a product, you can use the Fairphone. (www.fairphone. com/en/impact).

STEP 1: Familiarize Yourself with the SDGs and Their Targets
Time Estimate: 15–30 Minutes

Read the short descriptions of all 17 Sustainable Development Goals on the SDG Compass website (https:// sdgcompass.org). (For any SDGs that are unclear, or that you want more information on, read the sections "The role of business" and "Key business themes addressed by this SDG.")

For SDGs that you find particularly relevant to your product or service, read the section "The SDG targets." This provides more concrete actionable goals. For example, one of the targets for the "life below water" SDG is 14.5, "By 2020, conserve at least 10 percent of coastal and marine areas, consistent with national and international law and based on best available scientific information."

STEP 2: Tag Your Product's Positive and Negative Impacts on SDGs
Time Estimate: 30–60 Minutes

Use the diagram below or make your own map of where your product, accompanying services, and related aspects of your company either benefit SDGs to accelerate progress toward the goals, or hurt SDGs by impeding progress or causing social/environmental damage. Don't include all SDGs, just the ones most relevant to the impacts your product or service has on the world.

POSITIVE IMPACTS (supporting the SDG)

Product Life Cycle

Include considerations such as:

- Which SDG are most relevant and important to your product, product category, and industry?

- Which SDGs are relevant to each life cycle stage? If you're designing a service or software, not a product, remember the physical hardware your software/service uses and the energy that it requires, including communications bandwidth. Also remember related physical products (e.g., if you're making an online fashion store, include impacts of the clothes sold).

- Which SDGs and targets are currently being hurt? Are there opportunities to mitigate your product or company's negative impacts?

- Which SDGs and targets are currently being helped? Are there ways to increase your positive impact?

- Which SDGs and targets are not addressed but should be? Is there a way to start addressing them?

For details of your product or service you need but don't know, research to find them. For example, look at:

- The claims/data your company publishes online or in reports.

- Industry reports on typical details for similar companies.

- Reports or scores by watchdog groups (e.g. Global Witness, Greenpeace, Ethos ESG).

- Other reports on the product, product category, or industry.

Note the details you have found, with references to the sources.

STEP 3: Set Goals for Improvement
Time Estimate: 10–20 Minutes

Based on the map or list you have just made, choose your top one or two opportunities for improvement, and set improvement goals.

Include considerations such as:

- Why did you choose this area to focus on? (Maximizing already-positive impacts, mitigating negative impacts, or other?)

- What indicators and metrics should we use to drive improvements? Refer to relevant targets in the SDG Compass website's "the SDG targets" section for that SDG.

- What's the current baseline score of that indicator/metric, for your product and for the industry?

- What's a good target for improvement in that indicator/metric? Is there a short-term and longer-term target?

STEP 4: Create an Action Plan
Time Estimate: 10–30 Minutes

Based on the improvement goals you've set, how would you go about reaching those goals?

Include considerations such as:

- What aspects of the business need to be involved in this change (e.g., material suppliers, clients, your managing executive, etc.)? How to enroll them?

- Are there systemic changes needed outside your company (e.g., government regulations, social norms, etc.)? How can your product/company/team contribute to those larger changes? Are there existing efforts in this area to support?

- Are there partnerships to pursue? With others in your industry? With experts? With environmental or social groups? With governments?

STEP 5: Summarize Your Findings and Your Plan
Time Estimate: 15–30 Minutes

Use the following table to summarize your work. If you chose more than one SDG to work on, make additional columns to list each SDG in its own column.

SDG Which did you choose and why? What part of the life cycle does it mainly apply to?
Problem/Benefit What's the current problem to be mitigated, or benefit to be expanded?
Goal What metric will you measure success by? What's the current (baseline) score? What score do you want to achieve, by when?
Plan How will you work on your goal? Who do you need to enroll in the process?

Checklist for Self-Assessment

To score your success on this exercise, see if you...

☐   *Created a map or list of your product's positive and negative impact on the SDGs, over your product's life cycle. (For digital/services, include energy use and related physical things.)*

☐   *Did additional research, as needed, to understand your product or industry's impacts on the SDGs. Cited sources for findings.*

☐   *Chose goals, metrics, and targets for improvements using clear and appropriate criteria.*

☐   *Articulated a basic plan to work toward your goals and targets.*

☐   *Summarized your findings and plan in a quick reference table.*

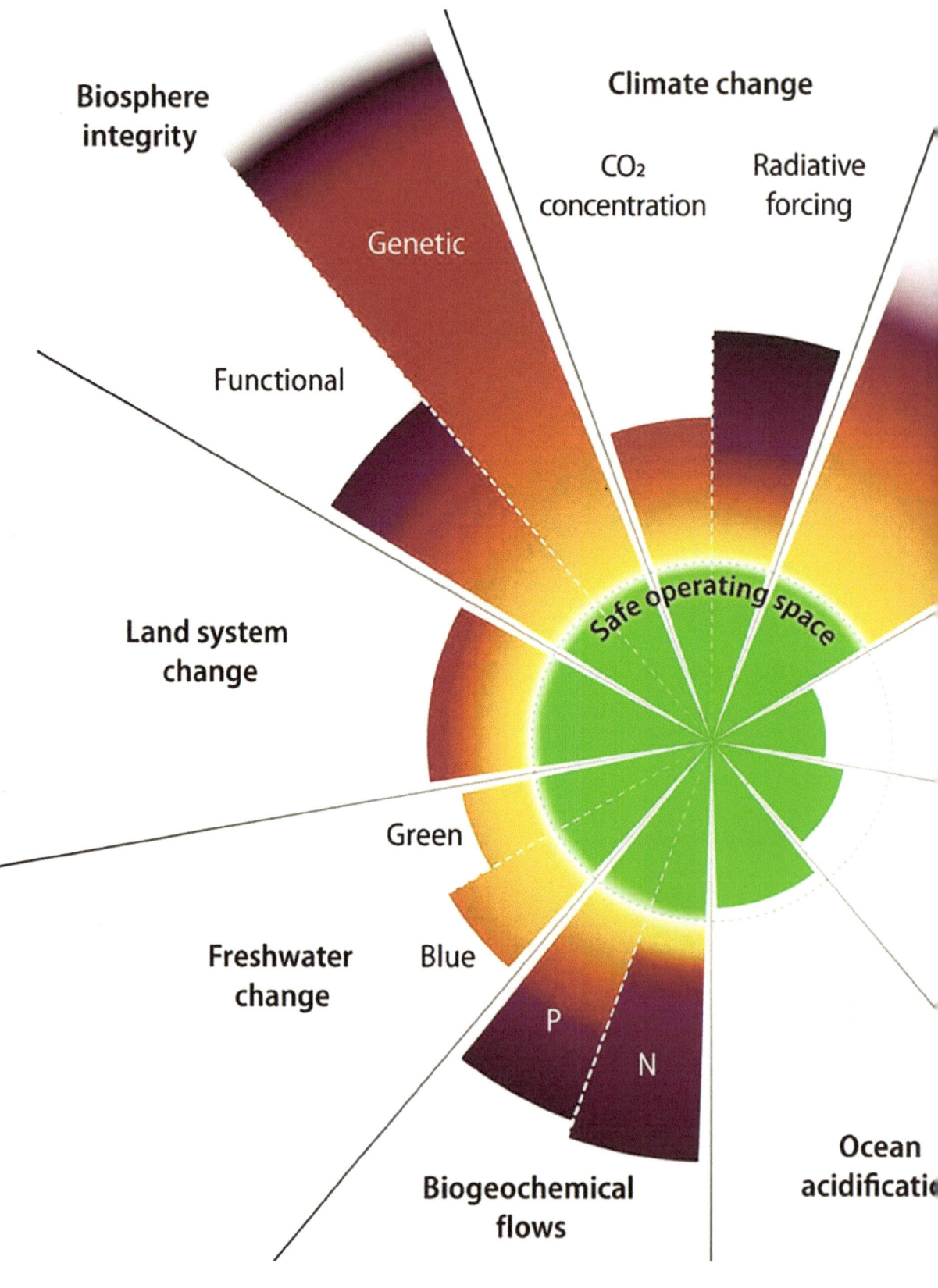

Novel entities

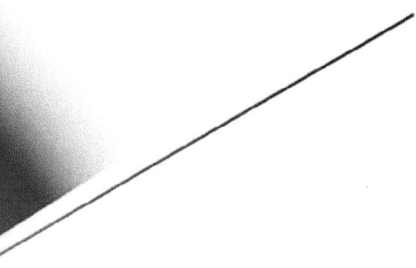

Stratospheric ozone
depletion

Atmospheric
aerosol loading

CHAPTER 2
Planetary
Boundaries

*Conny Bakker, Jeremy
Faludi, and Ruud
Balkenende*

Goals

- Identify the nine planetary boundaries and understand how they are interlinked
- Compare and contrast the different representations of the planetary boundaries (the Planetary Boundaries model, the SDG Wedding Cake, the Doughnut framework)

DOI: 10.4324/9781003504672-3

Why It Matters

What kinds of environmental impacts are serious problems that you should consider in your designs? The Planetary Boundaries model quantifies the environmental sustainability metrics (with large uncertainties) on a planetary scale to answer these questions. It is a risk model, showing how we are increasing the risk of reaching potential ecological boundaries or "tipping points." Once we persistently move beyond these boundaries, the Earth system may tip over into a very different state, one much less hospitable to the development of human societies. Thus, our designs must work to restore ecosystems to back within the planetary boundaries.

Summary

- The Planetary Boundaries model depicts the proposed safe operating space for humanity, for nine planetary systems. Central in the model is the green "safe" zone, where there is very little risk of eroding the earth system.

- Six planetary boundaries are "in the red," notably the two core boundaries of Climate Change and Biosphere Integrity.

- "Doughnut Economics" uses the Planetary Boundaries and adds a "social foundation" which helps remind us that addressing environmental problems should not result in social injustice.

Earth came out of its last ice age some 12,000 years ago. The geologic period after this, called the Holocene, was characterized by a stable climate that has allowed human civilizations to rise and flourish. However, the past 200 years have dramatically changed this, with the Industrial Revolution. Human actions are now the main drivers of global environmental change. Thus, we risk pushing the Earth system outside the stable environmental state of the Holocene, with potentially catastrophic consequences for human societies as well as many other species.

For example, the Earth has a natural carbon cycle, where carbon atoms continually and repeatedly go from the atmosphere to the land and back into the atmosphere (Figure 2.1). Carbon is stored in a variety of reservoirs, including plants (e.g., stems and leaves) and animals (e.g., bones and other tissues). In the atmosphere, carbon is stored as gases, such as carbon dioxide. In oceans, the vast majority of carbon is stored dissolved in water as carbonate; some is in the bodies of marine organisms, such as clam shells or coral skeletons. Carbon is also contained within rocks, minerals (like oil), and other sediment buried beneath the ocean and land's surface geology.

Carbon moves through ecosystems over short and long time scales. Short time scales (years or decades) happen when carbon stored in plants or animals is released back to the soil and the atmosphere when they die and decay, or when a volcano erupts. Long-term scales (millions of years) happen when carbon is removed from

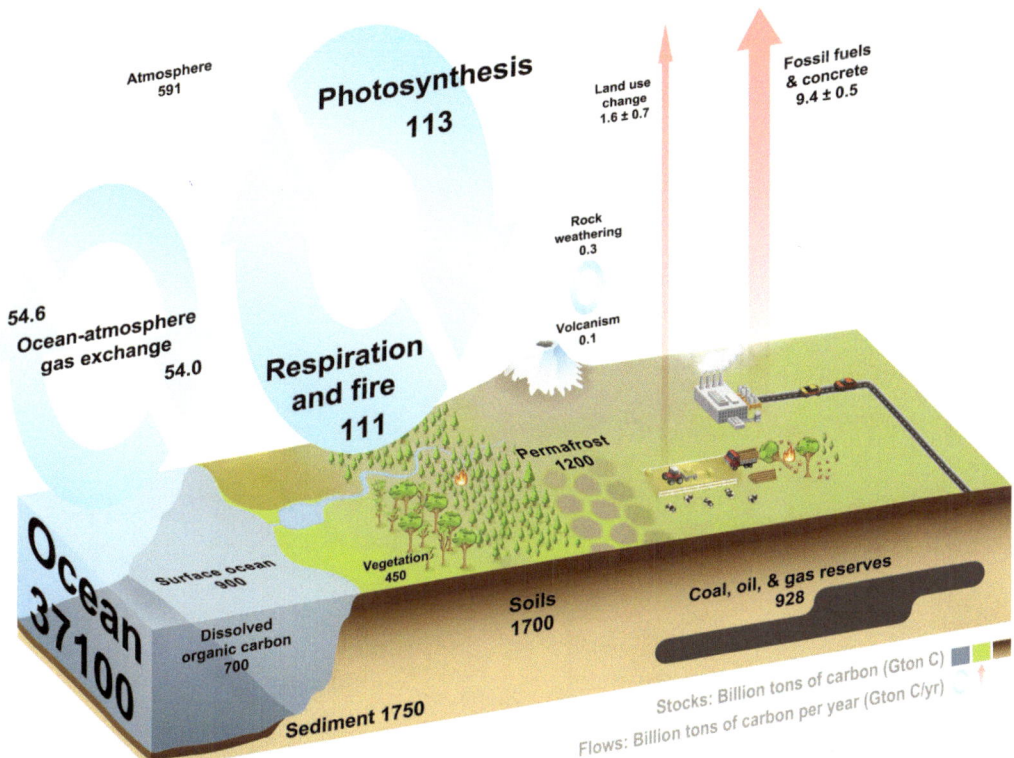

Figure 2.1 A simplified version of the natural carbon cycle, plus anthropogenic emissions. For a detailed version, see the IPCC's Sixth Assessment (IPCC, 2023).

Source: Image based on IPCC Fifth Assessment (2013) with Sixth Assessment numbers.

short-term circulation ("sequestered") by rock weathering, or plants and animals becoming fossilized into coal or oil underground, or removed from seawater by shells and bones of dead marine animals and plankton collecting on the sea floor and geologically transforming into limestone.

The carbon cycle is in a dynamic equilibrium, meaning that the amount of carbon naturally released from reservoirs is equal to the amount that is naturally absorbed by reservoirs (Figure 2.1's blue circles). Humans have upset this balance by mining and burning fossilized carbon ("fossil fuels") and cutting down vegetation, both shown

in red arrows in Figure 2.1. However, any carbon released at the same rate it is being sequestered is "carbon neutral." For example, if trees are cut down and burned for fuel no faster than new trees grow to full size, that is carbon neutral. And if more new trees grow to full size faster than they are cut and burned, that is "carbon negative," or regenerative. Because industry has overshot the natural equilibrium so much, we need decades or centuries of carbon negative industry to restore balance to the system's natural cycle.

How do we regain ecosystem stability? In 2009, a team led by Johan Rockström and Will Steffen assessed what factors were

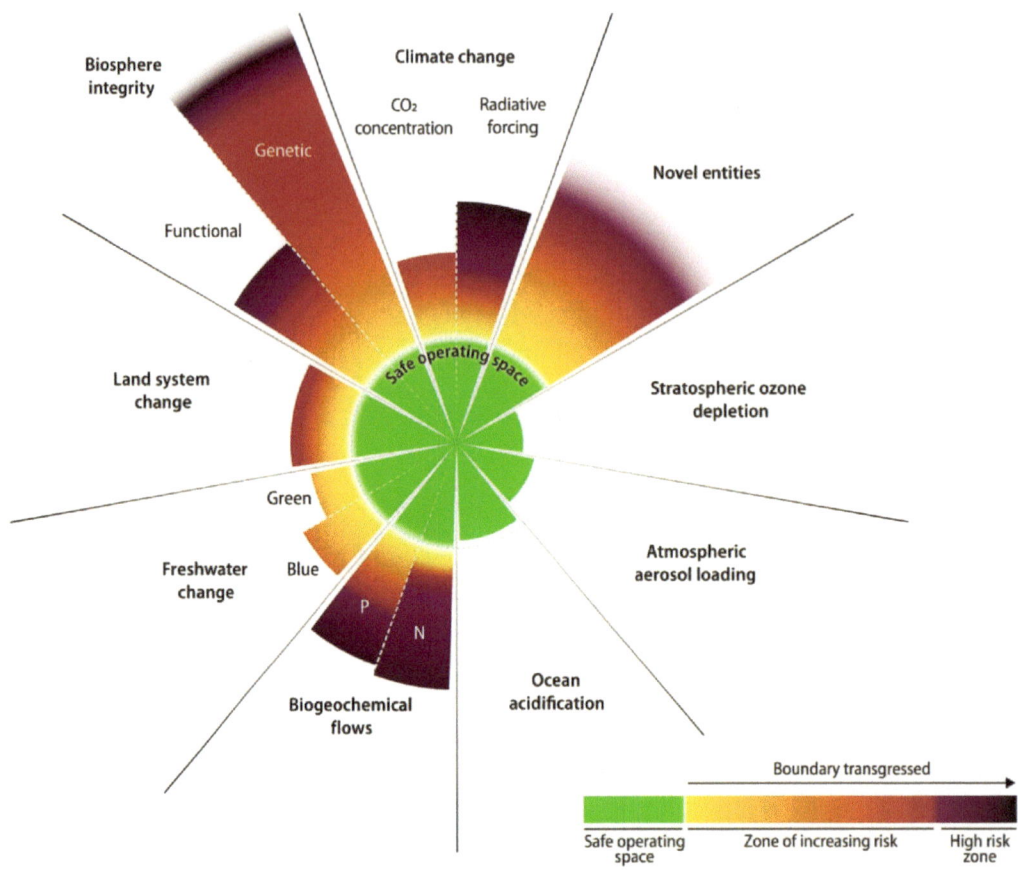

Figure 2.2 Planetary boundaries model. Within the green circle (the Safe Operating Space), damage is still reversible; beyond it in the yellow to red, there is increasing risk of irreversible disruptive ecosystem change; purple is known high risk.

Source: Richardson et al. (2023).

disrupting global ecosystems and determined "a safe operating space for humanity" within certain planetary boundaries (Rockström et al., 2009), see Figure 2.2. They found that we should not exceed certain levels of nine Earth system processes: climate change, freshwater change, stratospheric ozone depletion, atmospheric aerosol loading, ocean acidification, biogeochemical flows (of nitrogen and phosphorous), novel entities, land-system change, and biosphere integrity.

How do these Earth system processes stabilize or disrupt ecosystems?

- **Climate change** results from increased CO_2 and other greenhouse gases in the atmosphere due to human activities, such as burning fossil fuels and deforestation. It is divided into two parts because the large majority of it is due to CO_2, but other greenhouse gases also cause "radiative forcing" (the mechanism causing climate change). Radiative forcing is where visible and other light from the sun passes through the atmosphere and hits the Earth, is absorbed and turns into heat, which is re-radiated as infrared light, but some gases are not so transparent to infrared as to visible light, so they reflect some of the infrared back down to Earth, causing warming. Hence the term "greenhouse gas." Some amount of warmth is good and healthy for ecosystems (Figure 2.1 shows natural CO_2 levels), but too much disrupts life on Earth. The effects are global: rising average global temperatures (not evenly distributed), sea-level rise, and more extreme fluctuations (e.g., extreme hot and cold, extreme rainfall or drought). In the 2015 Paris Agreement on Climate Change, 196 countries pledged to limit global warming to 1.5 degrees Celsius. In 2023, a UN climate report concluded that carbon emissions in the past decade are higher than ever, and we are on a pathway to global warming of more than double the 1.5-degree limit (IPCC, 2023). As a result, the global climate is at very high risk of irreversibly shifting into a new state, with different locations and sizes of forests, deserts, coastlines, species habitats, and more.

- **Freshwater change** affects the functioning of global freshwater systems. The Planetary Boundaries model distinguishes between green and blue water. Green water is the water available to plants (rain, soil moisture). Blue water is fresh water in rivers and lakes. Again like carbon, water circulates through global ecosystems in natural cycles, but these cycles have been disrupted by human industry. However, for water there is not too much emission but too much depletion, mostly for irrigating crops but

also for manufacturing buildings, roads, and products. As soils are drying out due to droughts, the green water boundary has been transgressed (Stockholm Resilience Centre, 2022).

- **Stratospheric ozone** shields the Earth's surface from much of the sun's harmful ultraviolet (UV) radiation. A hole in the ozone layer was discovered in the late 1970s, caused by industrial chemicals like chlorofluorocarbons (CFCs) in refrigerants, propellants for aerosols, blowing agents, and such. Worldwide concerns over cancer risks and other harmful effects led to the Montreal Protocol in 1989, which banned a range of ozone-depleting substances, including CFCs. Ozone levels have been recovering since then, an enormous environmental success, which is why this Earth system process is in the green.

- **Atmospheric aerosol loading** is ultra-fine particulate matter floating in the air. Dust from desert storms is a natural aerosol. Smoke, soot, and other pollution caused by human activity have detrimental effects, such as lung cancer in humans and animals, or damaging plant health. Some aerosol particles, like black carbon, make climate change worse, because they absorb heat. Others, like sulphur particles, have a cooling effect because they reflect sunlight away from the surface. This Earth system is within the boundary.

- **Ocean acidification** means the pH of sea water decreases due to the increased uptake of carbon dioxide (CO_2) from the atmosphere. Just like in carbonated drinks, carbonic acid is part of the process of CO_2 gas turning into liquid carbonate and bicarbonate. Ocean water is naturally slightly basic/alkaline, but in the past 200 years its acidity has increased 30% (from pH 8.2 to 8.1). As the ocean becomes more acidic, it dissolves calcium carbonate from sea shells and other calcifying organisms, and makes less carbonate available to build and maintain their shells, skeletons, and similar structures. That harms many marine species, but this Earth system is within the boundary.

- **Biogeochemical flows** (also called "eutrophication") refers to nitrogen ("N") and phosphorus ("P"), both nutrients for plant growth and used in fertilizers. Like carbon and water, these elements circulate through global ecosystems in natural cycles, but these cycles have been disrupted by industry over-fertilizing land for crops. Part of the fertilizer is not taken up by plants and ends up in the water and the soils. Too many nutrients in water cause algae blooms, which hurt ecosystems in two ways: First, large algae blooms shade out other aquatic plants. Second, and more important, when the algae die, their decomposition pulls oxygen of the water, suffocating fish and other aquatic life. Locations with large algae blooms are called "dead zones."

- **Novel entities** is the burden of synthetic chemicals and other novel entities on the environment. There are an estimated 350,000 different types of manufactured chemicals on the global market. These include plastics, pesticides, industrial chemicals, chemicals in consumer products, nanoparticles, antibiotics and other pharmaceuticals. These are all novel entities, created by human activities with largely unknown effects on the Earth system. Many have known harmful impacts on climate change, ozone depletion, and other boundaries, like the CFCs mentioned above, but the majority have never had their toxicity or other impacts properly assessed, much less the impacts from combinations of them. In 2022, a team of scientists concluded that humanity has exceeded the novel entities safe planetary boundary because the increasing production volumes and high number of them exceed the ability of society to assess and monitor their safety risks (Persson et al., 2022).

- **Land-system change** relates to deforestation and other natural habitat destruction to convert land to agriculture or other uses. It is one of the main forces behind the serious reductions in biodiversity. It also affects climate and ecosystems by changing rainfall patterns, soil erosion into waterways, absorption of the sun's heat, and other factors.

- **Biosphere integrity** consists of two aspects: functional and genetic biodiversity loss. Humans are causing a sixth global extinction wave, which many scientists argue could become as impactful as the asteroid that killed the dinosaurs, but there is still time to prevent that level of damage. Genetic biodiversity loss is measured by the rate at which species go extinct; specifically, number of extinctions per million species-years (E/MSY). The latest reports show that species are going extinct globally at rates unprecedented since the end of the dinosaurs (IPBES, 2019). The normal extinction rate of mammals is about 2 extinctions per 10,000 species per 100 years (2 E/MSY); current mammal extinction rates are 40–120 E/MSY (other animals, like insects, are far higher). The boundary is 10 E/MSY. Functional biodiversity loss is measured as the percentage of the Earth's biomass growth being appropriated/reduced from ecosystems by industry. The baseline pre-industrial rate of biomass growth for any specific location can be measured geologically, in gigatons of carbon per year pulled into living plants through photosynthesis, and then into animals, etc. Then today's "human appropriation" of biomass for that same location combines how many gigatons of carbon biomass per year are used by humans (e.g., food or materials in products) and how many gigatons never grew because of ecosystem damage (e.g., overfishing, deforestation). The boundary is for less than 10% appropriation of preindustrial biomass, meaning over 90% remaining for biosphere function. Current values for most ecosystems are around 30% appropriation.

2.2 Understanding the Planetary Boundaries Model

For most of the Earth system processes, the authors of the model were able to define a safe operating space and an "area of increasing risk." For instance, climate change has its safe planetary boundary at 350 ppm CO_2 (shown in green in Figure 2.2), with an area of uncertainty of 350–450 ppm (Steffen et al., 2015). In 2022, CO_2 measured 420 ppm. Climate change is increasingly moving into a high-risk area (shown in red/purple in Figure 2.2). It means that we are one step closer to the predicted irreversible disruptive effects climate change will have on our human societies.

It is important to realize how interlinked all the Earth system processes are. Biosphere integrity, for instance, is driven by species extinction, which is caused by crossing other

boundaries (e.g., coral reefs being killed by ocean acidification, land use change killing endemic species, climate change making it harder for species to survive, algae blooms from eutrophication killing fish, etc.). Burning fossil fuels drives climate change and ocean acidification through CO_2, and also drives atmospheric aerosol (particulate) pollution. This means that we do not have the luxury to concentrate our efforts on any one of them in isolation from the others.

Also be aware that you shouldn't compare the red/purple bars with each other. It would be incorrect to conclude that the novel entities boundary is "worse" than climate change because it has a bigger bar – each of these processes has its own set of boundaries and metrics. All Earth system processes that have moved into the red zone warrant our close attention. Even those that have not are still worth our attention—atmospheric aerosols are in the "safe" zone, not at high risk, but roughly 3.4 million people per year die from particulate pollution (Yim et al., 2024), which could be prevented.

To set priorities between bars, the model's authors identified two of the nine boundaries as "core boundaries": climate change and biosphere integrity. They are core because they "have the potential to drive the Earth System into a new state should they be substantially and persistently transgressed" (Steffen et al., 2015). Climate change and biosphere integrity are closely integrated and connected to all the other planetary boundaries. While the other boundaries, once crossed, may cause massive problems, they would not result in a new state of the whole Earth system. Climate change and biosphere integrity are, however, on a different level— they are crucial to Earth system functioning. Exceeding their boundaries could irreversibly shift the world from its current relatively stable ecosystems into radically different ecosystems that today's life is not adapted to—it would be a disaster for humanity and other species worldwide.

2.3 Mitigation and Adaptation

Responding to climate change and other boundaries involves two possible approaches: mitigation and adaptation. Mitigation is reducing and stabilizing impacts, adaptation is changing infrastructure or behavior to adapt to changes already in the pipeline. For example, approximately 3.3 to 3.6 billion people live in contexts that are highly vulnerable to climate change, and many species are vulnerable to climate change as well. Mitigation measures include a rapid phase-out of CO_2 and other greenhouse gases and improving the quality of the sinks that accumulate and store greenhouse gases, such as oceans, forests, and soil. Examples of adaptation are the building of levees or dikes, enhancing natural water retention by restoring wetlands and rivers, and the development of drought-resistant crops. Responses to biodiversity loss often go together with climate measures, and include the protection of vulnerable habitats, and active nature restoration, for instance, through reforestation and rewilding.

You can design significant mitigations and adaptations. Some scientists (Bresseler, 2021) have calculated that every extra 4400 tons of CO_2 emitted kills one more person by the end of the century. This implies that, for example, a company's smartphone production whose manufacturing, use, and end of life emits 80 kg CO_2 per phone, with 230 million phones/yr produced, will cause more than 4000 additional people to die by 2100 for every year of phone production, from climate change alone. This is obviously not precise, but if you as a designer or engineer could reduce those emissions by 10%, you would save hundreds of lives every year, and thousands of lives across your career.

2.4. Planet First?

Economies depend on the society they are built on, and all societies depend on the planet we live on. A collapsing environment also causes societies to collapse, which causes economies to collapse. When discussing the Sustainable Development Goals, Rockström argued that the SDGs should be visualized differently—not as a set of goals that are all equally important, but as a "wedding cake" of layers depending on the ones below (Figure 2.3). The biosphere SDG targets are related to the Planetary Boundaries.

The wedding cake figure is a plea to give more weight to the biosphere targets in the SDGs, which currently seem to be relatively underrepresented, compared to the many societal and economic goals. It argues that the Triple Bottom Line's circles of environment, society, and economy are not really a Venn diagram of equally important circles overlapping, as shown in Chapter 1, but that the Prosperity circle is completely within the People circle, which is completely within the Planet circle. The wedding cake's reframing of priorities asks us to take a long view–the needs of ecosystems require a long-term commitment, which is easy to

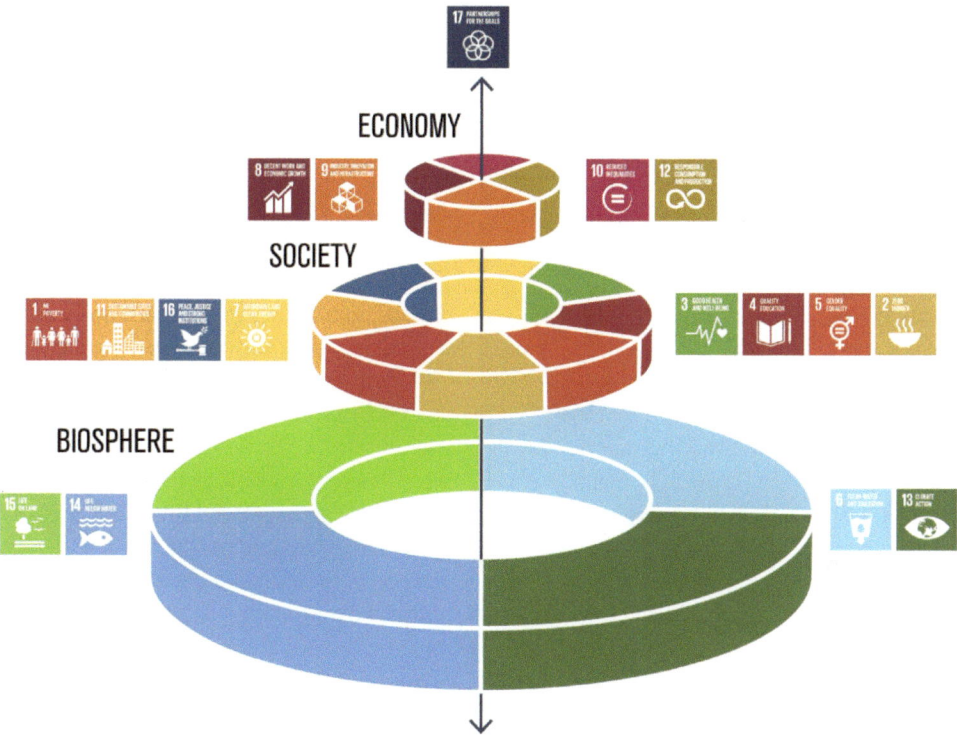

Figure 2.3 The SDG wedding cake: a new way of viewing the sustainable development goals

Source: Azote for Stockholm Resilience Centre, Stockholm University.

forget or ignore when dealing with short-term societal and economic crises. You could even argue that it gives more importance to ecological concerns than human development. In other words: planet first?

2.5 Criticism of the Planetary Boundaries Model

The Planetary Boundaries model has been (and still is) very successful: it has raised awareness of the state of the Earth's ecosystems across the world and its thinking has been incorporated into government

regulations and industry sustainability visions. The model has also drawn a fair amount of criticism. Biermann and Kim (2020) created a comprehensive overview, from which we summarize a few points: One of the critiques is the possibility that the boundaries may be misused by policymakers to "justify prolonged degradation of the environment up to the point of no return." This is especially concerning since troubles are usually not apparent until after ecological overshoot has happened.

A second point of concern is that the model is expert-driven and technocratic. Who gets to decide on the precise values of the boundaries that are to be protected?

Should this be left to experts from the natural sciences who are predominantly based in wealthy industrialized countries? The implementation of the planetary boundaries might constrain economic growth and potentially the development prospects of the Global South. Many question whether this is acceptable, and whether establishing planetary boundaries shouldn't be a much more democratic process, with stakeholder consultation and participation.

Linked to this point is the critique that the model lacks attention to basic human needs and values. It led Kate Raworth to develop a "Doughnut" framework, which suggests both a safe and just operating space for humanity.

2.6 Doughnut Economics

The "Doughnut Economics" framework of sustainability targets, by Kate Raworth (2017), combines planetary boundaries as an environmental outer limit (ceiling) with a social inner limit (foundation) of basic human needs and values, such as housing, social equity, and education (Figure 2.4). It says that industry and society must not overshoot the nine Planetary Boundaries on the outside of the doughnut, but at the same time must not fall short of the 12 social targets of the UN SDGs on the inside of the doughnut. Doughnut Economics, as the name implies, is an economic mindset, and was intended for use in governments. However, it can also be used by company executives and managers.

To use Doughnut Economics or Planetary Boundaries in design, use them to guide your vision for what mitigations and adaptations you will build to help the world thrive. Pursuing your vision can take many forms, but the two most common are (1) choosing a boundary to invent a new product or service around, or (2) taking an existing product or service and improving its performance on the boundaries.

Choosing a boundary to target, for example, could be the climate change boundary, or the gender equality boundary. You might invent a carbon capture process, or start a woman-owned business. You may even be able to target multiple boundaries to improve together. For example, LanzaTech is a company using bacteria to convert industrial carbon emissions into packaging and other materials, and their CEO is a woman, as are half their board of directors. Inventing around a boundary is entrepreneurship, even if it's within an existing company, because you need to build a new business around the new design.

Improving an existing product or service is more typical for most designers. If you already work at a company that already makes a certain product or service, benchmark your performance on the boundaries, and see where you can best improve it. That could be finding the worst boundary violation and reducing it, or finding the best boundary restoration that's already happening and magnifying it, or some combination of improvements to several boundaries. As mentioned in the earlier smartphone example, even a 10% impact reduction for some products in very large-scale production could save hundreds of lives per year around the world.

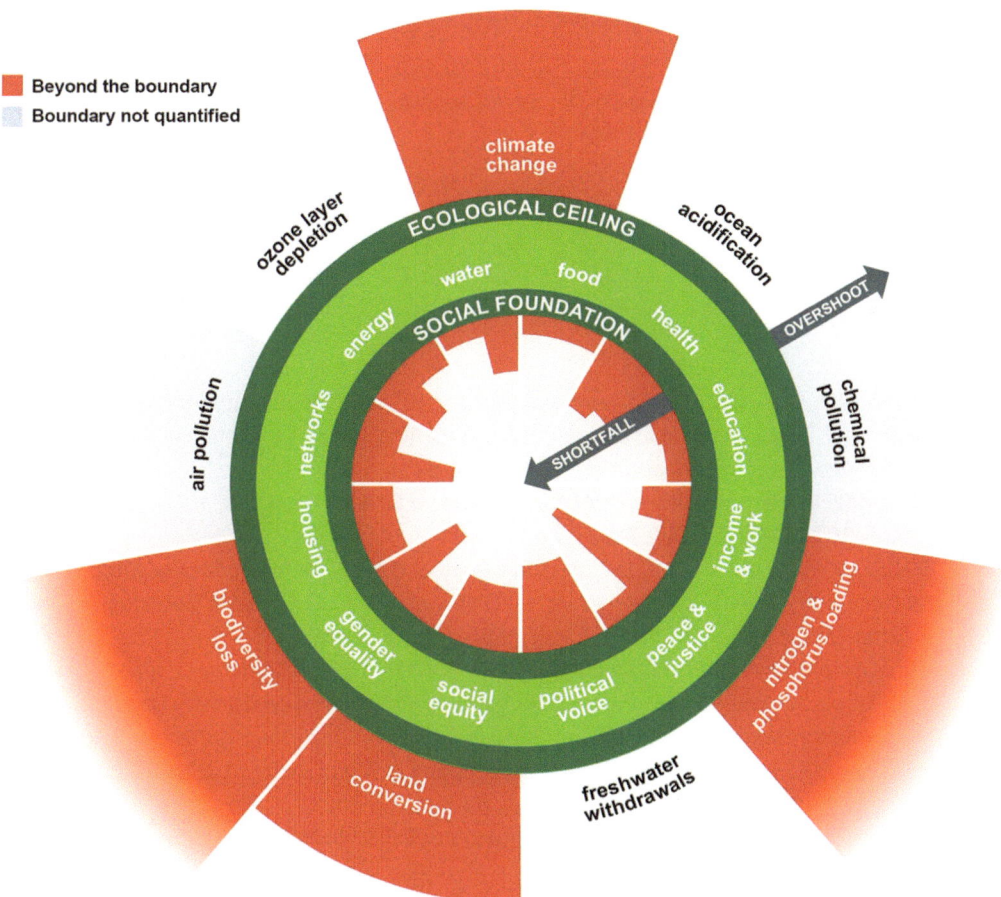

Figure 2.4 Doughnut economics model

Source: Doughnuts Economics Action Lab; Doughnuteconomics.org.

Resources and References

Resources for Further Study

- Steffen, W., Richardson, K., Rockström, J., Cornell, S. E., Fetzer, I., Bennett, E. M., Biggs, R., Carpenter, S. R., de Vries, W., de Wit, C. A., Folke, C., Gerten, D., Heinke, J., Mace, G. M., Persson, L. M., Ramanathan, V., Reyers, B., & Sörlin, S. (2015, February 13). Planetary boundaries: Guiding human development on a changing planet. *Science*, 347(6223).

- Raworth, K. (2017). *Doughnut economics: Seven ways to think like a 21st-century economist*. Random House Business.

- Project Drawdown. Available at: https://drawdown.org, lists top-priority climate interventions with calculations of expected carbon reductions. It has many opportunities for design and entrepreneurship.

References

Biermann, F., & Kim, R. E. (2020). The boundaries of the Planetary Boundary Framework: A critical appraisal of approaches to define a "safe operating space" for humanity. *Annual Review of Environment and Resources*, 45(1), 497–521.

Bressler, R. D. (2021). The mortality cost of carbon. *Nature Communications*, 12(1), 4467. https://doi.org/10.1038/s41467-021-24487-w

IPBES. (2019) Global assessment report on biodiversity and ecosystem services of the Intergovernmental Science-Policy Platform on Biodiversity and Ecosystem Services. (edited by E. S. Brondizio, J. Settele, S. Díaz, & H. T. Ngo). IPBES Secretariat.

IPCC. (2023). *Climate change 2023: Synthesis report: Contribution of Working Groups I, II and III to the Sixth Assessment Report of the Intergovernmental Panel on Climate Change*. Intergovernmental Panel on Climate Change.

Persson, L., et al. (2022) Outside the safe operating space of the planetary boundary for novel entities. *Environmental Science & Technology*, 56(3), 1510–1521.

Raworth, K. (2017). *Doughnut economics: Seven ways to think like a 21st-century economist*. Random House Business.

Richardson, K., Steffen, W., Lucht, W., Bendtsen, J., Cornell, S. E., Donges, J. F., Drüke, M., Fetzer, I., Bala, G., von Bloh, W., Feulner, G., Fiedler, S., Gerten, D., Gleeson, T., Hofmann, M., Huiskamp, W., Kummu, M., Mohan, C., Nogués-Bravo, D., … Rockström, J. (2023). Earth beyond six of nine planetary boundaries. *Science Advances*, 9(37), eadh2458.

Rockström, J., Steffen, W., Noone, K., Persson, S., Chapin, F. S., Lambin, E. F., Lenton, T. M., Scheffer, M., Folke, C., Schellnhuber, H. J., Nykvist, B., de Wit, C. A., Hughes, T., van der Leeuw, S., Rodhe, H., Sörlin, S., Snyder, P. K., Costanza, R., Svedin, U., . . . Foley, J. A. (2009). A safe operating space for humanity. *Nature*, 461, 472–475.

Steffen, W., Richardson, K., Rockström, J., Cornell, S.E., Fetzer, I., Bennett, E.M., Biggs, R., Carpenter, S.R., de Vries, W., de Wit,C.A., Folke, c., Gerten, D., Heinke, J., Mace, G.M.,

Persson, L. M., Ramanathan, V., Reyers, B., & Sörlin, S. (2015, February 13). Planetary boundaries: Guiding human development on a changing planet. *Science*, 347(6223).

Stockholm Resilience Centre. (2022, April 26). Freshwater boundary exceeds safe limits. Available at: https://www. stockholmresilience.org/research/research-news/2022–04–26-freshwater-boundary-exceeds-safe-limits.html

Wang-Erlandsson, L., Tobian, A., van der Ent, R. J., Fetzer, I., te Wierik, S., Porkka, M., Staal, A., Jaramillo, F., Dahlmann, H., Singh, C., Greve, P., Gerten, D., Keys, P. W., Gleeson, T., Cornell, S. E., Steffen, W., Bai, X., & Rockström, J. (2022). A planetary boundary for green water. *Nature Reviews Earth & Environment*, 3, 380–392.

Yim, S. H. L., Li, Y., Huang, T., Lim, J. T., Lee, H. F., Chotirmall, S. H., Dong, G. H., Abisheganaden, J., Wedzicha, J. A., Schuster, S. C., Horton, B. P., & Sung, J. J. Y. (2024). Global health impacts of ambient fine particulate pollution associated with climate variability. *Environment International*, 186, 108587.

How to Apply #2: Assess Your Product's Impacts Using Doughnut Economics
Time Estimate: 40 Minutes–2.5 Hours

Qualitatively assess your product or service using the Doughnut economics model, both in terms of the environmental ceiling and the social foundation. You could also assess a whole company or industry, even a government or community program. (Optional: See the Doughnut Economics Action Lab (DEAL) for more related tools and resources: https://Doughnuteconomics.org/tools.)

STEP 1: List Impacts on the Ecological Ceiling
Time Estimate: 15–60 Minutes

Make a list of the impact(s) your product/service has on each of the categories of the ecological ceiling in the Doughnut economics model (see Figure 2.4). For "chemical pollution," generalize it to the planetary boundary "novel entities"; for "excessive fertilizer use," use the planetary boundary "biogeochemical flows"; for "air pollution," use the planetary boundary "atmospheric aerosol loading." Remember to consider all life cycle stages of your product or service, from material extraction to manufacturing to customer use to end of life. Write down:

- Which ecological ceiling categories are most impacted by your product/service?

- Which categories are least affected?

- Are any categories positively impacted?

STEP 2: List Impacts on the Social Foundation
Time Estimate: 15–60 Minutes

Make a list of the impact(s) your product/service has on each of the categories of the social foundation in the Doughnut economics model (see Figure 2.4). Almost all categories match the UN Sustainable Development Goals, so you may want to look back at the exercise for SDGs in Chapter 1. The Doughnut's "Social Equity" is SDG 10, reduced inequalities. "Housing" is SDG11, sustainable cities and communities. The social categories not perfectly matching SDGs are "Networks," defined as "access to networks of transport, of communications, and of community support"; and "Political Voice," defined as "ensure people have voice in, and influence over, decisions that affect their lives." For more information on the categories, see the Doughnut Economics Action Lab's "Dimensions of the Doughnut." Remember to consider all life cycle stages of your product or service. Write down:

- Which social foundation categories are most impacted by your product/service?

- Which categories are least affected?

- Are any categories positively impacted?

STEP 3: Decide Which Factor to Focus on in Design
Time Estimate: 10–20 Minutes

Integrate the understanding you've gained from Step 1 and Step 2 to decide what you most want to improve in your product or service. You don't need to ideate actual design solutions, just list potential goals for improvement. Write down:

- One or two overall goals, considering all environmental and social aspects.

- What tensions exist between your environmental and social goals? How might you navigate or resolve those tensions?
- Are there synergies, where your social and ecological goals reinforce each other?

BONUS: Ideate actual design changes you can make to your product, its life cycle, the business model, or other aspects to bring everything within the ecological ceiling and social foundation. (We'll do this in depth in later chapters.)

Checklist for Self-Assessment

To score your success on this exercise, see if you…

- ☐ *Listed your product/service/company's impacts on the Ecological Ceiling categories.*

- ☐ *Listed your product/service/company's impacts on the Social Foundation categories.*

- ☐ *Listed overall goal(s) integrating environmental and social concerns.*

- ☐ *Listed tensions and/or synergies between your environmental and social goals.*

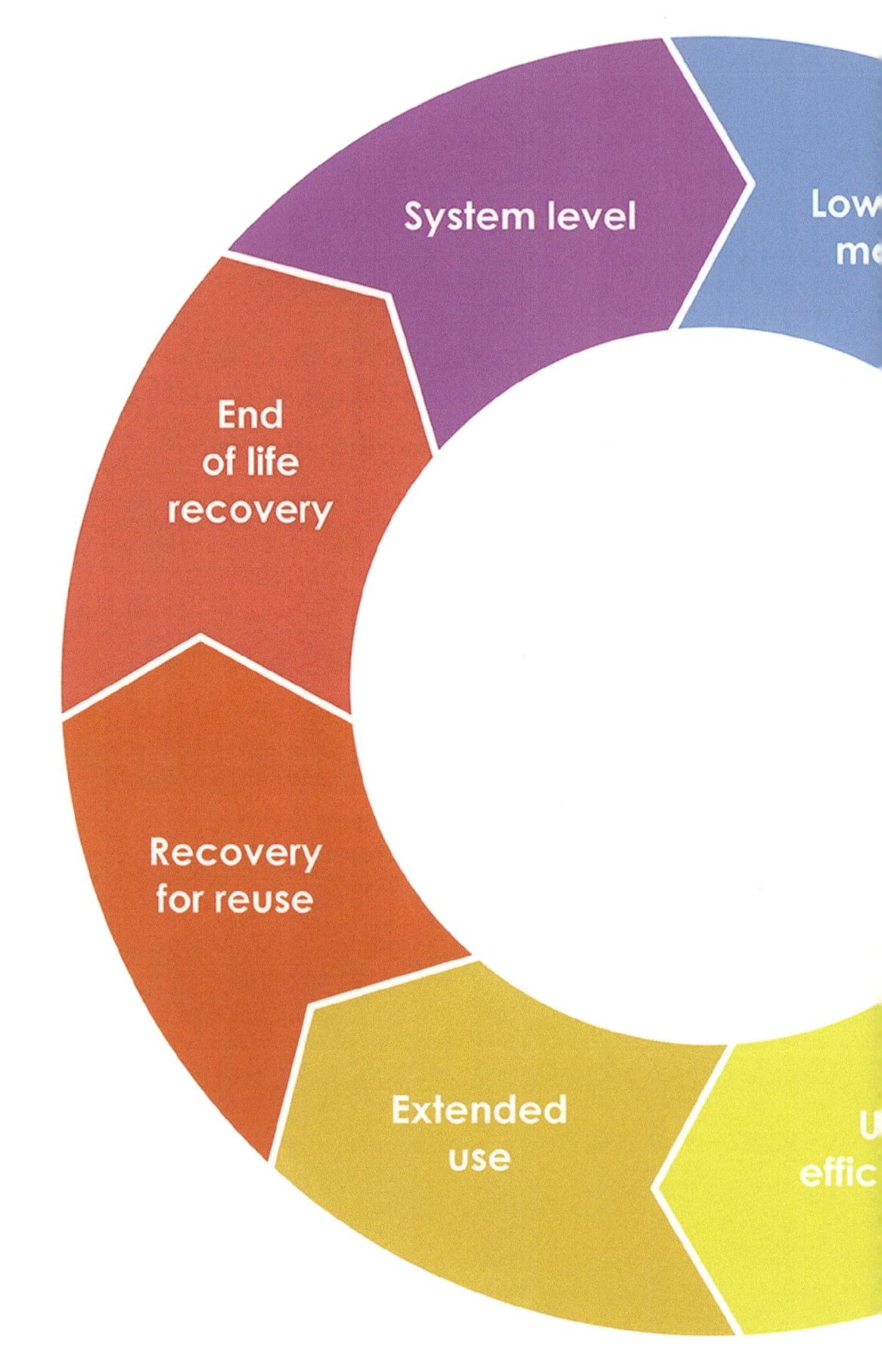

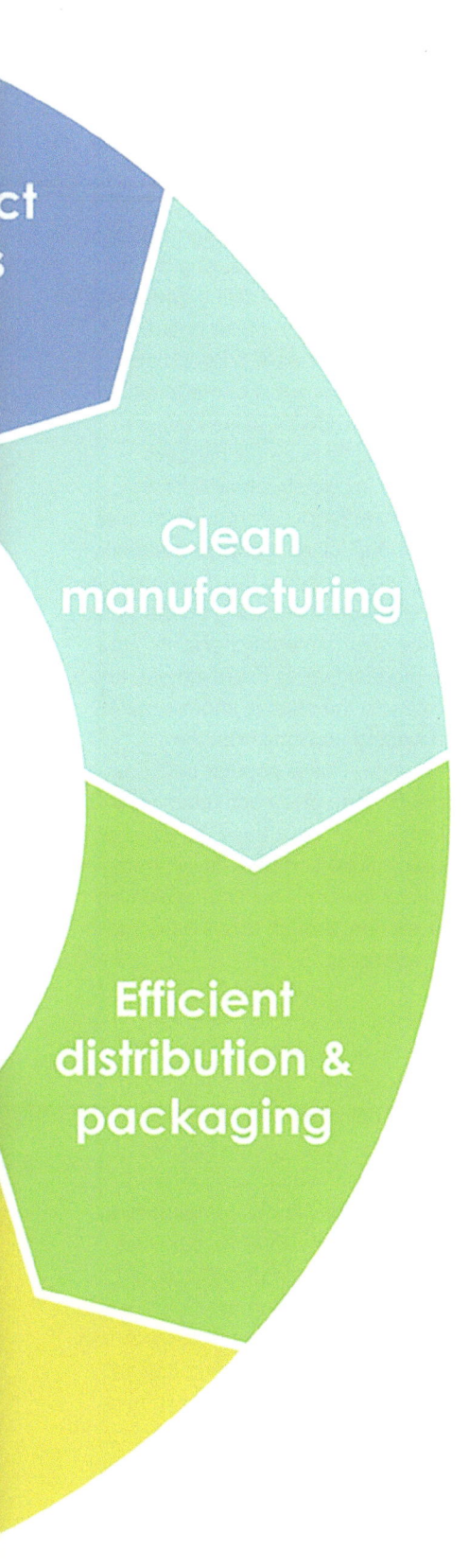

CHAPTER 3

Ecodesign

*Conny Bakker,
Jeremy Faludi, and
Ruud Balkenende*

Goals

- Use the Ecodesign Strategy Wheel to ideate sustainable product redesign
- Use the Ecodesign Strategy Wheel to qualitatively compare sustainability of different designs

DOI: 10.4324/9781003504672-4

Why It Matters

The concept of life cycle thinking underpins most sustainable product design assessments and ideation. Mapping the life cycle of a product, from raw materials sourcing to recycling, allows you to qualitatively judge the resource demands and waste at each stage and consider design interventions for them. The Ecodesign Strategy Wheel is a design tool to both judge impacts and suggest design strategies for each life cycle stage.

Summary

- The product life cycle is the basis for environmental assessments, both qualitative and quantitative.

- The Ecodesign Strategy Wheel is a tool that helps you consider all stages in a product's life cycle.

- While the Ecodesign Strategy Wheel is a helpful tool for making relatively quick judgment calls about how and where to improve a product's environmental profile, there are several drawbacks, like trade-offs and lack of data, that need to be taken into account.

3.1 Ecodesign

Ecodesign is a sustainable design approach with the explicit goal of minimizing the environmental impact of existing and newly developed products along their entire life cycle (van Doorsselaer & Koopmans, 2021). The core concept is eco-efficiency: maximizing the product's value to users while minimizing its potential negative environmental impact. In other words, companies try to produce products or offer services with the lowest possible materials and energy while keeping the same function. It especially suggests focusing on the worst impacts (the environmental "hotspots") to make the biggest overall improvement, and tries to estimate the relative improvement of different design options from the baseline design. This is popular because eco-efficiency often gives both economic and environmental. advantage, a "win-win." It should also include company-level strategies like producing fewer products to avoid user overconsumption, though this is harder to argue economically.

In the mid-1990s, several authors proposed extreme forms of eco-efficiency which they called the "factor" approaches. Von Weizsäcker, Lovins, and Lovins (1995) launched "Factor Four" whose goal was to cut resource use down to one quarter of the baseline design without compromising usability or profitability. In 2010, the Rocky Mountain Institute created the "Factor 10 Engineering" (10xE) design principles, which argued that cutting energy and resource use down to one-tenth of the baseline design can be very profitable—but it does require transformational design. Most companies that embrace ecodesign tend to take a less radical approach, using it to shave off a few

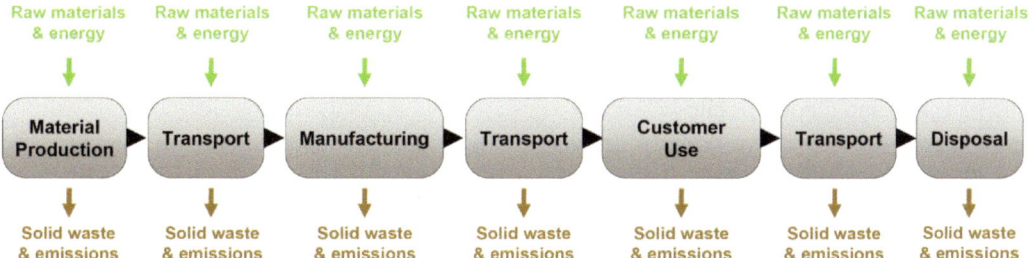

Figure 3.1 The life cycle of a product

percent of material or energy consumption, thus incrementally improving product footprints. However, constantly improving incremental savings can compound over the years into radically large savings. In computers, Moore's Law has not only improved speed but efficiency, improving the number of computations per kilowatt-hour by ten times every decade, so that computers in 2010 were 100 billion times more energy-efficient than the computers in 1950 (Johnsson, 2012).

3.2 Life Cycle Thinking

Product life cycles are complex: raw materials are mined or grown and processed across the world, parts are produced and assembled in different locations. Products are packaged, shipped, stored, sometimes repackaged, sold, and shipped again before they reach their end-use destination. After customer use, products enter an equally or more complex logistics network—for instance, to return unwanted products, to return broken products for repair or refurbishment, to redistribute and remarket products for second-hand sales, or to transport waste products to their next destination (be it landfill, incineration, or recycling). All this doesn't even mention the product's packaging, which has a lifecycle of

its own. Or in case of a digitally connected product, the extensive communication network infrastructure with its energy use and continuous work to keep the software up to date. Software and other digital products are not exempt from this—they all run on physical hardware and use energy. In order to map this complexity and get a sense of the associated environmental impacts, we use a (simplified) representation of a product life cycle (Figure 3.1).

The idea of using a product life cycle as the basis for environmental assessment comes from the theory of industrial metabolism. The metabolism of industry is defined as the whole integrated collection of physical processes that convert raw materials and energy, plus labor, into finished products and wastes in a (more or less) steady-state condition (Ayres, 1989). Robert Ayres uses the term "metabolism" because of the remarkable parallels between biological and industrial activities, "both are materials processing systems driven by a flow of free energy."

For each process stage, materials and energy are needed as input, for instance, for the production of polyethylene, oil needs to be drilled and refined (needing input of all kinds of machinery and energy as well as human labor). This leads to output of the plastic itself, as well as different kinds of waste (heat, catalysts, emissions to air, water, etc.). Mapping the product life cycle and its inputs and outputs in this way is the starting point for quantitative assessments like life cycle assessment (LCA), covered in following chapters, but it can also be used to do a qualitative assessment (life cycle thinking).

3.3 The Ecodesign Strategy Wheel

The Ecodesign Strategy Wheel was first developed at TU Delft in the 1990s (Brezet & Hemel, 1997), also popularized in the Okala Practitioner guide (White et al., 2013). What

follows is an updated version. It helps you consider all stages in a product's life cycle. It can be used in different ways, for instance:

- To suggest ecodesign strategies for ideation.

- To assess a product on its environmental merits, and identify improvement options.

- To compare a product's environmental merits before and after redesign, or compare design options.

The Ecodesign Strategy Wheel presents eight strategies along the spokes of the wheel, following the life cycle stages of the product (see Figure 3.2). To use it as an ideation tool, simply brainstorm based on the life cycle strategy you are focusing on

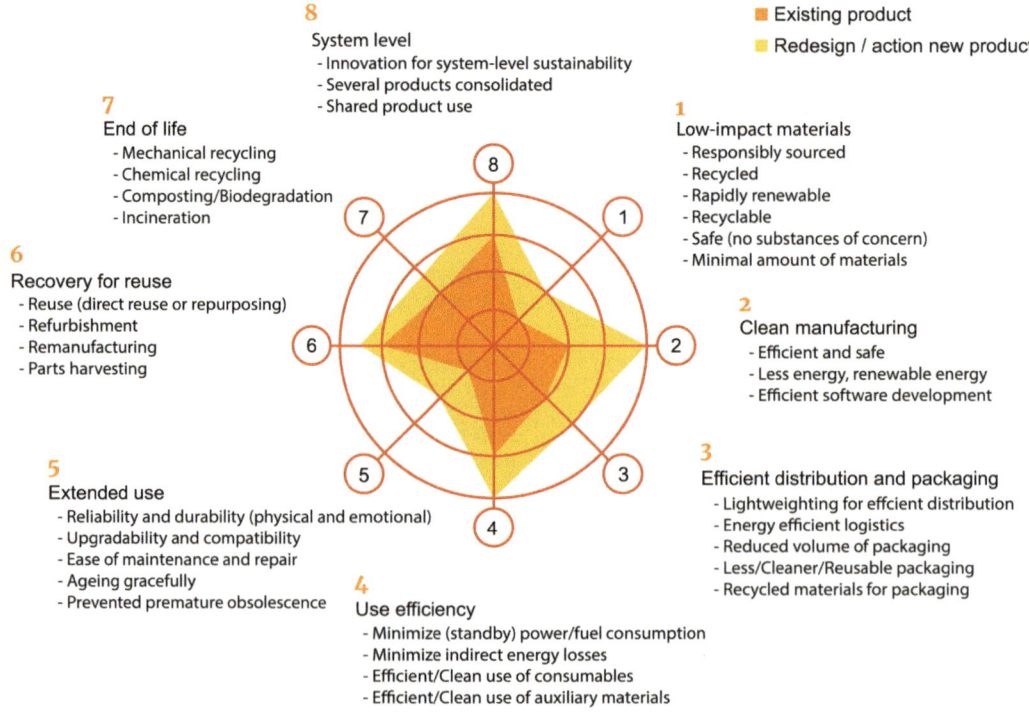

Figure 3.2 The Ecodesign Strategy Wheel

Source: Brezet & van Hemel (1997).

(e.g., "low-impact materials"), especially brainstorming the sub-strategies listed (e.g., responsibly sourced, recycled and recyclable, and safe). To help choose which life cycle stage to focus on, you may also use the wheel to qualitatively guess your priorities by assigning a score to each strategy. The better a product meets the strategy (or list of sub-strategies), the higher the score, with the best scores marked on the outer ring of the wheel and the worst scores close to the center. As all scores are based on educated guesses, and are not calculated, the strategy wheel cannot be used to determine the actual ecological impact of a product.

Each strategy includes a list of sub-strategies. These are explained below. A product doesn't have to use all of these to get a positive score. Also, if a strategy is not applicable, you can skip it.

1. Low-impact materials

   • Choose responsibly sourced materials—ones that at least minimize environmental degradation or social injustice in their extraction or harvesting, and ideally drive environmental restoration and social flourishing.

   • Choose recycled materials. Recycled plastics and metals often have somewhat different properties than new ones, but are suitable for many applications.

   • Choose recyclable materials. Don't use composite or blended materials if it can be avoided.

   • Sustainably harvested biomaterials, harvested only as quickly as they grow back. This is easier with rapidly renewable materials (e.g., cork, bamboo, wool).

   • Choose safe materials: avoid "substances of concern"—chemicals which may have hazardous effects on human health and the environment.

   • Minimize amount of material use.

2. Clean manufacturing

   • Ensure efficient and safe processing, manufacturing, and assembly.

   • Minimize production waste and ensure recycling of production waste.

   • Use less energy and/or renewable energy for manufacturing.

   • Minimize the resource needs of software development (e.g., AI training).

3. Efficient distribution and packaging

   • Lightweighting for efficient distribution.

   • Energy-efficient logistics.

   • Reduce volume of packaging, for example, concentrated products use less packaging.

   • Reusable or refillable packaging.

   • Packaging made from recycled materials.

4. Use efficiency

   • Minimize energy consumption (electricity or fuel), both when in use and in standby modes.

   • Minimize indirect energy losses. This is relevant to, for instance, building insulation products and windows.

- Efficient/clean consumables, for example, water and detergents in washing machines, ink in printers. Eliminate, minimize, and/or use more sustainable ones.

- Efficient/clean auxiliary materials, for example. lubricants or refrigerant gases. Eliminate, minimize, and/or use more sustainable ones.

5. Extended use

- Reliability and durability—both physical and emotional durability are important to extend a product's useful life.

- Upgradability (software and hardware) and compatibility. Enable the user to keep the product up to date. Ensure backward and forward compatibility. This means that the design is compatible with previous or future versions of itself.

- Ease and cost of repair and maintenance (e.g., faster disassembly time, available spare parts).

- Aging gracefully. Choose materials that retain their aesthetic value as they wear over time.

- Prevent premature obsolescence. Create products and associated services that can continue to serve emotional, functional, and technical needs over time, and over multiple product generations.

6. Recovery for reuse

- Reuse of the product when the original user no longer wants it. This can be direct reuse, or repurposing. With repurposing, the product gets a different purpose than originally intended (e.g., using old tires as playground equipment).

- Refurbishment: restore a product to an acceptable level of functionality (e.g., a computer taken back, cleaned, and its battery replaced, then resold at a discount).

- Remanufacturing: restore a product to as good or better than a new product (e.g., a computer completely disassembled, all components cleaned and quality tested, replacing any not performing at 100 percent, reassembled, then resold).

- Parts harvesting: retrieval of components, modules, or parts from obsolete products, to use them as spare parts for repair, maintenance, remanufacture, etc.

7. End of life

- Mechanical recycling: shredding the product and sorting materials, then grinding up and melting materials (usually metal, glass, or plastic) for a new life in a new product.

- Chemical recycling: breaking down a material (usually plastic) to its basic chemical building blocks, and reconstructing into new material. This uses more energy and chemicals than mechanical recycling, but matches primary materials in quality.

- Composting/Biodegradation: breakdown of organic materials or biodegradable plastics. Composting requires aerobic conditions (with oxygen); anaerobic conditions (without oxygen) create methane, which can be burned as fuel.

- Incineration: burning materials, usually with energy recovery (using the heat and/or generating electricity with it). All material value is lost, but this is better than landfill (burying in the ground).

8. System level

- Innovation for system-level sustainability benefits. For example, rethinking how to meet the user needs without a physical product is usually more innovative and beneficial than changing product architecture or features.

- Consolidate several products into one by meeting related needs. Smartphones are a good example.

- Share products among multiple users, for example, rental tools or library books.

- Use software instead of hardware ("dematerialize"). Any electronics used to run the software must have lower impacts than the hardware they replace.

3.4 Limitations of the Ecodesign Strategy Wheel

The Ecodesign Strategy Wheel comes with a few drawbacks and warnings. Be aware of these when you start working with the method:

- There will be trade-offs. When you choose to lightweight your product by using less material, you might reduce material and manufacturing impacts but shorten the product's life. Trade-offs are unavoidable, and should not alarm you, but they should be identified and considered carefully.

- Assessing a product with educated guesses is not easy. Design for Sustainability has developed into a domain with specialized knowledge. Without that knowledge (for instance, about recycling technology, ecodesign regulations, design for repair, etc.), the outcomes of the qualitative assessment could be rather meaningless. The strategy wheel will simply confirm your pre-conceived biases. Work with experts if you need to, or use quantitative tools like LCA to make sure you're on the right track.

- The wheel was developed with consumer durables in mind. If you want to use it for other products (single-use items, or software, or services), you may need to skip a number of strategies.

- The wheel has a bias toward environmental sustainability. It hardly considers business or social sustainability. If you want to work with these as well, you need to use different tools, for instance, the UN SDGs (specifically the "targets and indicators" for each goal) or the Global Reporting Initiative standards (GRI, 2024).

- Finally, the first question you should always ask before starting with the Ecodesign Strategy Wheel is: do we really need this product? How could we make users happy with less stuff?

Resources and References

Resources for Further Study

- Brezet, H., & van Hemel, C. (1997). *Ecodesign: A promising approach to sustainable production and consumption*. United Nations Environment Programme.

- White, P., Belletire, S., & Pierre, L.S. (2013). *Okala practitioner: Integrating ecological design*. IDSA.

References

Ayres, R. (1989). Industrial metabolism. In J. Ausubel & H. Sladovich (Eds.), *Technology and environment* (pp. 23–49). National Academy Press.

Brezet, H., & van Hemel, C. (1997). *Ecodesign: A promising approach to sustainable production and consumption*. United Nations Environment Programme.

Global Reporting Initiative (2024). The GRI Standards. Available at: https://www.globalreporting.org/standards/

Johnsson, L. (2012). Overview of data centers energy efficiency evolution. In I. Ahmed & S. Ranka (Eds.), *Handbook of Energy-Aware and Green Computing,* (vol. 2). Chapman and Hall/CRC.

Rocky Mountain Institute (2010) Factor 10 engineering design principles. Available at: https://rmi.org/wp-content/uploads/2017/05/2010-10_10XEPrinciples-1.pdf

Van Doorsselaer, K., & Koopman, R. (2021) *Ecodesign; A life cycle approach for a sustainable future*. Hanser Publications.

Von Weizsäcker, E., Lovins, A., & Lovins, H. (1995) *Factor Vier; Doppelter Wohlstand—halbierter Naturverbrauch; der neue Bericht an den Club of Rome*. Droemer Knaur.

White, P., Belletire, S., & Pierre, L.S. (2013). *Okala practitioner: Integrating ecological design*. IDSA.

How to Apply #3: Brainstorm Using the Ecodesign Strategy Wheel
Time Estimate: 1.5–2.5 Hours

Use the Ecodesign Strategy Wheel to brainstorm improvements on the product or service you're working on. You can do this for a green redesign of something existing, or if creating something new, you can imagine what your impacts and priorities would likely be.

STEP 1 (Optional): Choose an Ecodesign Strategy to Focus On
Time Estimate: 10–45 Minutes

Set your priorities for redesign by scoring your product or service on the Ecodesign Strategy Wheel's list of eight overall strategies in the various life cycle stages plus the overall system.

This step is optional because it's better to use evidence-based metrics such as life cycle assessment (discussed in later chapters) or eco-certification scorecards to set priorities, but if you don't yet have those skills, or don't have time, you can do this.

To choose which strategy/strategies to focus on:

- Score your product on its performance for all eight ecodesign strategies, using the spiderweb graph of Figure 3.2. Use the wheel's sub-strategies to inform your scoring, and use data you have on your product and comparable products. Score on a scale of 0 to 4, where 0 means it's terrible at fulfilling that strategy, and 4 means it's excellent at fulfilling that strategy. This is not precise or empirical, like life cycle assessment in later chapters, but make your best guess.

- Briefly list the reasons for your scores (< 10 words per strategy).

- Wherever you're unsure about your product's performance, do some research and describe what you learn in footnotes to your list of reasons for your ecodesign strategy scores. Cite your sources. Note where questions remain or more research is needed.

- If your product or service is software, remember the physical hardware your software runs on and the energy it uses, including communications bandwidth. Also remember related physical products (e.g., if you're making an online fashion store, include impacts of the clothes sold).

Based on your list above, decide what ecodesign strategy/strategies to focus on improving. Choose the worst-performing strategy/strategies to get the most improvement, and/or choose other strategies you think have greater leverage to give the best overall improvement.

STEP 2: Brainstorm Alternatives and Improvements
Time Estimate: 30–60 Minutes

For your chosen ecodesign strategy, brainstorm ways to reduce environmental impacts.

- Generate at least 10 ideas for each sub-strategy listed in the chapter, ideally 50+ ideas total. (For example, if you chose the "recovery for reuse" strategy, have 10 ideas for reuse, 10 ideas for refurbishment, 10 for remanufacturing, and 10 for parts harvesting, plus more ideas.)

- Be specific and concrete, it gets you more new ideas from the same general concepts; have wild ideas, too, it's a brainstorm.

- Use your notes and research from the previous step for ideas and inspiration.

If you chose more than one ecodesign strategy to brainstorm on, repeat this for each strategy.

STEP 3: Decide What Idea(s) to Act on
Time Estimate: 15–30 Minutes

Choose one or more winning ideas to move forward with, by qualitatively scoring the likely outcome of your newly brainstormed ideas.

- First, to save time by not scoring all your new ideas, review your new ideas and choose your top three to five favorites (roughly). You might do this with dot-voting or other methods. You might choose based on potential for impact, ease of implementation, user experience, or other criteria.

- For each favorite new idea, estimate the score your product would have in each ecodesign strategy with the new idea. Use the same spiderweb graph and same rating scale as before.

- *Note: your new idea might only improve the one strategy you brainstormed on, but it might also improve or worsen scores in other strategies. (For example, creating a takeback program to collect and refurbish 90 percent of your products for resale to new customers could arguably improve scores in "recovery for reuse" and "low-impact materials" and "clean manufacturing.")*

Checklist for Self-Assessment

To score your success on this exercise, see if you...

☐ *Scored your product/service on all eight ecodesign strategies.*

☐ *Listed succinct notes for why the scores were what they were.*

☐ *Listed what research you did to fill in knowledge gaps, and listed outstanding questions.*

☐ *Listed what ecodesign strategy you focused on.*

☐ *Brainstormed 10+ ideas for each sub-strategy, with 50+ ideas total.*

☐ *Scored your top few newly-brainstormed ideas on all eight ecodesign strategies.*

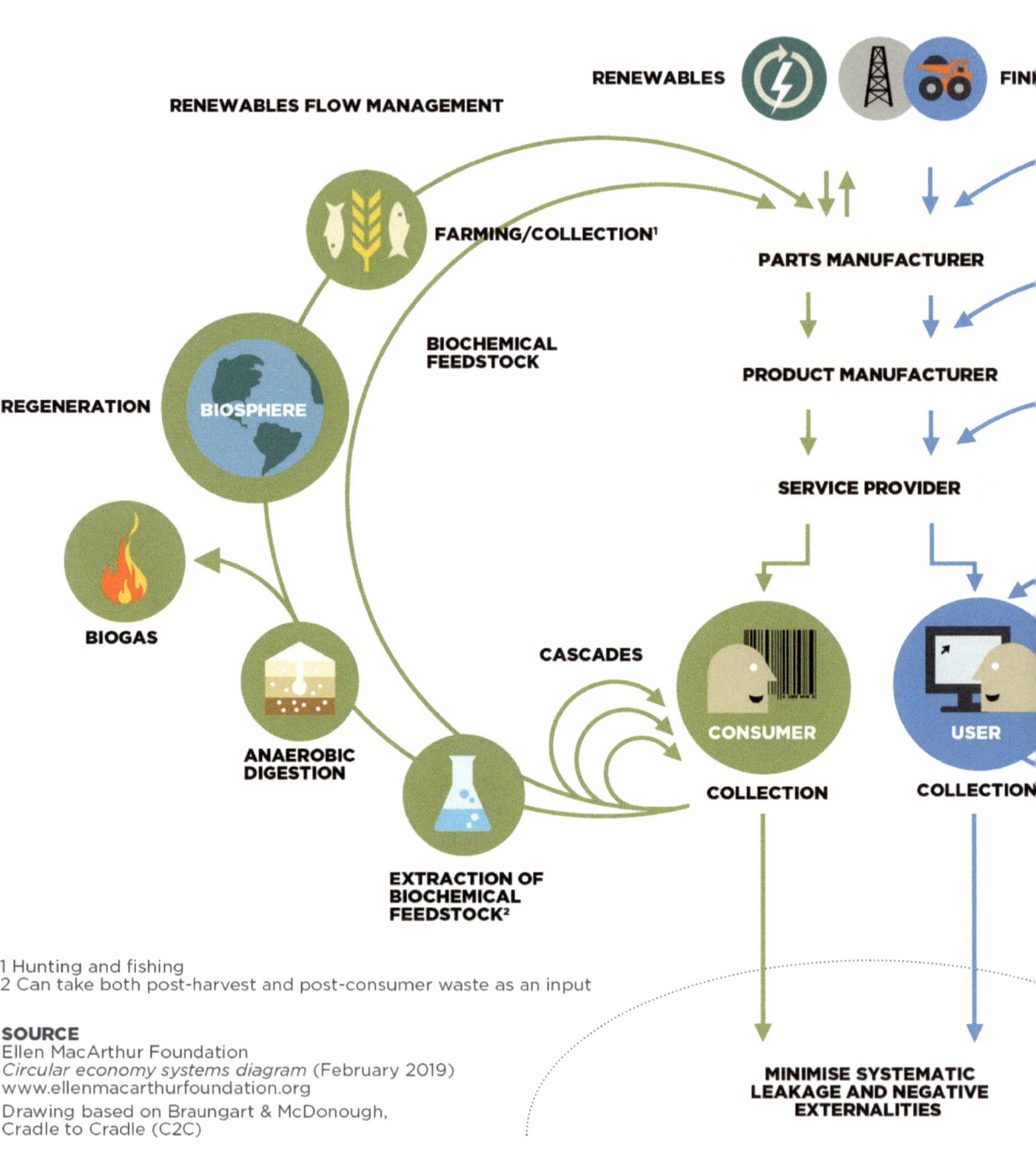

RENEWABLES FLOW MANAGEMENT

RENEWABLES

FIN

FARMING/COLLECTION[1]

PARTS MANUFACTURER

BIOCHEMICAL
FEEDSTOCK

PRODUCT MANUFACTURER

REGENERATION

BIOSPHERE

SERVICE PROVIDER

BIOGAS

CASCADES

CONSUMER

USER

ANAEROBIC
DIGESTION

COLLECTION

COLLECTION

EXTRACTION OF
BIOCHEMICAL
FEEDSTOCK[2]

1 Hunting and fishing
2 Can take both post-harvest and post-consumer waste as an input

MINIMISE SYSTEMATIC
LEAKAGE AND NEGATIVE
EXTERNALITIES

SOURCE
Ellen MacArthur Foundation
Circular economy systems diagram (February 2019)
www.ellenmacarthurfoundation.org

Drawing based on Braungart & McDonough,
Cradle to Cradle (C2C)

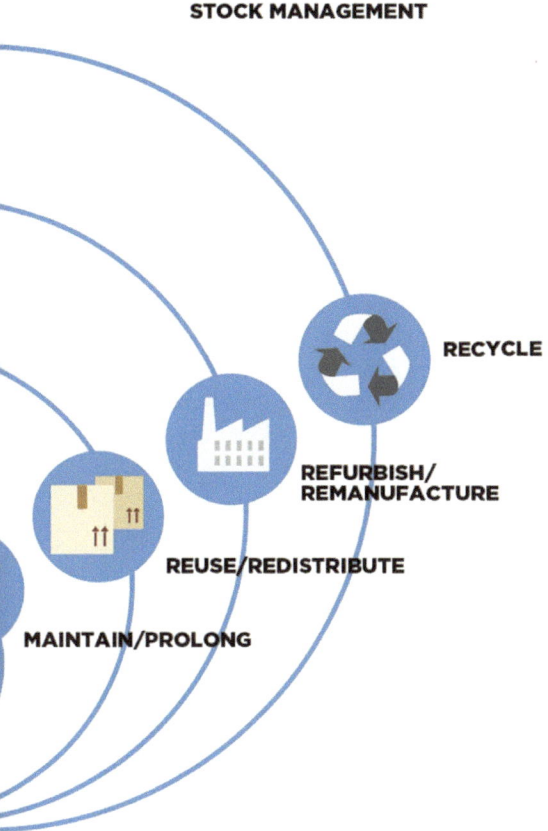

STOCK MANAGEMENT

RECYCLE

REFURBISH/
REMANUFACTURE

REUSE/REDISTRIBUTE

MAINTAIN/PROLONG

ELLEN MACARTHUR
FOUNDATION

Source: Ellen MacArthur Foundation (2019).

CHAPTER 4
Cradle to Cradle and the Circular Economy

*Conny Bakker,
Jeremy Faludi,
and Ruud Balkenende*

Goals

- Analyze how the Cradle to Cradle concept can be used in design
- Analyze the Circular Economy concept, its advantages and disadvantages, and its importance for sustainable design

DOI: 10.4324/9781003504672-5

Why It Matters

The Circular Economy is currently one of the most popular ways to design for sustainability. To understand the Circular Economy, and its advantages and disadvantages, we should first understand the Cradle to Cradle (C2C) concept. The Circular Economy was built on C2C's foundations, expanding it to consider the economics that drive design decisions.

Summary

- The Cradle to Cradle (C2C) concept takes natural systems as a source of inspiration, striving toward a positive vision where human society and industry regard waste as "food" for new cycles of use, and are powered entirely by renewable energy.

- Cradle to Cradle is not about efficiency but about effectiveness—waste need not be minimized if it is food for other cycles, providing positive regenerative impacts. However, this is difficult to achieve in practice.

- The Circular Economy concept elaborates on the Cradle to Cradle model, prioritizing longer product lifetime and recovering whole products over material recovery through recycling.

- The Circular Economy also combines ecological thinking with economic thinking, making a business argument for sustainability.

- Both concepts have been very influential in shaping the way we think about design for sustainability.

4.1 Cradle to Cradle

The Cradle to Cradle concept was popularized by architect William McDonough and chemist Michael Braungart in their book, *Cradle to Cradle: Remaking the Way We Make Things* (McDonough & Braungart, 2002). The book was an instant hit among designers and architects, and inspired many companies to change the way they designed buildings and products.

At its core, C2C has a radical proposition: it argues that doing "less bad" by creating efficient products and processes (as suggested by ecodesign) is simply not going to bring us sustainability. Instead, we should focus on doing "good" from the outset. Beneficial or regenerative design is the starting point of Cradle to Cradle. It emphasizes effectiveness, not efficiency: "doing the right things" versus "doing things right," because the latter can lead to optimizing fundamentally bad systems. They use the metaphor of a cherry tree: it manufactures its wood, flowers, leaves, and cherries out of air, sunlight, and local soil without depleting resources. Its blossoms and fruit feed animals and its branches provide shade and habitat. Even when the blossoms and fruits fall to the ground unused, they decompose to give their nutrients to microorganisms, plants, animals, and the soil. Even the "waste" is not pollution, it all contributes to a healthy ecosystem.

The three guiding C2C principles are:

1. **Waste equals food** (materials are nutrients that feed biological or technological cycles).

2. **Use current solar income** (use clean renewable energy, as natural life uses sunlight).

3. **Respect diversity** (take local contexts, resources, and culture into account when designing).

Once these principles are in place, the authors argue, and our products and materials are powered by renewable energy, are safe, and can be regenerated endlessly, there is no reason why we shouldn't have an abundance of them. It is no surprise that companies and designers like Cradle to Cradle!

4.1.1 Biological and Technological Cycles

The idea of "waste equals food" is the foundation of Cradle to Cradle. McDonough and Braungart based this principle on natural cycles of nutrient flows. In nature, these are closed loop cycles, meaning that there is, generally, no waste. Plants and animals are broken down after they die, releasing nutrients for new growth, etc., in endless cycles, powered by the sun. Cradle to Cradle asks industries to mimic these natural processes by creating products that regenerate endlessly, in either biological or technical cycles, see Figure 4.1.

The biological cycle is for organic materials (e.g., derived from plants or animals), used to produce so-called "consumption products" that can, after use, safely return to the biosphere (e.g., as compost). The technical cycle is for synthetic or inorganic materials (used in "service products") that cannot regenerate in the biocycle. Technical cycle materials also need to be safe and non-toxic, and it should be possible to recycle them in perpetuity without loss of material integrity or quality. This allows materials to be used over and over again instead of being "downcycled" into lesser products, ultimately becoming waste.

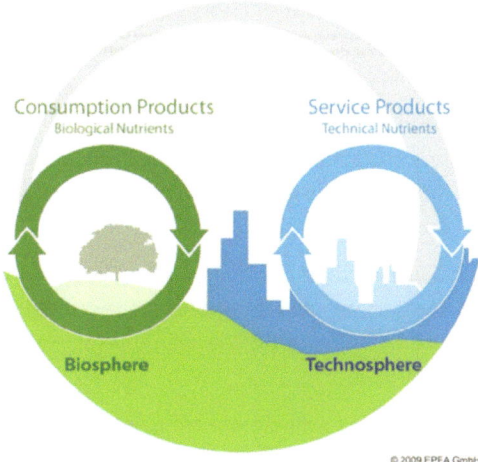

Figure 4.1 Cradle to Cradle's bio-cycle and technical cycle

Designers should create products made from either "technical" or "biological" materials and not mix the two, as this will result in "monstrous hybrids" that cannot become food in either cycle, according to the authors. The cost and difficulty of separation are some of the largest barriers to recycling and composting.

4.1.2 C2C Certification

Like the UN Sustainable Development Goals (SDGs), Cradle to Cradle's three principles are just goals; you need tools and strategies to achieve them. Also like the SDGs, the rest of this book's chapters provide such support. Besides this, companies can pay for a Cradle to Cradle eco-label certificate, which acts as both design guidance and metrics for success. Products and company operations are assessed to achieve different levels of certification based on five criteria: material health, product circularity, clean air and climate protection, water and soil stewardship, and social fairness.

- **Material health** ensures that materials are safe for humans and the environment, avoiding toxicity hazards.

- **Product circularity** drives reuse, remanufacturing, recycling, composting, etc., for regenerative product and process design.

- **Clean air and climate protection** drives companies toward net-positive renewable energy and carbon sequestration in materials.

- **Water and soil stewardship** safeguards clean water and ground around company and supplier factories.

- **Social fairness** drives human rights and a fair and equitable society around company and supplier factories, workplaces, and communities.

The certification includes many quantitative metrics, like a circularity metric for the percentage of material recovered, the percentage of heat and electricity from renewable generation, and the percentage of materials assessed for toxicity. It also includes qualitative metrics, like whether or not the company has a plan to switch to renewable energy or greener materials, and whether a social sustainability assessment has been performed. Certification is done through an independent non-profit, the Cradle to Cradle Products Innovation Institute. C2C certification criteria have been regularly updated over the years. It's interesting to see how the original C2C concept has been translated into five wide-ranging criteria that have some resemblance to the Sustainable Development Goals.

4.1.3 Limits of Cradle to Cradle

While the waste equals food principle is inspiring, it is very difficult for designers to realize this in practice. The recycling of plastics or paper, for instance, almost always results in degradation of material properties, which means that some form of downcycling over time is inevitable.

The idea of regenerative design also tends to misguide designers into thinking that renewable biomaterials can and should always be absorbed in the biosphere: many renewable materials are dyed, coated, mixed, blended, or otherwise treated in ways which can make them resistant to decomposition in the biological cycle, or worse, leave toxic contamination.

This brings up the question of how to deal with "monstrous hybrids." A cotton-elastane garment is a good example: while cotton is a renewable material (if not dyed), blending it with the synthetic material elastane (a polyether-polyurea copolymer) makes it unfit for the biological cycle; such garments should be treated in the technocycle. Unfortunately, elastane is notoriously difficult to recycle, which results in the garments being downcycled immediately. The same is true for renewable materials treated with toxic or non-biodegradable chemicals, as mentioned above. McDonough and Braungart would argue that cotton-elastane garments are monstrous hybrids and should be avoided— which is a fair point, except that the elastane enables the creation of high-quality, durable and comfortable garments that may last much longer than biocycle-only garments. How to deal with such tensions is not discussed in the C2C philosophy.

Finally, C2C's basic tenet is often contested, which is that as long as a product is regenerative and fulfills the C2C criteria, there is no limit to how much of it we can have. This is sometimes referred to as the "abundance" argument. Its attraction is the idea of no limits (endless consumption!), as opposed to the rather pessimistic "limits to growth" argument. It is a false idea, however—even perfectly safe and regenerative products that are powered by renewable energy rely on an extensive infrastructure (renewable energy production, transportation, etc.), which in its turn relies on large amounts of materials and energy. From a systemic perspective, limits to growth are very real; while we would certainly benefit from C2C products, it doesn't absolve us from having to address the very basis of our consumerist society.

4.2 The Circular Economy

While the concept of a Circular Economy is not new, it was successfully reinvigorated by the UK-based Ellen MacArthur Foundation in 2013, with their publication of "Towards the Circular Economy: An economic and business rationale for an accelerated transition" (Ellen MacArthur Foundation, 2013). The work of the Foundation found widespread uptake, especially in industry, leading to the European Commission developing a Circular Economy Action Plan to guide their sustainable product policy (European Commission, 2020).

According to the Ellen MacArthur Foundation, the Circular Economy challenges our linear "take-make-use-waste" system, and is based on three principles:

1. Eliminate waste and pollution.

2. Circulate products and materials (at their highest value).

3. Regenerate nature.

A circular economy should be built on renewable energy and materials, and its focus should be the decoupling of economic activity from the consumption of finite resources. According to the Ellen MacArthur Foundation, a successful circular economy is driven by design, as we urgently need creativity and innovation to transition away from the wasteful linear economy.

4.2.1 Circular Economy Principles

1. **Eliminate waste and pollution.** "Waste is a design choice," the Ellen MacArthur Foundation argues:

> There is no waste in nature, it is a concept we have introduced. From tiny, short-lived products, like crisp packets, all the way up to seemingly permanent structures like buildings and roads, the economy is filled with things that have been designed without asking: What happens to this at the end of its life?
>
> (Ellen MacArthur Foundation, 2022)

Products and materials can be looped back into the economy through maintenance, repair, reuse, refurbishment, remanufacture and recycling. Organic materials should go back to the biosphere, to become nutrients for new organic materials. It is easy to see the parallels to Cradle to Cradle.

2. **Circulate products and materials at their highest value.** An interesting distinction between Cradle to Cradle and the Circular Economy is the focus on high economic value loops. Where C2C is mainly concerned with the cycling of materials as "nutrients" for the biosphere or technosphere, the Circular Economy adds that these nutrients are also worth money, and they have much more value as products or components than as raw materials. A mobile phone worth

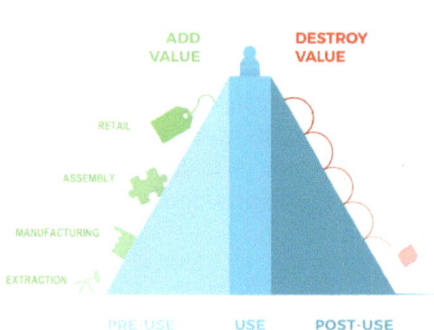 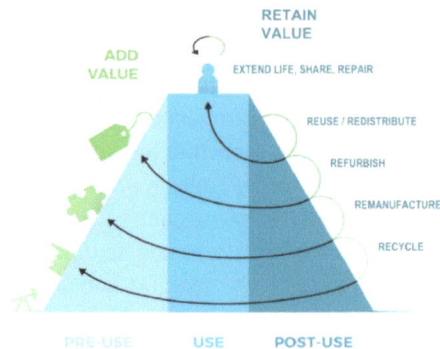

Figure 4.2 The Value Hill. Value is retained by keeping products in use for longer (at the top of the hill) and looping products and materials back through reuse, repair, refurbishment, remanufacture, and recycling.

Source: Elisa Achterberg (Circle Economy & Sustainable Finance Lab), Jeroen Hinfelaar (Nuovalente), and Nancy Bocken (TU Delft); (Achterberg et al., 2016).

€1000 might only contain €5–20 in gold, copper, lithium, etc. The vast majority of its financial value comes from all the manufacturing and quality control, etc., and while the financial cost only loosely correlates with the environmental impacts of manufacturing, there is certainly an economic incentive to avoid losing so much value. Roughly speaking, the more the whole product's integrity is maintained, the more economic value is also retained. Longer-lived products are best, followed by repair, remanufacturing, etc., and recycling is the last resort. The Value Hill (see Figure 4.2) illustrates this principle.

3. **Regenerate nature**. This principle moves away from products/materials and toward natural systems. According to the Ellen MacArthur Foundation, "regenerate nature" is about employing practices that allow nature to rebuild soils and ecosystems, recover biodiversity, and return biological materials to the earth. Food production is a case in point. Our global food system runs on synthetic fertilizers, pesticides, fossil fuels, and water. Soil health is ignored, excess

fertilizer runs off into the sea and causes dead zones through algae blooms, pesticides kill more than just the harmful insects, and climate change results from fuel use, tilled soil outgassing, and livestock. Regenerative farming practices can fix these problems by reducing reliance on synthetic inputs and by building healthy soils that absorb rather than release carbon. There is a long way to go, as regenerative farming practices currently make up a small percentage of total agriculture.

4.2.2 Decoupling

Decoupling is one of the goals of a circular economy. It means that, once a circular economy is in place, the economy would be able to grow without corresponding increases in environmental pressure. In a linear economy, increasing production (and thus growth in GDP) usually lead to increasing resource and energy use, and increasing waste and pollution. In such a linear economy, growth and environmental pressure are coupled, or linked. In a perfect circular economy, where waste no longer exists and products and materials are constantly looped back through reuse and recycling powered by renewable energy, decoupling becomes a possibility.

You may recognize the similarity with the "abundance" argument of Cradle to Cradle. Decouple, and you can grow. It is, however, not easy, in spite of many claims otherwise. An extensive review of the evidence of decoupling by Haberl et al. (2020) concludes: "large rapid absolute reductions of resource use and greenhouse gas emissions cannot be achieved through observed decoupling rates, hence decoupling needs to be complemented by sufficiency-oriented strategies and strict enforcement of absolute reduction targets." In other words: we need

to use and consume less. And while a circular economy will certainly help to reduce environmental pressure, it is no guarantee that it will help us reduce consumption and growth.

4.2.3 The Butterfly Diagram

The most-used visualization of the circular economy is the "butterfly diagram" (Figure 4.3). It combines C2C's concept of biological and technical nutrient cycles with prioritizing different levels of cycles, to retain more environmental and economic value. It is more complicated than meets the eye. In the middle of Figure 4.3, from top to bottom, is the product lifecycle: from raw materials production to end of life. The

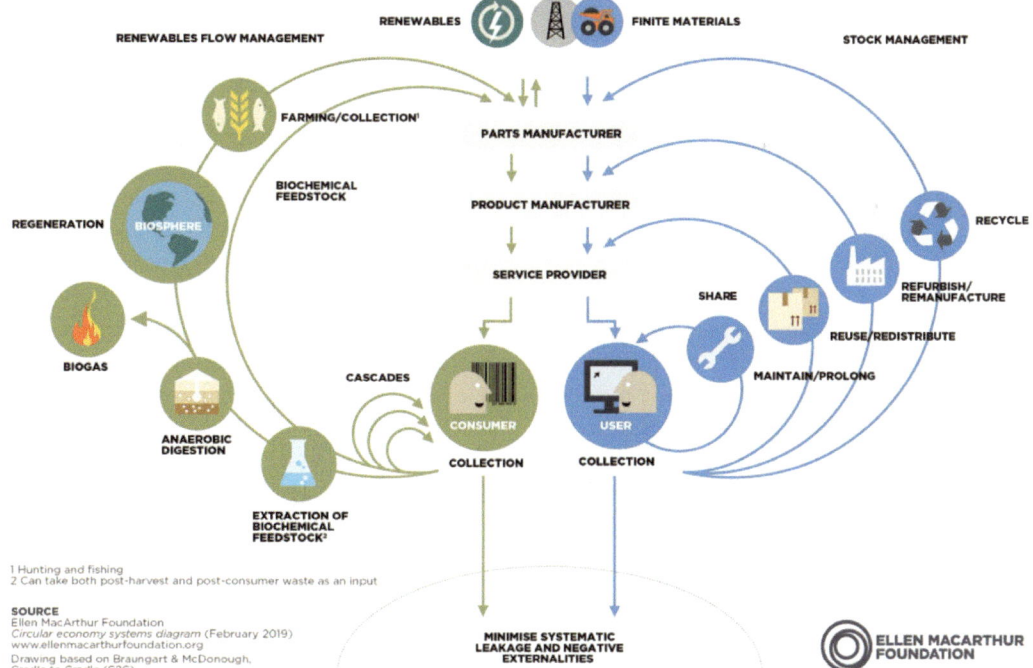

Figure 4.3 The butterfly diagram of the Circular Economy with its biological cycles ("renewables flow management") and technological cycles ("stock management")

Source: https://www.ellenmacarthurfoundation.org.

biocycle, in green, starts with renewable materials; the technocycle, in blue, deals with finite materials, just as in C2C. At end of life there is no "waste." The model asks you to minimize leakage from the system and avoid negative "externalities." Negative externalities are when the production or consumption of a product results in a cost to a third party that is not paid by the company or the customer (hence external). For example, cheap "fast fashion" satisfies customer desires and company profits, but the workers work in sweatshops for exploitative wages, and pollution and waste harm the people, plants, and animals living near the factories and waste dumps.

The butterfly diagram furthermore distinguishes between consumers and users. The idea is that in the techocycle, people are product "users"—they are temporary owners of the thing and get to use it for their benefit. After they're done with it, it has to be returned (collected) for reuse in the technocycle. Biocycle products are not used, they are consumed. Food, or biodegradable plastic bags are examples. After use, these products (or what's left of them) are collected to be processed in the biocycle.

Both cycles have multiple loops. In the technocycle it's easy to see the priority order: the smaller the loop, the higher the product value and the lower the impact of continued cycling. Sharing, maintaining, and prolonging the life of a product are preferred over the other loops because they are tighter loops. When they're no longer an option, then design for reusing components through refurbishment, remanufacturing, and parts harvesting. Finally, the longest loop is the last resort option: recovering raw materials through recycling.

In the biocycle, the loops are called "cascades." The idea is that a renewable material can go through several transformations before it ends up in (an) aerobic digestion or chemical recycling. For example, a bio-based polymer textile could be made from orange peels, but with completely different mechanical and aesthetic properties from an orange peel; unlike recycled steel or plastic, which aim to be as similar to the original as possible.

The biocycle, finally, has a few lesser-known processes. Aerobic digestion, or composting, happens when an organic material is broken down by micro-organisms under oxygen-rich conditions. Composting can be done at elevated temperature through an industrial process, or under ambient (normal) conditions in the environment. If composting is successful, biodegradable plastics, for instance, degrade into CO_2 and water. This is called "regeneration" in the butterfly diagram. Anaerobic digestion is the breakdown of organic materials in the absence of oxygen; here, materials break down into CO, CO_2, methane, and water. In an industrial process, methane can be captured and burned as biogas to replace fossil fuels. Finally, there are a number of thermochemical recycling processes (e.g., pyrolysis, gasification) that can yield gas and/or oil ("extraction of biochemical feedstock"). The feedstock can be used to create new polymers, or as energy ("biogas").

Overall, compared to Cradle the Cradle, the butterfly diagram creates a much more detailed (and economically viable) picture of how the biocycle and technocycle could be made to work in real life. It shows how the circular economy has taken the ideas of C2C and built upon them to create this new interpretation.

4.2.4 Criticism of the Circular Economy

Like all highly visible concepts, the Circular Economy has drawn its share of criticism. We've already addressed the difficulty of achieving decoupling, even if we could have a perfect circular economy. Many argue that instead of perpetual growth, we need to shrink economic activity in developed countries, aiming for "degrowth" (Schmelzer et al., 2022).

Other critiques focus on the technocratic nature of the Circular Economy, with its strong focus on material flows and without consideration of global equity and social sustainability issues. Opponents also find the concept of Circular Economy too open to multiple interpretations, some of which are too focused on economics. As long as this remains unclear, there is the risk of simplistic interpretations and greenwash. Many companies have jumped on the CE bandwagon by claiming that they use "recyclable materials," which are a pale shadow of real circular economy goals, because the ability to be recycled says nothing about whether they actually will be recycled.

4.3 R-Strategies

While not part of the original Ellen MacArthur descriptions of the Circular Economy, later academics and some European governments have combined eco-efficiency principles with the Circular Economy's eco-effectiveness for a more comprehensive list of sustainability strategies. Originally called the "three R's" (Reduce, Reuse, Recycle), these grew to "9 R's" or "10 R's" and are usually just called "R-strategies" to not bother with the exact number. Their exact definitions vary by source, but the 10 R's listed by the Netherlands Environmental Assessment Agency (Potting et al., 2017) are:

0. Refuse
1. Rethink
2. Reduce
3. Reuse
4. Repair
5. Refurbish
6. Remanufacture
7. Repurpose
8. Recycle
9. Recover

Strategies 3–8 are mostly the same as the Circular Economy strategies in the butterfly diagram, and in the same order. However, before those, the strategy "refuse" is a recommendation to avoid overconsumption, meeting human needs in different ways, either with existing products or without products. "Rethink" means redesign the product not for individual use, but to be shared by many users, or have one product perform the functions of several previous products, as in the Ecodesign Strategy Wheel's system stage. "Reduce" is designing for less material and energy use, as in the Ecodesign Strategy Wheel's first through fourth stages. "Repurpose" is also from Ecodesign. Finally, the last strategy "recover" is incinerating materials for energy recovery. This is even more of a last resort than recycling, but still better than incinerating without energy recovery, or landfilling.

Resources and References

Resources for Further Study

- Ellen MacArthur Foundation. (2013). Towards the circular economy Vol. 1: an economic and business rationale for an accelerated transition. Available at: https://www.ellenmacarthurfoundation.org/ towards-the-circular-economy-vol-1-an-economic-and-business-rationale-for-an

- McDonough, W., & Braungart, M. (2002). *Cradle to cradle: Remaking the way we make Things* (1st ed.). North Point Press.

- Cradle to Cradle certification tutorial on VentureWell Tools for Design and Sustainability. Available at: https://venturewell.org/tools_for_design/measuring-sustainability/cradle-to-cradle/

References

Achterberg, E., Hinfelaar, J., Bocken, N., Heideveld, A., Kerkhof, J., Fischer, A. & Ahsmann, B. (2016). Master circular business with the value hill. Available at: https://www.circonl.nl/ resources/uploads/2019/11/value-hill-white-paper.pdf

Ellen MacArthur Foundation. (2013). Ellen MacArthur Foundation, Towards the circular economy, Vol. 1: an economic and business rationale for an accelerated transition. Available at: https://www.ellenmacarthurfoundation.org/ towards-the-circular-economy-vol-1-an-economic-and-business-rationale-for-an

Ellen Macarthur Foundation. (2019, February). Circular Economy Systems Diagram. Ellen Macarthur Foundation. Available at: https://www.ellenmacarthurfoundation.org/ circular-economy-diagram

Ellen MacArthur Foundation. (2022, February 16). Eliminate waste and pollution. Available at: https://ellenmacarthurfoundation.org/eliminate-waste-and-pollution

European Commission (2020). *Circular economy action plan: For a cleaner and more competitive Europe*. Publications Office of the European Union. https://data.europa.eu/ doi/10.2779/05068

Haberl, H., et al. (2020) A systematic review of the evidence on decoupling of GDP, resource use and GHG emissions, part II: Synthesizing the insights. *Environmental Research Letters*, 15(6).

McDonough, W., & Braungart, M. (2002). *Cradle to cradle: Remaking the way we make things* (1st ed.). North Point Press.

Potting, J., Hekkert, M. P., Worrell, E., & Hanemaaijer, A. (2017). *Circular economy: Measuring innovation in the product chain.* Planbureau Voor de Leefomgeving, 2544.

Schmelzer, M., Vetter, A., & Vansintjan, A. (2022). *The future is degrowth: A guide to a world beyond capitalism*. Verso Books.

How to Apply #4: Brainstorm Using Circular Economy Principles
Time Estimate: 1–2.5 Hours

Work with the circular economy Value Hill diagram to brainstorm design, process, or business model changes for your product. The diagram creates a simple template for ten brainstorms.

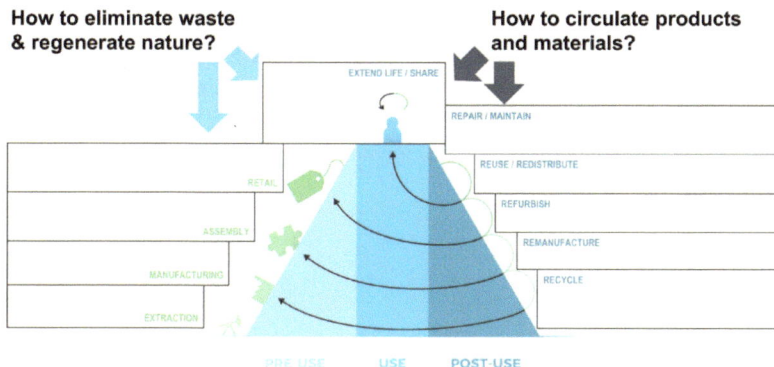

STEP 1: Generate Use Phase Concepts
Time Estimate: 10–15 Minutes

Brainstorm how to lengthen life, share, and otherwise extend your product's use, slowing its journey to its end of life, and write down the ideas that you generate. Strive for 10 or more ideas; the more, the better. Be specific and concrete, it gets you more solutions from the same general ideas. However, defer judgment while brainstorming—don't be afraid of proposing wild ideas too. Your ideas may not all fit in the box in the figure; if not, just list them somewhere clearly connected to the Use box.

STEP 2: Generate Post-Use Phase Concepts
Time Estimate: 10–30 Minutes

Brainstorm how to tighten the circularity loops in the five boxes of your product's "post-use" phase: repair, reuse, refurbishment, etc. How can you circulate the whole product, its components, or its materials to reduce waste and maximize the value they retain, perhaps even increasing value? Think about how the post-use phase corresponds with pre-use stages through the value loops proposed in the diagram (e.g. Refurbish to Retail).

Strive for five or more ideas per box, with five or more ideas circulating not just materials but components or the whole product. Again, the more ideas, the better; be specific and concrete. But again, defer judgment and encourage wild ideas. If your ideas don't all fit in the boxes, just list them somewhere clearly connected to each box.

STEP 3: Generate Pre-Use Phase Concepts
Time Estimate: 10–30 Minutes

In addition to your post-use recovery ideas, brainstorm new ideas in all of the four boxes of your product's "pre-use" phase: extraction, manufacturing, etc. How can the production of your product or service add value to what's now waste, recovering or repurposing materials, components, or whole products, or regenerating nature through its production?

Strive for five or more ideas per box, with five or more ideas regenerating nature or adding value to waste. These can build on your post-use phase ideas, or they can simply be about production processes. Again, the more ideas, the better; be specific and concrete, but defer judgment. If your ideas don't all fit in the boxes, just list them somewhere clearly connected to each box.

STEP 4: Choose Your Favorite Idea to Develop More
Time Estimate: 20–45 Minutes

Choose an idea, or set of ideas that work together, and develop it more. Are changes needed in your product's materials, architecture, user experience, manufacturing, or business model? Sketch or briefly write the design changes and how they would impact factors like cost, production, and consumer behavior.

Checklist for Self-Assessment

To score your success on this exercise, see if you…

- ☐　*Had 10+ ideas for the use phase.*
- ☐　*Had 5+ ideas for each box in the post-use phase.*
- ☐　*Had 5+ ideas for each box in the pre-use phase.*
- ☐　*Had 5+ ideas circulating not just materials but components or the whole product.*
- ☐　*Had 5+ ideas regenerating nature or adding value to waste.*
- ☐　*Connected ideas across the pre-use and post-use phases.*
- ☐　*Chose one or more idea to develop more.*
- ☐　*Sketched or described design changes and their estimated impacts on cost, production, and use.*

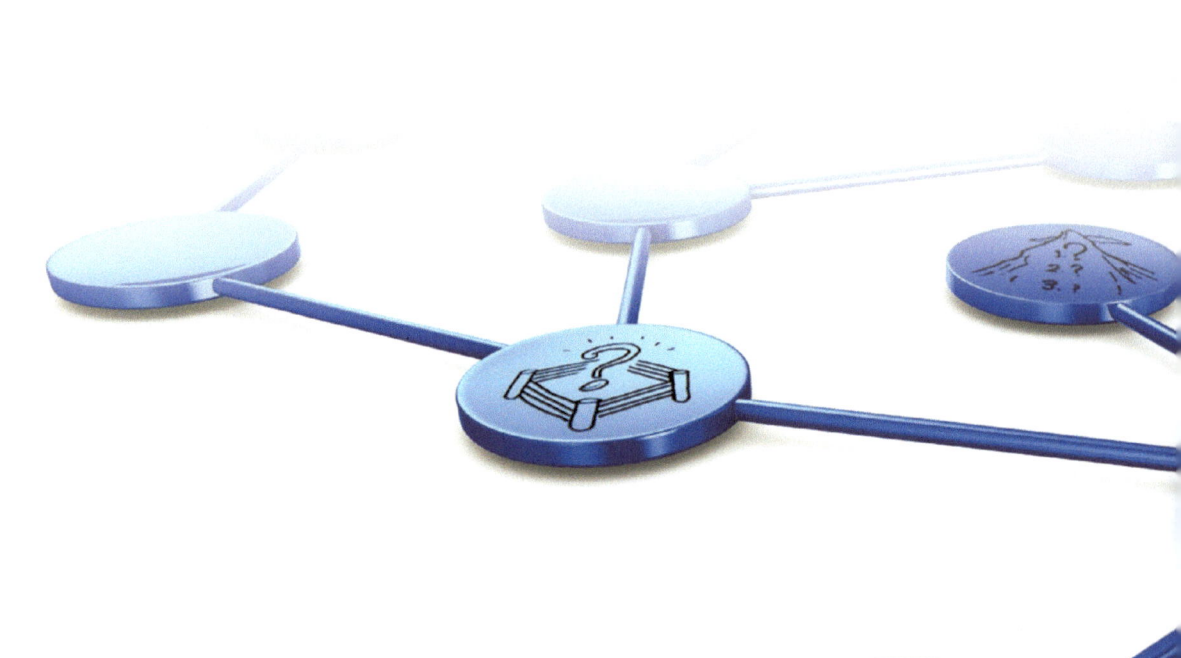

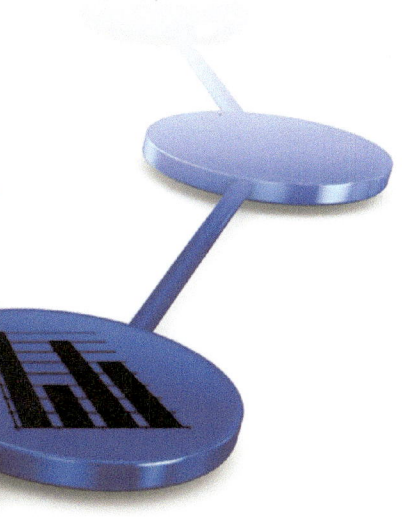

CHAPTER 5
Systems

Introduction

*Jeremy Faludi,
Ruud Balkenende,
and Conny Bakker*

Goals

- Describe the basic characteristics of a system
- Draw a system map
- Set priorities using evidence-based assessment
- Ideate solutions on the system map
- Decide on the best ideas using the earlier priorities

DOI: 10.4324/9781003504672-6

Why It Matters

Systems thinking helps you step back and look at the big picture to make sure you're solving the right problems. Whole System Mapping is one of the simplest methods for systems thinking, making it easier and faster to use. It also integrates into the design process better than many methods, because it not only provides analysis but integrates the analysis into better brainstorming and decision-making. It can also help serve as a framework for the design process, including other tools and methods within it.

Summary

- At their most basic, systems have parts, relationships, and boundaries. They can get much more complex.

- Systems thinking can mean design of a system, designing something while considering the system around it, or designing something to change a larger system.

- Whole System Mapping (a simple method to integrate systems thinking into design) is a four-step process: (1) draw the system map, (2) set priorities, (3) brainstorm on the system map, and (4) decide.

- Whole System Mapping's brainstorming on the system map should include at least one idea for everything on the map, and several ideas to eliminate parts of the system.

5.1 What Are Systems?

Systems are integrated sets of things, people, processes, etc., that have some effect on the world beyond what the individual components do. Much of the value of the system is not in the individual components, but in the relationships between them. For example, one telephone is not useful, its value lies in its ability to connect a person to other people through other telephones. Systems can be nested, where the system of a product may be part of an encompassing system, and may have other systems within it. For example, a refrigerator is a system, and it is one part of the encompassing system of the kitchen, and part of the encompassing system of the regional electricity grid, and part of the encompassing systems of global society and environment. When you think in systems, you have to decide what scale of system(s) you're thinking about.

5.1.1 Parts of a System

Systems have multiple parts interacting with each other; these include physical parts, events, processes, people, and others. For example, a refrigerator's immediate product system has several physical components (doors, insulation, a thermostat, etc.). In its encompassing system, it also has events (the user putting food inside) and processes (manufacturing the refrigerator), people (the users, the purchaser, the factory workers), see Figure 5.1. Events or processes should include all lifecycle stages of the product: material extraction, manufacturing, transport, use, and end of life. You may not need to model people, but failure to include people (especially their irrational desires) is a common cause of failed systems thinking.

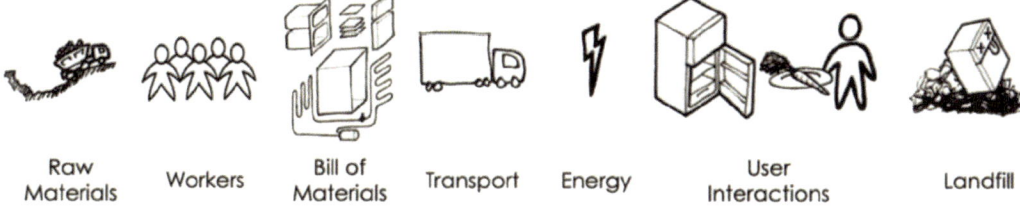

| Raw Materials | Workers | Bill of Materials | Transport | Energy | User Interactions | Landfill |

Figure 5.1 Some parts of a refrigerator's system, including hardware, events, and people

5.1.2 Relationships

Optimizing parts does not automatically optimize the whole; sometimes it does the opposite. The relationships between parts of a system can be more important than the parts themselves. As Figure 5.1 shows, relationships include cause and effect; nearness or distance in space and time; interpersonal relationships between people, or between a person and things or events; and more. A single part or event may have multiple causes and multiple effects. For example, for a mobile phone to make a call (one event), the phone needs power, the number of the person to call, and a connection to a cell tower (which is a whole network). If the system is missing any one of these components, it prevents the call. Likewise, having the user press the call button sets in motion a whole cascade of events in the phone software and hardware, as well as on the cellular network, in order to connect the call. In turn, the person being called may take actions based on the call, such as looking in the refrigerator for food the caller asked about.

When designing systems, set your performance objectives at the whole system level, not at an individual component level. If separate teams work on separate parts of the system, make sure their performance objectives serve the whole system, not just their parts. For example, ecodesign might suggest improving manufacturing impacts by making a refrigerator with cotton batting insulation instead of polyurethane spray foam. However, the vast majority of a typical refrigerator's total lifetime environmental impacts are actually from electricity use during its life, not manufacturing, and good polyurethane insulation has more than double the insulation value of cotton batting. Thus, while polyurethane foam is worse for manufacturing impacts, it lowers total lifetime environmental impacts greatly. The insulation team should first prioritize energy performance, and only after that is it worth their time and money to reduce the impacts of the insulation material.

5.1.3 Boundaries

Systems have boundaries—some things are part of the system, others are outside the system. If the system is "closed," the system does not depend on anything outside and does not affect or contribute to anything outside. However, for sustainable design, almost all systems you consider will be "open" systems (shown in Figure 5.1), requiring some kind of inputs from the outside and producing some kind of outputs (either beneficial or harmful) to the outside.

When you model systems, boundaries help determine how complicated and comprehensive your model is. Drawing the boundaries tighter makes it easier and faster to define your system, but you may miss important details. Drawing the boundaries wider can include more details but will increase modeling effort and uncertainty.

5.2 Modeling and Redesigning Systems

All models are wrong, but some models are useful. Different models are useful for different purposes. For example, a topological map of a landscape identifies peaks or valleys, and a street map navigates you there. To work with a system, deconstruct and define its parts, but then also reintegrate and synthesize them to understand the whole. Only looking at the whole is shallow, but only looking at parts misses the forest for the trees.

Many methods exist to analyze, build, and change systems. These fall into three categories, according to Mieke van der Bijl-Brouwer: (1) design of technical systems—designing a self-contained complex system, e.g., an airplane or a national electrical grid, (2) system-conscious design—designing a product or service to benefit an encompassing system, e.g., a refrigerator that benefits the environment and society, and (3) system-shifting design— designing a product or service to change an encompassing system, e.g., something that helps transition industrial agriculture to regenerative healthy food systems. Different system design methods are used for these. For example, this chapter's Whole System Mapping primarily applies to points (1) and (2). Chapter 6 mentions "systems engineering." usually just used for point

(1); the Systemic Design Toolkit (Jones & Van Ael, 2022), usually just for point (3); and System Dynamics/"12 Leverage Points" (Meadows, 1999), for all three.

The fastest and easiest way to model systems is by drawing qualitative system maps. More sophisticated methods are covered in Chapter 6. Qualitative maps allow you to simultaneously see the whole, the components, and their connections. Even the simplest maps help you understand, strategize, and communicate the system's parts, relationships, functions, etc. Mapping is especially effective when you involve multiple stakeholders with expertise from different parts of the system (e.g. mechanical engineer, electrical engineer, designer, marketer, supply chain manager, etc.). Involving multiple stakeholders helps everyone understand the entire system better by sharing their different points of view and aligning everyone in their understanding of shared goals. Qualitative visual maps like Figure 5.2 are used by the Whole System Mapping method, and others like "Gigamapping" (Sevaldson, 2011), to first understand problems and then to brainstorm solutions.

5.3 Whole System Mapping

Whole System Mapping is a very simple method, so it is relatively fast and easy to use, but it acts as a foundation to build deeper and more complex practices on top of it, whether they are systems-related or other sustainable design tools and methods. While most systems thinking methods only provide analysis, Whole System Mapping integrates analysis with ideation and decision-making.

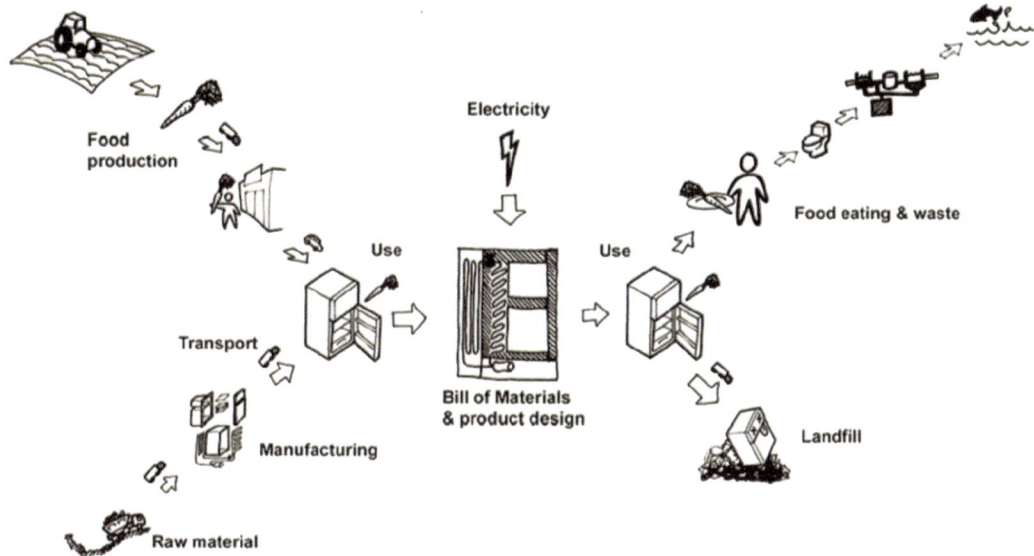

Figure 5.2 A qualitative system map of a refrigerator

Thus, industry designers, engineers, and managers value it for both sustainability and innovation, saying it broadens their creative scope, focuses thoughts on higher-leverage details, enhances collaboration across specialties, and helps converge on relevant solutions (Faludi et al., 2020). This also lets it serve as an overall framework for the design process, including other tools and methods within it (not only other systems thinking tools, but tools about materials, energy, etc.).

However, other systems thinking tools are better for some circumstances. Whole System Mapping usually focuses on a physical product, the user, and the product life cycle; not stakeholder relationships, socio-cultural concerns, feedback loops, paradigms, or leverage points. You can add these to your map to integrate 12 Leverage Points or the Systemic Design Toolkit, but they add complexity. Such tools help when you're not designing a product but a service or organizational change.

Whole System Mapping has four steps:

1. Draw a whole system map.

2. Set priorities based on a quantitative sustainability metric (usually LCA) plus business metrics.

3. Brainstorm solutions on your system map, for more thorough and radical ideas.

4. Decide on the winning idea(s) based on your priorities.

These steps are often iterative, especially steps two and three, with different brainstorms based on different priorities, sometimes redefining priorities or parts of the system map.

5.3.1 Step 1: Draw a Whole System Map

For product design, drawing the system map includes the product's main components, the product life cycle, the user interactions, other things the product is used with, and the connections between these. You don't need to include every nut and bolt, but at least all major subassemblies; you also don't need every event or process, but all major life cycle stages. There is no "right answer" to how much depth you should go into—tune your level of depth to your goals, your ability to redesign the parts in question, and how much time you have.

For service design or software design, the map is similar, with more focus on the user experience and other actors; it may also include relevant policies, training, and organization. Physical products and consumables, such as energy used by the services or software must be included, because most impacts arise from them. For example, a clothing retail app should include the clothes—they may cause the largest impacts in the system.

For product design, consider a refrigerator. The system map needs all the major subassemblies (Figure 5.3): structural frame, doors, exterior and interior skins, shelves, insulation, cooling coils inside and heat exhaust coils outside, driven by a compressor and valve that are controlled by a thermostat. You might consider that there is a cool section and freezer section.

Drawing the whole life cycle means summarizing the raw material mining/growing, manufacturing into the product, transporting materials to the factory and the product to the user, energy use during the fridge's life, and the end of life—landfill? Recycling? Other? This helps you consider the environmental or social impacts at all stages (Figure 5.4).

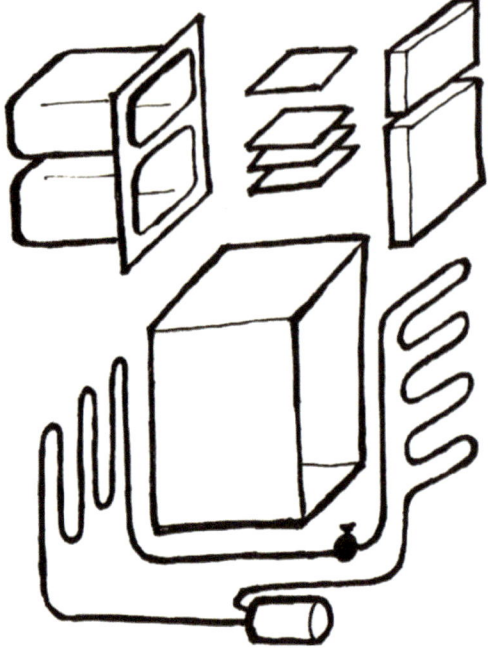

Figure 5.3 Basic parts (subassemblies) of a refrigerator

Next, draw the user interaction(s), to show the user needs. A refrigerator's purpose is to preserve the user's food. It does this by keeping food cold. The user interaction is the user opening and closing the door to put food in, or take it out. Opening the door results in a loss of cold air, requiring more energy to cool back down.

Next, draw other things your product is used with. With a refrigerator, it's the food. These other products can sometimes be more impactful than your product itself, especially if you're designing software or services. Finally, draw the connections between everything in the map. Figure 5.4 shows the flow of time with arrows, a good default way to connect everything. (You choose your level of detail.)

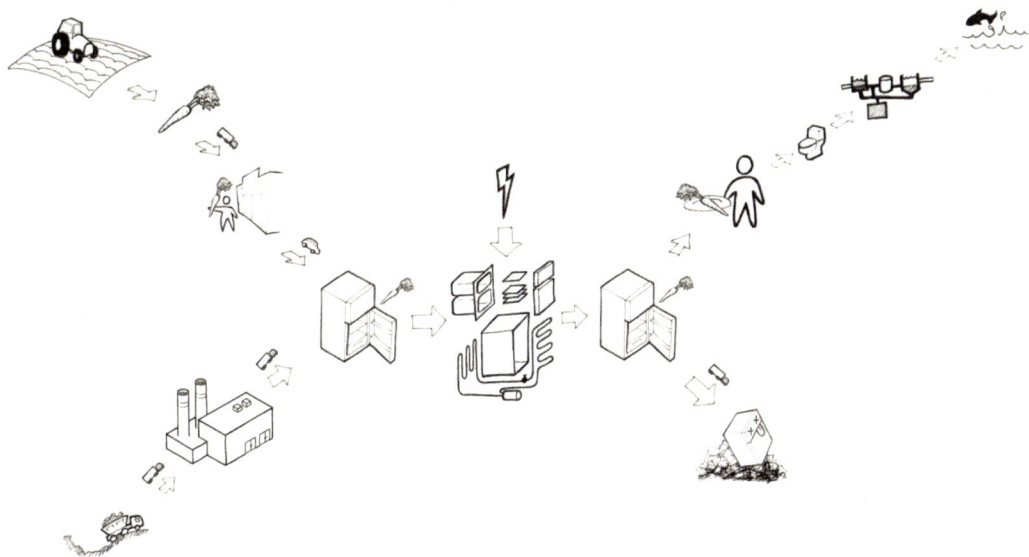

Figure 5.4 Refrigerator extended system map, with the life cycle of the food it preserves

Note, while drawing a system map, you might think of new ideas for improvement. That's fine, but set them aside for now, and bring them back in step 3 (ideation). For step 1, focus on the system *as it is now*. If you wish to integrate deeper system modeling, such as feedback loops or actor mapping layers, do that here.

5.3.2 Step 2: Prioritize Objectives by Assessing Impacts

Now that you see the system in the map, identify your greatest leverage points for improvement. This usually means identifying the biggest impacts and minimizing them, or seeing how they can become positive impacts. Include both sustainability priorities and business priorities. Determine your priorities not through guesswork, but quantitative analysis like life cycle assessment (LCA) and financial metrics. The

list can be as long and detailed as you want, but more focused problem statements make better brainstorms—two or three priorities are better than five. You can always focus on other priorities in future iterations.

Ideally, assess the impacts of your whole system using quantitative evidence-based metrics. For most products, LCA is the most objective and rigorous method, but it has its limits, so you might choose social impact metrics, or an eco-label certification, etc. You might compare against a benchmark: what's the best theoretically possible energy use for a refrigerator, or what's a natural organism whose energy use actually benefits the ecosystem, like a tree?

When assessing, you might decide some things in the system map can be ignored, because you have no influence on them. Figure 5.5 shows boundaries drawn so the

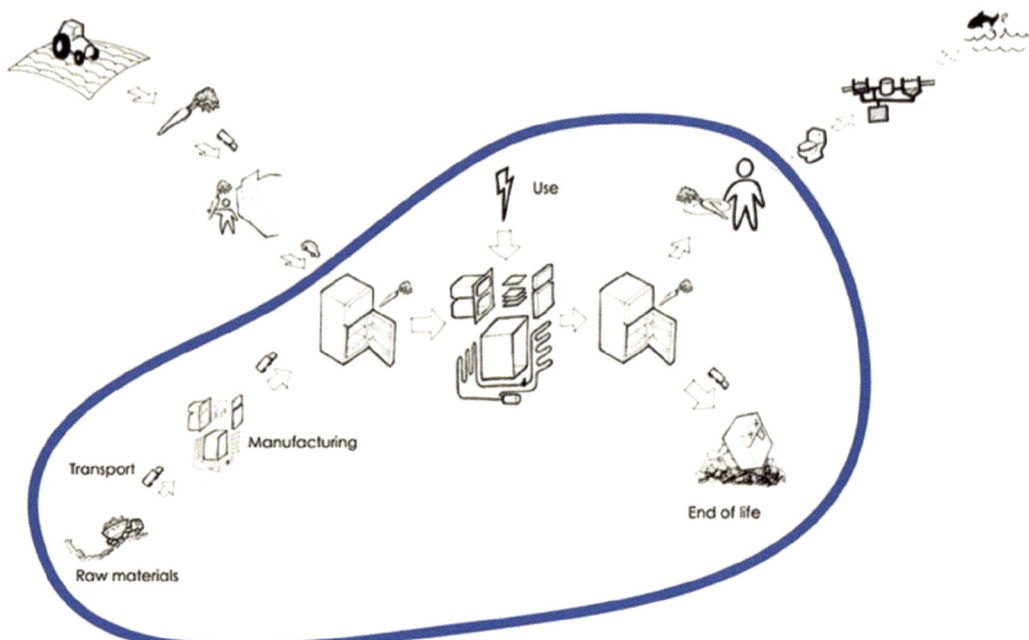

Figure 5.5 Refrigerator system map with boundaries drawn to show what is and is not assessed in the LCA

LCA will not count food impacts, only fridge impacts, because your company doesn't make the food. However, considering bigger systems can help you be bolder and make bigger change.

Figure 5.6's LCA shows that the biggest impact in a refrigerator's life cycle comes from energy consumption during use. Thus, the top priority is to improve the impacts of lifetime energy use. This becomes the main goal of step 3's brainstorm. You could also broaden your goal to "minimize overall LCA score," but it's usually better to focus the brainstorm more.

Include business or user need priorities, because a product or service that nobody buys is not sustainable. These help frame your brainstorm. For example, how much more expensive could an eco-refrigerator be and still have most

people buy it—10%? Zero? Also, what key user need must not be sacrificed to sustainability? Convenience?

Your final priorities might be:

1. Minimize energy impacts.

2. No added cost.

3. Same user convenience.

These priorities become the brainstorm question for step 3. In this case, "How might we minimize energy impacts without adding cost or user inconvenience?" You might even set thresholds for success, like cutting energy impacts 50%, zero cost change, and perhaps zero added time to access food. The point of adding such metrics is to drive yourself to bolder goals, not to settle for improvements of a few percent.

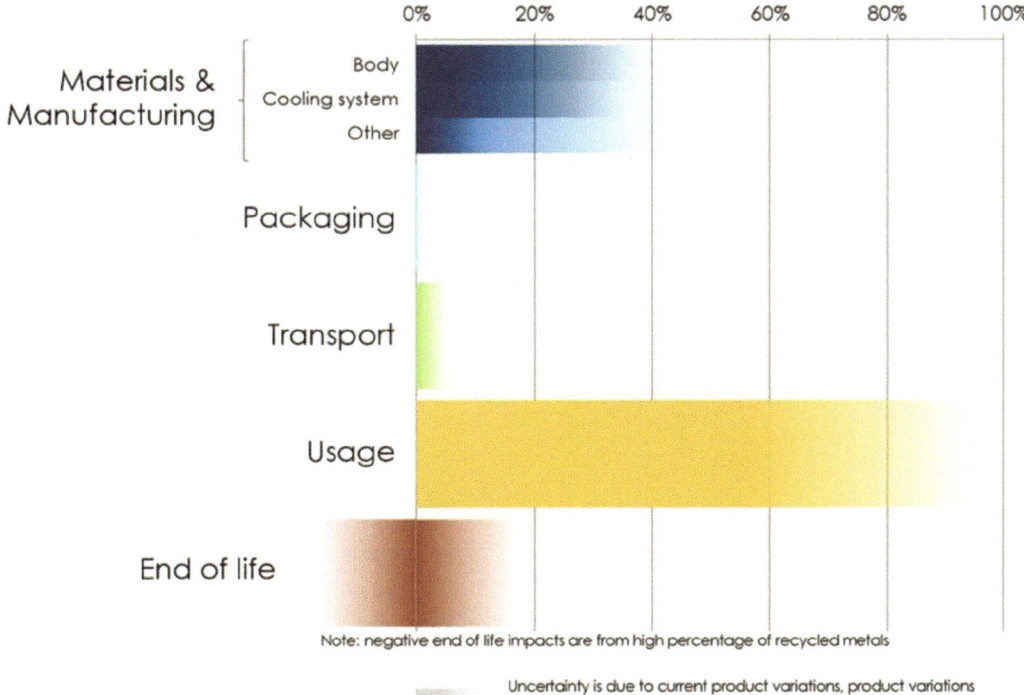

Lifetime impact by percent

Note: negative end of life impacts are from high percentage of recycled metals

Uncertainty is due to current product variations, product variations over time, and data limitations. When uncertainties overlap, consider addressing both impacts

Figure 5.6 Refrigerator LCA, showing large uncertainties (gradient blurs) because it averages several published LCAs

Use critical thinking—don't be a slave to assessment metrics, but use them to gain insight. Also beware of phrasing priorities in ways that limit your ideation: above, we did not say "reduce energy use" because that's just one possible solution. Sourcing renewable energy would reduce impacts without reducing the amount of energy used. Keeping your priority descriptions broad and centered around impacts, not solutions, means that your brainstorm in step 3 can be broader and more creative. As in step 1, you might have new ideas for improvement while formulating your priorities (like "double the amount of insulation in the fridge"). That's fine, just don't confuse them for priorities,

recognize them as solutions and set them aside until step 3.

5.3.3 Step 3: Brainstorm on System Map

Now you'll problem-solve! Brainstorming on the system map combines the big picture with specific details, both broadening and focusing your ideation. It avoids you fixating on one small part of the system, which is typical in most brainstorms. You force yourself out of that fixation by making sure you have new ideas affecting everything in the system map. Then you push for more radical ideas by having ideas that eliminate parts of the system map.

To brainstorm on the system map, first write your problem statement from step 2 somewhere on the system map you drew in step 1. As mentioned above, for a refrigerator it could be "How might we minimize energy impacts without adding cost or user inconvenience?" So, either digitally or using sticky notes, write or sketch your new ideas and place them on the system map next to the relevant part of the existing system.

Figure 5.7 shows the beginning of a brainstorm on a system map. It has ideas for some parts of the system but not others. Keep brainstorming until you have new ideas for every part of the system.

Try to have new ideas that affect multiple parts of the system, for example, changing the refrigerator's skin material might improve its insulation value for energy savings. Figure 5.8 shows ideas for reusing components or recycling materials, which avoid new material use. These don't appear to help with your top priority—energy use—but they could if they enable higher performance insulation that's more expensive but recovered at end of life. You can obviously spend more time on areas you think will have more leverage, but don't ignore "irrelevant" areas, the best ideas might surprise you.

When brainstorming, get specific about implementation to have more ideas—for example, rather than one idea "better insulation," think of several specific ideas: "polyurethane spray foam insulation,"

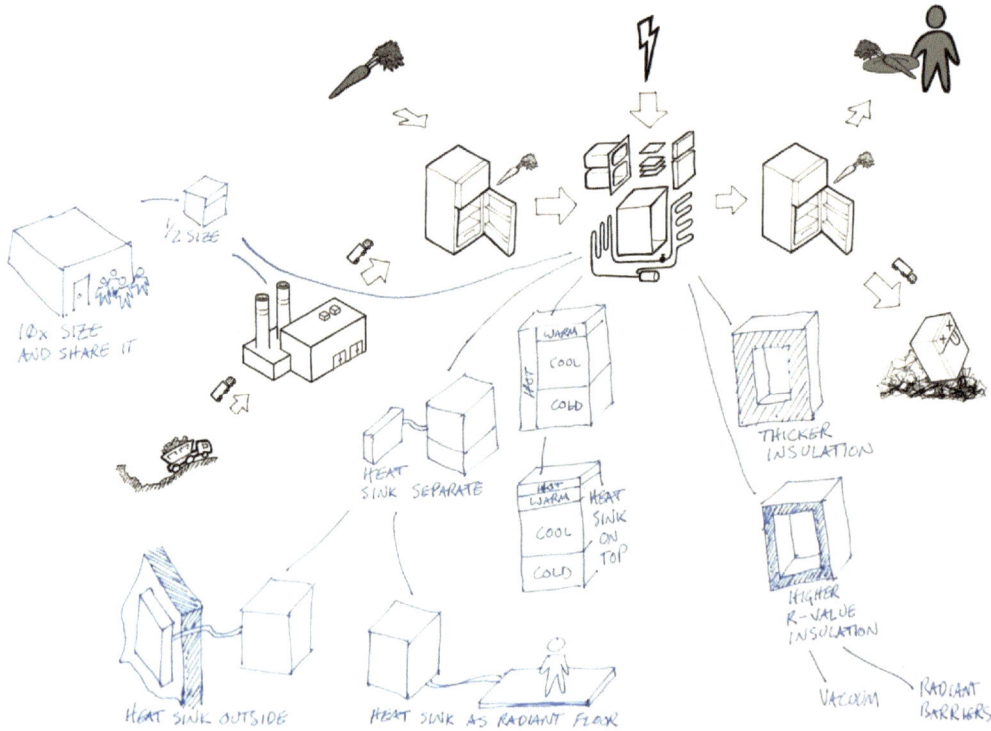

Figure 5.7 Brainstorming on system map, with new ideas for some parts but not others

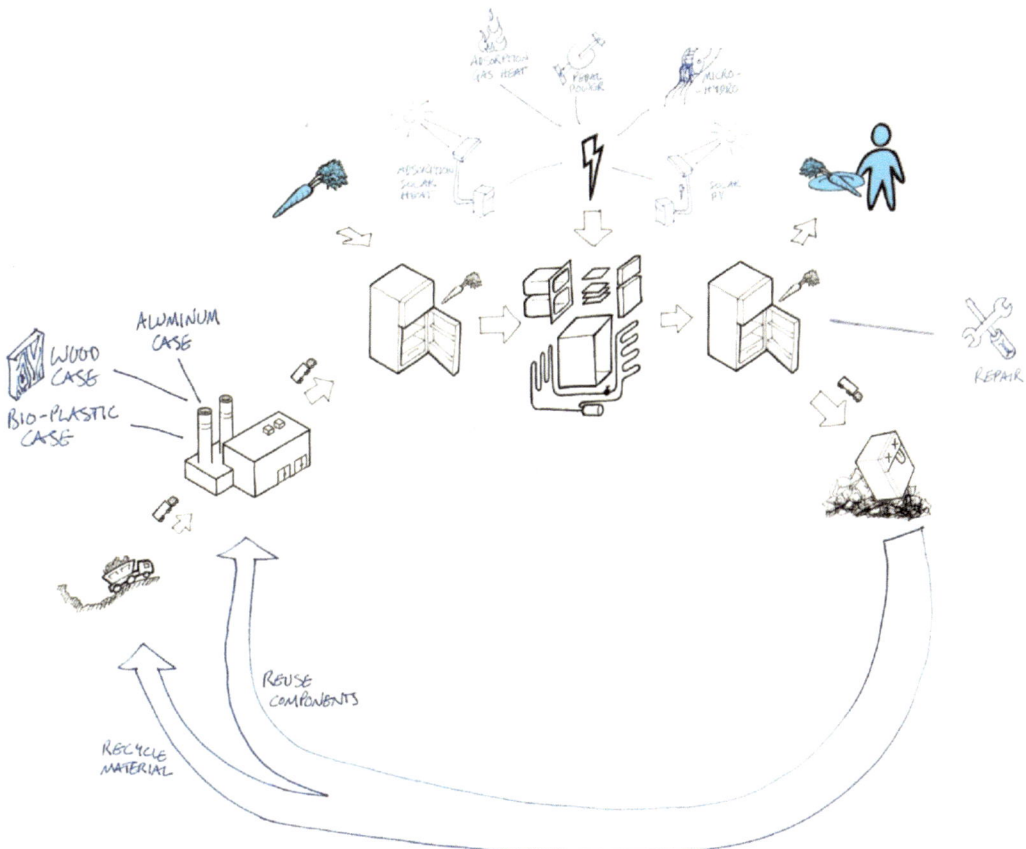

Figure 5.8 Continued brainstorming on system map, to cover more parts of the system

"recycled jeans insulation," "vacuum panel insulation," etc.

In addition to having new ideas for all parts of the system, have some ideas that *eliminate parts of the system*. Figure 5.9 shows growing food inside the fridge or preserving food by salting, drying, etc.

Eliminating parts of the system usually gives more radical innovations; they can also save money while saving materials, energy, or other impacts. For example, drying or salting food eliminates the entire product—this saves money for the user and saves

massive amounts of energy and materials. However, a refrigerator manufacturer would need to invent a new business model. Eliminating parts of the system can also help you think about reducing user consumption habits.

If desired, you can also integrate other ideation tools and methods here, such as the ecodesign strategy wheel, biomimicry, checklist-based prompts or card decks, or other systems thinking methods such as Factor Ten Engineering or 12 Leverage Points. They can take your ideation to the next level.

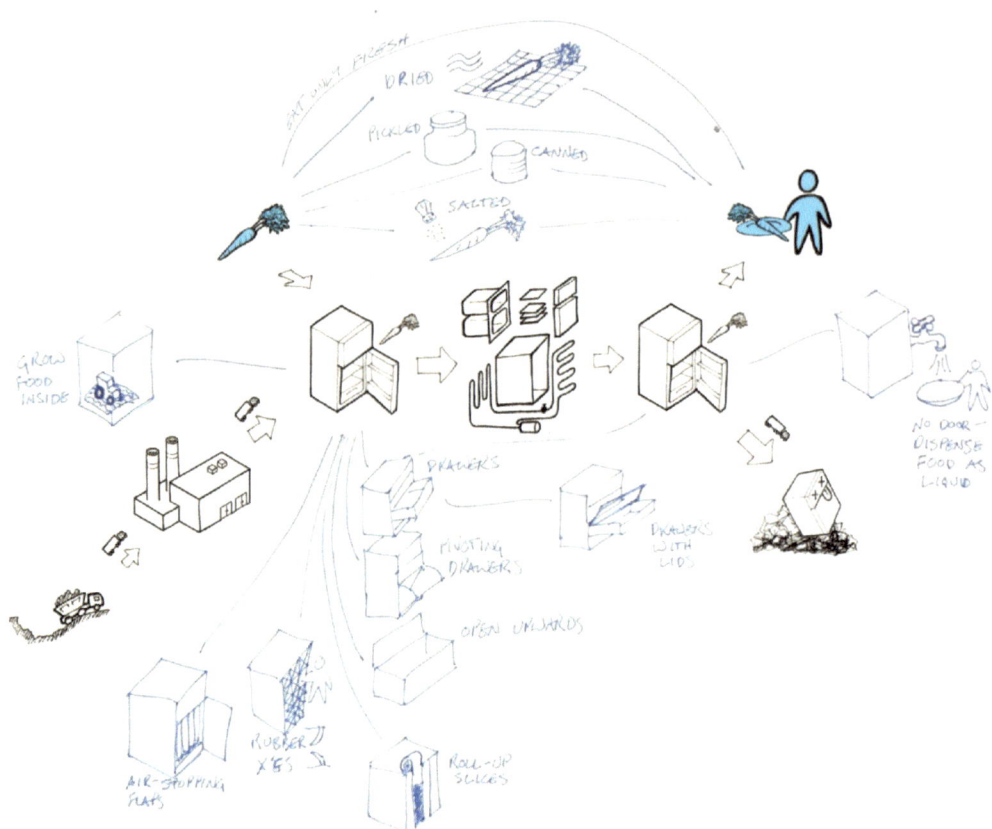

Figure 5.9 Brainstorming to eliminate parts of the system. Several ideas eliminate the product entirely

Remember the first rule of brainstorming: defer judgment. Write down even your craziest, stupidest ideas, and keep having more ideas; there are no bad ideas during brainstorms. After, in step 4, you decide between good and bad ideas. You might also iterate—reframe the problem, and redo your priorities and brainstorm, maybe even redo the system map.

5.3.4 Step 4: Decide by Priorities

Once you have lots of solutions, how do you choose the best? Decide based on the priorities you established in step 2. This keeps you on target for real improvement of the whole system, not greenwashing, if your priorities are evidence-based. Take only the best-scoring idea(s) into the rest of your product development process.

Scoring every brainstorm idea would take too long, and in any brainstorm, most ideas are bad (which is okay), so first use dot voting to narrow down to your favorite ideas. Dot voting means everyone votes three times with three dots. They can put one dot each on their favorite three ideas, or put all three dots on one idea. Either way, the ideas with the most dots "win." Narrow it down to three

Objective	Weight	Ideas			
		Mushroom insulation	Glass door	Drawers	Solar
Energy impacts	5	3	2	4	5
Financial cost	4	3	2	2	1
User behavior	4	3	5	4	3
Total Score		39	38	44	41

Figure 5.10 A decision matrix, scoring four brainstorm ideas according to three priorities

to six ideas, roughly. While dot voting, you may notice overlapping ideas and combine or differentiate them before voting.

Score your top ideas on all priorities from step 2, weighing the scores by the importance of priorities.

If you used LCA in step 2, do an estimated LCA on each new idea. Also estimate scores for your business priorities. One good way is to make a decision matrix like Figure 5.10, which shows a decision matrix with four new ideas, each scored on priorities from step 2. Each column is an idea, each row is a priority. The column "weight" shows how important each priority is. Note that weights are not ranking—a 1 weighs 1/5th what a 5 weighs, so would that really count as a priority at all? If you need refinement, you can use a scale of 1–10. In this table, the "Drawers" idea wins because it scores 5x4 + 4x2 + 4x4 = 44. Thus, the decision matrix helps you juggle many different measures of success at the same time.

Of course, don't be a slave to the decision matrix—critical thinking is still key. But don't use intuition alone. The decision matrix helps you consider and discuss all priorities in your decisions. In the end, decide which idea(s) to carry forward into product development only when they're best for the whole system, not just a part of it.

5.3.5 Step X: Iterate as Needed

No design process is ever this linear, but implementing these steps all together in one workshop or over the course of months in a larger design process should help you see the bigger picture, target the most important problem(s), break out of your normal boxes to find creative new ideas, and pick the best ones to pursue further. Especially iterate if you feel the need for greater depth, perhaps adding other systems thinking tools for a more complex system map, or adding other design methods to the ideation step.

For example, step 1's system map can include more advanced system modeling (see Chapter 6), or a Customer Journey Map for a service, or other methods. Step 2's priorities can be done with LCA, circularity scorecards, point-based eco-certifications (e.g., Cradle to Cradle or EPEAT), or others. Step 3's brainstorming can be enhanced by additional ideation methods mentioned above. Step 4's deciding should always use the same assessments as in the Priorities step, but can also add more.

You might do weeks of iterations, and do it at several points throughout the design process. It helps you integrate evidence-based decision-making into your early-stage design, and helps you connect low-level design decisions with big-picture priorities.

Resources and References

Resources for Further Study

- Meadows, D. (1999). Leverage points: Places to intervene in a system. The Sustainability Institute. Available at: https://donellameadows.org/archives/leverage-points-places-to-intervene-in-a-system/

- Rocky Mountain Institute (RMI). (2010). Factor Ten Engineering Design Principles. Available at: https://rmi.org/insight/ factor-ten-engineering-design-principles/

References

Faludi, J., Yiu, F., & Agogino, A. (2020). Where do professionals find sustainability and innovation value? Empirical tests of three sustainable design methods. *Design Science*, 6.

Jones, P. H., & Van Ael, K. (2022). *Design journeys through complex systems: Practice tools for systemic design*. Bis Publishers.

Meadows, D. (1999). Leverage points: Places to intervene in a system. The Sustainability Institute. Available at: https://donellameadows.org/archives/leverage-points-places-to-intervene-in-a-system/

Sevaldson, B. (2011). GIGA-Mapping: Visualisation for complexity and systems thinking in design. *Nordes*, 4.

How to Apply #5: Apply Whole System Mapping to a Product Redesign
Time Estimate: 2–4 Hours

STEP 1: Create a Whole System Map of Your Product
Time Estimate: 30–60 Minutes

Sketch a mind-map of your product's (or service's) whole system. The map needs to include the following features:

- The product's major components and how they connect (a full Bill of Materials, or at least major subassemblies; if a service or digital product, what hardware is used?).

- The product's full life cycle (raw material, manufacturing, transport, use, disposal).

- When/why/how the user uses the product or service.

- How the system's nodes are connected—lines or arrows to show the flow of time or material or logic.
- If other products or infrastructure are always used with your product or service (like clothes for a dryer, or roads for a car), include them in the map, too.

- Sticky notes on a whiteboard or big sheet of paper are best, or their digital equivalent, so that you can move things around as you build the map.

STEP 2: Set Priorities Based on LCA and Business
Time Estimate: 5–10 Minutes

To set priorities, quantitatively measure the biggest opportunities to improve your product. This exercise uses a basic estimated life cycle assessment (LCA). Learning LCA takes time, so here are some alternatives:

- Find an existing relevant LCA, such as those at https://ProductDesign.green.
- Explore certification systems (like Cradle to Cradle or EPEAT).
- Use one of the graphs below to fake it for this exercise.

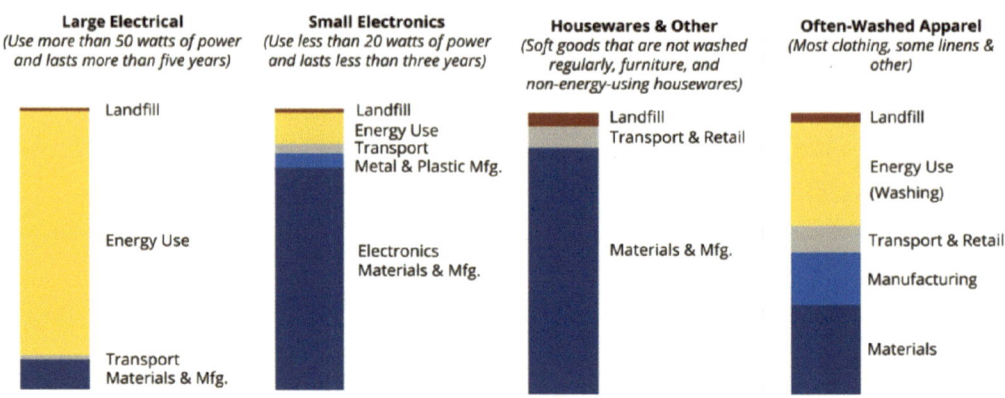

STEP 3: Brainstorm Solutions on Your System Map
Time Estimate: 20–40 Minutes

Now the fun part! Brainstorm on your Whole System Map from Step 1, to solve your top priority. For example, if your LCA shows that your product's energy use causes the most environmental damage and your business prioritizes low cost and visual appeal, your brainstorm's goal might be "Reduce lifetime energy impacts without compromising cost or aesthetics."

Ideally, brainstorm using a different color of sticky note than you did to make the system map, so you can tell your new ideas from the system components. Follow the rules of brainstorming to generate ideas: don't say no to any ideas, but also stay focused on your topic.

WELCOME WILD IDEAS

HEADLINE

GO FOR QUANTITY
WITHHOLD JUDGMENT

BUILD ON OTHERS' IDEAS

BE VISUAL

Use the whole system map to do the following:

1. Brainstorm at least one new idea for everything in the system map--every subassembly or step in the process (each sticky note). If that part of the system doesn't seem relevant to your priorities, have an idea that makes it relevant, connecting with other parts of the system. This gives you a more thorough brainstorm. You should have several ideas for the most relevant parts of the system.
2. Brainstorm ways to eliminate nodes in the system map. This process of elimination gives you more radical ideas: the more nodes you skip, the more radical the idea usually becomes.

If you want to get more advanced, you can integrate other systems-thinking methods into your brainstorm, or even use them to enhance your system map. Two particularly good ones are 12 Leverage Points by Donella Meadows and Factor Ten Engineering (10xE) by Rocky Mountain Institute.

STEP 4: Choose Winning Idea(S) Based on Your Priorities
Time Estimate: 10–40 Minutes

First, take your brainstormed list of ideas and narrow it down to 5–8 best options, considering what would best accomplish the priorities and delight users.

Next, measure each of your top options against your Step 2 top priorities for sustainability and business, using a decision matrix to weigh the different scores in different priorities. If you have LCA skills, do an estimated LCA on each for the environmental score. (Remember these results will have huge uncertainties, so only trust huge improvements!) As described in the chapter, don't mindlessly take the decision matrix's scores as final, but use it as a tool to help your critical thinking.

Finally, write down the winning idea (or combination of ideas). You could even illustrate it if you want. Now that you've finished the whole design method, you should have a clearer idea of the problem(s) you're solving and their context, a clearer set of priorities, lots of new ideas, and a decision of what to pursue further.

Checklist for Self-Assessment

To score your success on this exercise, see if you…

☐   *Illustrated all the product's major components/subassemblies and how they are connected.*

☐   *Illustrated the product's whole life cycle.*

☐   *Illustrated when/why the user uses the product or service.*

☐   *Illustrated what other things are always used with it.*

☐   *Illustrated the system with visual clarity (ideally even aesthetically).*

☐   *Quantitatively estimated the top sustainability priority (using LCA or other scorecard).*

☐   *Listed the top sustainability priority with business priorities, in order of importance.*

☐   *Generated new ideas for every node in the system map.*

☐   *Generated new ideas that eliminate nodes from the system map.*

☐   *Quantitatively estimated how well the new ideas achieve your listed priorities.*

☐   *Decided on idea(s) to move forward with, and described it/them succinctly.*

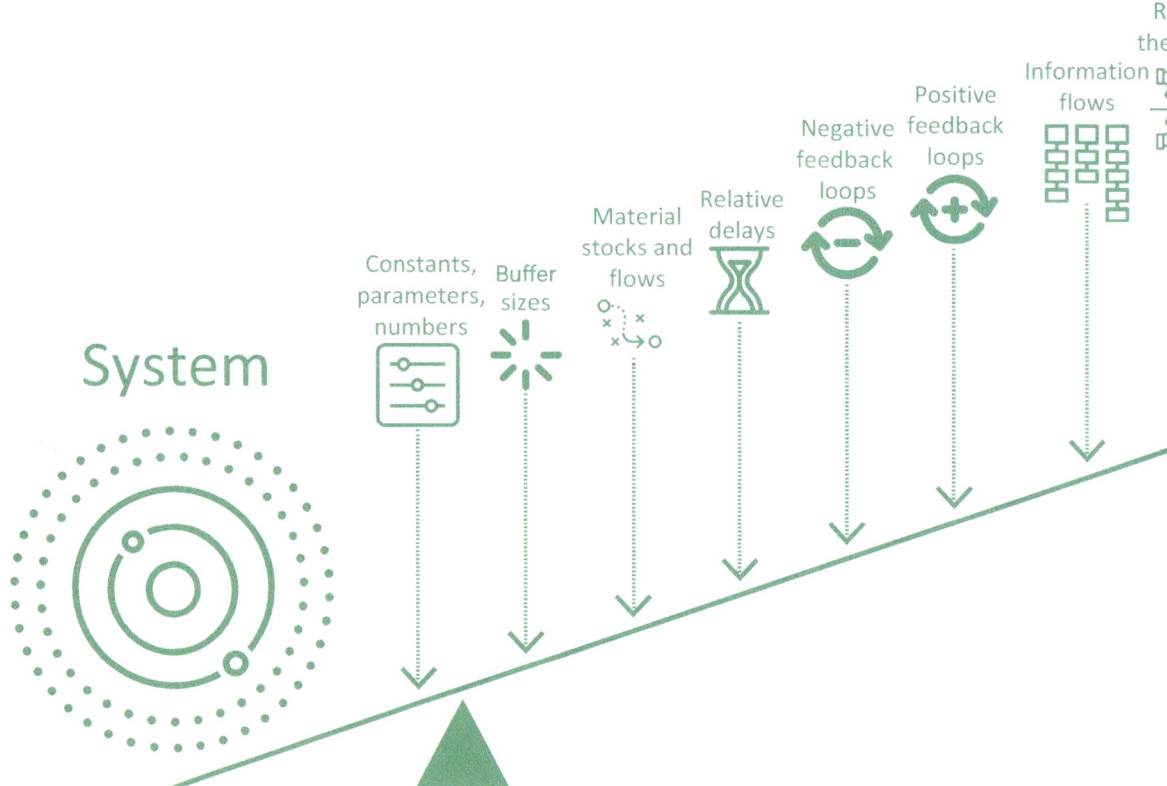

System

Constants, parameters, numbers

Buffer sizes

Material stocks and flows

Relative delays

Negative feedback loops

Positive feedback loops

Information flows

R●
the

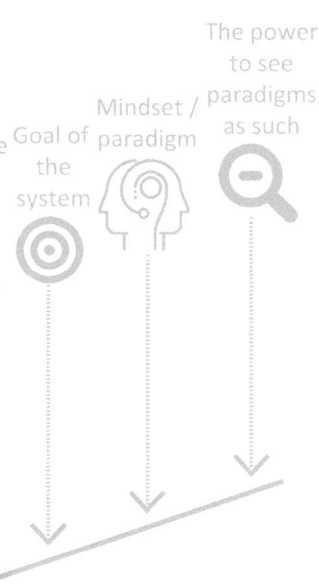

The power to see paradigms as such

Mindset / paradigm

Goal of the system

Source: Angheloiu & Tennant (2020).

CHAPTER 6
Systems
Complexity

Jeremy Faludi

Goals

- Differentiate stabilizing feedback from multiplying feedback

- List different levels on which systems can be understood and intervened in

- List which traits make systems resilient

- Ideate system solutions with the 12 Leverage Points

DOI: 10.4324/9781003504672-7

Why It Matters

Understanding systems at deeper levels can let you diagnose their problems better, and let you make deeper changes to them. Understanding dynamics of systems lets you understand how they will change (or not) over time. Understanding emergent properties and levels can help you avoid unintended consequences. Understanding resilience can help you create more resilient systems, or break undesired resilience.

Summary

- Dynamic systems have feedback loops (stabilizing and/or multiplying).

- Systems are greater than the sum of their parts, with "emergent" properties; thus, they can be modeled at different levels.

- One level of a system is the underlying paradigm motivating it; changing the paradigm can create a whole new system.

- Systems have inputs and outputs. Resilient systems keep giving the same output despite disruptions, due to their redundancy and/or flexibility. However, these traits may make them less efficient.

- Twelve Leverage Points is a method for designing deep system change.

6.1 Systems Deep Dive

Simple systems can be understood with simple system maps as shown in Chapter 5, but many sustainability issues are "wicked problems." Such problems have no one right answer because of complex interdependencies, often with contradictory or changing requirements that can't be fully characterized (Rittel & Webber, 1973). However, good and better solutions are possible, with cleverness and creativity. Such problems occur in complex systems. Complex systems not only have the parts, relationships, and boundaries of simple systems, but also dynamics and patterns from feedback loops; paradigms; levels arising from emergent properties; plus inputs and outputs with some resilience (Figure 6.1). All these will be described below.

6.1.1 Dynamics and Patterns (Feedback Loops)

Feedback loops are critical when making dynamic system models (Meadows, 2008). There are two kinds of feedback: stabilizing (negative feedback) and multiplying (positive feedback). The "negative" and "positive" are *not* bad and good, merely "less of the same" and "more of the same."

Stabilizing feedback loops drive the system to a steady state (Figure 6.2). The classic example is a thermostat—it senses the difference between a desired temperature setpoint and the actual temperature—the greater the temperature difference, the more the thermostat tells a heater or air conditioner to reduce the temperature difference. The temperature likely overshoots and then comes back toward the setpoint from the

Systems include...

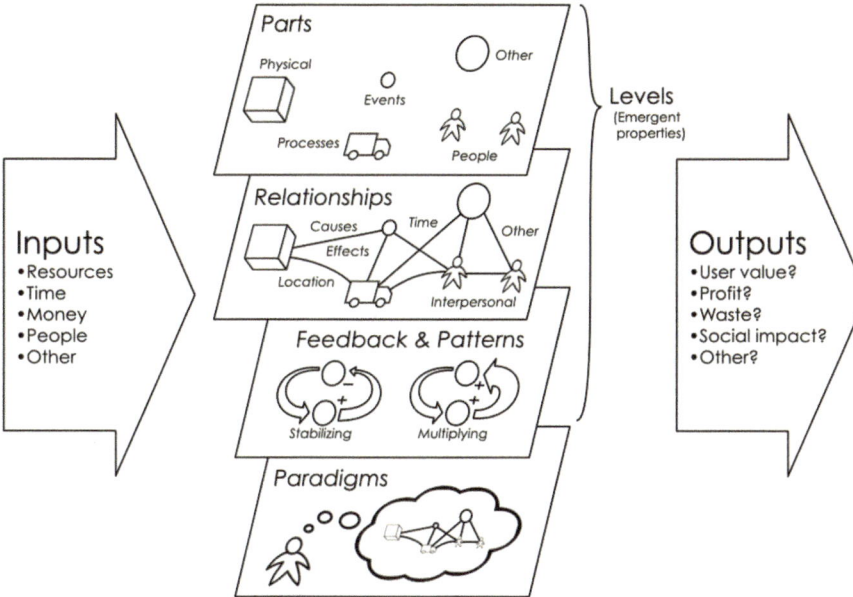

Figure 6.1 What complex open systems are made of

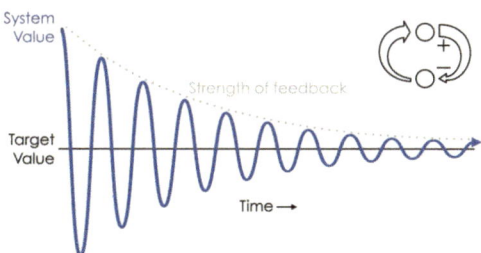

Figure 6.2 Stabilizing feedback

other direction, overshooting less with each cycle until it stabilizes, as Figure 6.2 shows.

For example, in economics, higher income taxes for people with higher incomes reduce the difference between high and low incomes to help stabilize the difference between rich versus poor. Such taxes may not have a consciously designed setpoint of desired income distribution, like a thermostat's

desired temperature setpoint, but they will still drive the system toward some steady state of income difference. Therefore, the intensity of feedback (how high the taxes are for what incomes) should be chosen carefully.

Multiplying feedback loops drive the system to do more of what it is already doing (Figure 6.3), hence the mathematical term "positive feedback," or the System Dynamics term "reinforcing feedback." For example, compound interest—the more money you have invested, the more interest you earn, which gives you

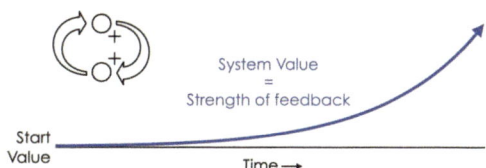

Figure 6.3 Multiplying feedback

more money, which earns more interest, etc. Multiplying feedback grows exponentially and runs away to infinity unless some other factor limits it. For example, a megaphone that hears and amplifies itself amplifying itself can emit a terrible shriek until you cover the microphone to dampen its input, bringing it back to a controlled state. In economics, tax breaks for the wealthy are a multiplying feedback loop that reinforce wealth inequality exponentially until outside forces intervene.

Again, the intensity of feedback matters. In megaphones, amplifier circuits have a resistor to tune their amplification to a reasonable level. In economics, an interest rate of 7% will double your money in 10 years, while 5% will double your money in 14 years. This "doubling time" can be a good measure of feedback strength.

6.1.2 Levels and Emergent Properties

What separates systems from collections of components is that the whole is greater than the sum of its parts. When the whole acts differently than you would predict from its components alone, that's an "emergent property"—a property the system has that the parts don't have. For example, a single ant is not very smart and not very strong by itself, wandering almost randomly. But ant colonies (Figure 6.4) can build complex mounds, organizing far-reaching trains to move large amounts of food back to their colony.

Because emergent properties usually can't be predicted by examining a system's parts, you don't necessarily need to know how

Figure 6.4 Ant colonies operate on a different level than individual ants

Source: Unsplash.

the parts work to understand the system. These are different "levels." Sometimes there are many levels. For example, human behavior when buying and using products emerges from the brain's network of neurons, emerging from each neuron's electrochemical switch, emerging from its biological structure, emerging from molecular chemistry, emerging from physics. The brain can be understood and intervened with on any of these levels. You can predict much human behavior without understanding neurons, much less chemistry or physics. But when you design a system, it may have emergent properties you never intended.

To account for unintended consequences in your system designs, prototype and simulate your system in as much detail as feasible, testing several options in both realistic and extreme scenarios to see what properties emerge. Then iterate based on the results, as many times as is feasible, to optimize your system's performance and resilience, discovering what you couldn't anticipate. This is the classic "design thinking" approach, used by hundreds of companies to develop thousands of products. For example, Nielsen Norman redesigned online interfaces for Danish Bank using four different tracks of design prototypes, with several iterative rounds of prototyping and user testing per track (Nielsen, 1993). One of the tracks improved customer satisfaction 41%, which sounds impressive, but their wide explorations with many iterations discovered a solution that improved satisfaction 242%!

System maps often only visualize a system at one level at a time, but considering the system at multiple levels can help acknowledge emergent properties, simplify complex systems, and help allocate tasks

to teams by subdividing the system. Two common approaches to levels are (1) systems engineering levels, and (2) business strategy "iceberg" levels.

6.1.2.1 Levels: Systems Engineering V Model

Companies creating complex products or technology systems, like airplanes or electricity grids, often use the systems engineering "V model" (Figure 6.5), with levels of concept, functional requirements, logical architecture, and physical architecture:

- The "concept" level defines the system's overall goals, and may define its inputs and outputs.
- The "functional requirements" level sets performance targets for user needs and other requirements and plans how they will be met.

- The "logical architecture" level defines how these requirements will be met in detail: what the inputs and outputs are; what tasks are done in what order and how they depend on each other; and otherwise defines how the system will function.

- The "physical architecture" level then plans how those functions will be performed by actual hardware, software, personnel, etc.

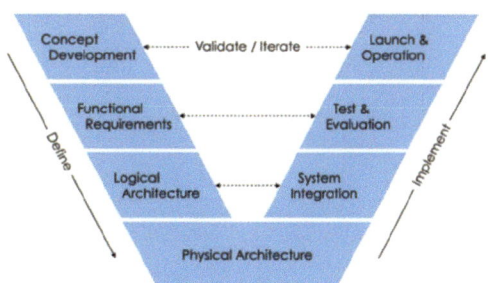

Figure 6.5 System engineering levels, shown in one version of the "V" model

Work begins at the concept level, flows down to the physical architecture level to define the system, then flows back up the levels to implement the plans (i.e., actually building and testing the system). Returning from the physical architecture level to the logical architecture level, the system's hardware and software are integrated to validate and possibly iterate the logical architecture level. Then climbing another level, the whole system is tested to validate (and possibly iterate) the functional requirements. Finally, the focus returns to the top level as the product or service is launched and commences operation; this step validates (and possibly iterates) the overall concept. For sustainable design, the launch and operation stage should also include maintenance, possible upgrade, and end-of-life considerations.

This is a top-down approach, favoring modular components; however, it can be iterative, as noted above. There are several variants of the "V" model; for details, see manuals by the United States government (U.S. Department of Transportation, 2007) and the German government (Graessler et al., 2018).

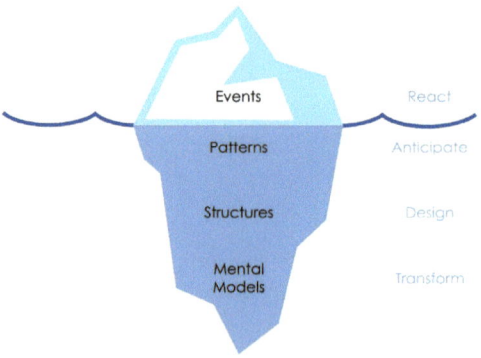

Figure 6.6 The iceberg model of systems

6.1.2.2 Levels: Iceberg Model

Policymakers, activists, and design consultants looking to redesign or influence existing systems often use Peter Senge's "Iceberg model" levels of Events, Patterns, Structures, and Mental Models (Senge, 1990). It's called an iceberg because only the top level is visible, with the rest below the surface at deeper levels of understanding (Figure 6.6). To understand this model, consider a factory:

• Events" are the visible observed phenomena or objects (e.g., water pollution from the factory).

• Patterns" are trends observed over time (e.g., increasing pollution from more factories).

• Structures" are the forces driving the patterns and relationships between parts of the system (e.g., demands for profit margins driving the factory to produce more waste than it can treat).

• Mental models" are the assumptions, values, and beliefs that keep the system locked in (e.g., the belief that company profit must be maximized without measuring costs to society and the environment).

Creative interventions can disrupt the patterns in any of these levels. For example, when pulling tomatoes from your refrigerator, if you hold the door open so long that the refrigerator dings at you, and you respond to the dinging by closing the door, you are *reacting* to an *event*. If you have learned how long you can keep the door open before the fridge dings and always shut it before then, you are *anticipating* a *pattern*. If you install strips of clear plastic inside the door so cold air doesn't leak out when you open the door, you *designed* a *structure*. If you decide that tomatoes are better eaten fresh and stored on the counter rather than being refrigerated at all, you *transformed* a *mental model*.

6.1.3 Mental Models

Mental models (paradigms) underlie all designed systems; we also understand natural systems through them. They are assumptions about how the world works, from empirically-measured phenomena like gravity to cultural assumptions like the belief that profits must be maximized. They include rules or laws (e.g., refrigerator industry safety and efficiency regulations), standards (electrical grid voltage), cultural factors (aesthetic values, morals, cuisine styles).

Figure 6.7 The iPhone did not create new technology but changed the paradigm of phones to include music playing, GPS directions, games, and other functions previously performed by separate products.

Source: Image by Matoo.Studio on Unsplash.

Paradigms are useful, often necessary, but they also box you in, preventing you from even thinking of, much less trying, some solutions. When you have a hammer, every problem looks like a nail. Switching between paradigms, or breaking out of them entirely, can expand creativity. For example, the original iPhone was a breakthrough because it broke the mental barriers between phones, music players, GPS mappers, and more (Figure 6.7).

Whole markets can be disrupted by new business strategies for products or services, and whole companies or industries can be disrupted through clever organizational change. This iceberg model can be added to the Whole System Mapping method as discussed in Chapter 5, with different system maps for the different levels, or a very complex map containing multiple levels at once.

6.1.4 Inputs, Outputs, and Resilience

Systems do things. Open systems produce outputs based on inputs. Closed systems have no inputs or outputs of the whole system, but different parts within the system have inputs and outputs you might want to measure and improve. Considering a tree as an open system, its inputs would include sunlight, CO_2, water, etc., and its outputs would include the production of leaves, shade, habitat for animals, perhaps fruits, etc. Considering a consumer product's life cycle as a closed system, it includes product hardware features, customer behavior, money, etc. as parts of the system driving other parts, and includes user benefits, raw materials use, energy use, waste, pollution, etc. as parts of the system being driven by the other parts. If you measure these parts of the system as outputs, some you'd like to keep or grow (user benefits), but some you'd like to reduce or eliminate (waste and pollution).

Resilient systems keep producing the same outputs even when there are changes in inputs, changes within the system, or changes in your desired outputs. This usually happens through redundancy (having multiple units, multiple pathways, or multiple methods to accomplish something) or flexibility (components of the system can change what they do in response to demand). You could have redundancy/flexibility of parts in the system, or redundancy/flexibility of relationships between parts. They can happen at any level, from physical hardware to information to mental models.

In sustainability, we often strive to build efficient systems, eliminating redundancy; however, this usually makes systems less resilient, more vulnerable to disruption (Figure 6.8). To have both efficiency and resiliency, you need adaptability and information.

For example, the fashion industry overproduces inventory just in case demand is high, throwing away clothes that go unsold. This redundancy provides resilient profits, but is terribly resource-inefficient (Figure 6.9). Clothing factories that can quickly switch between producing many different kinds of clothes, and have a fast accurate feedback loop between what's wanted and what's produced, could provide resilient profits without the waste. Such flexibility and information loops can create self-organizing

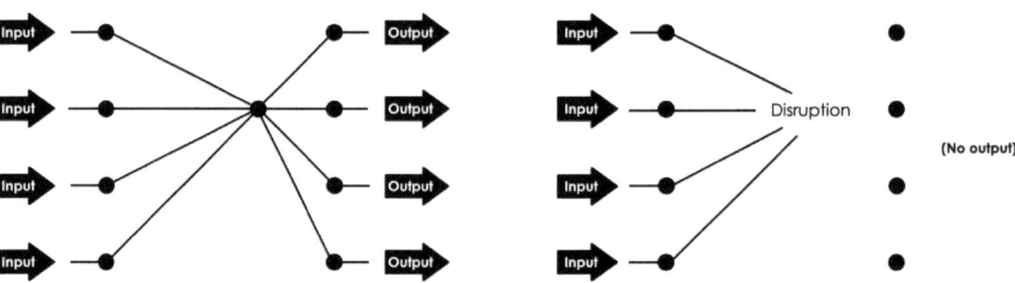

Figure 6.8 Centralized systems are efficient but easily disrupted

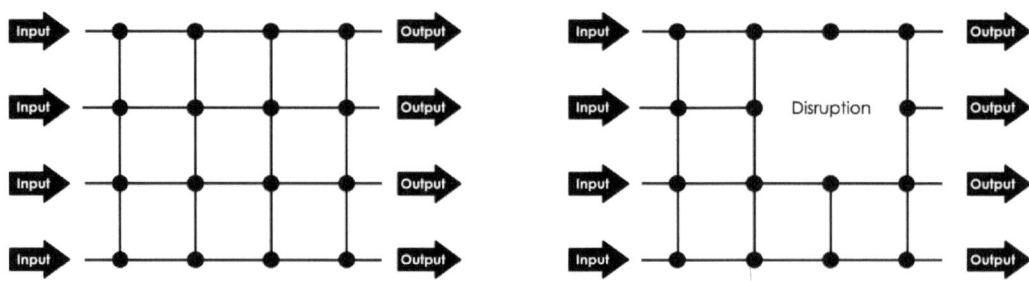

Figure 6.9 Redundant systems are inefficient but resilient

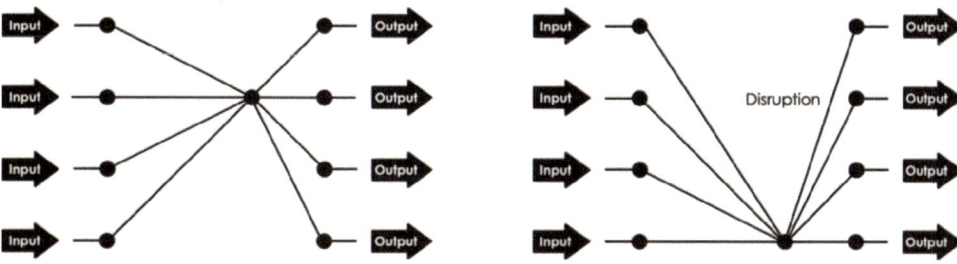

Figure 6.10 Self-organizing systems can be both efficient and resilient

systems (Figure 6.10), which can, in theory, be both efficient and resilient.

We want sustainable systems to be resilient, so they survive and thrive long after we've stopped investing time and money into building them. However, resilience is not always good—clothing overproduction with waste and exploited labor is currently quite resilient, as are other problematic systems such as racism, sexism, and wealth inequality. We need to break their resilience to replace them with more sustainable and just systems.

6.2 Advanced System Modeling

Qualitative static system maps, described in Chapter 5, are fast and easy. However, they don't capture system dynamics from feedback loops, and can't quantitatively predict the amount of change in outputs from changes in inputs or the system. You can add depth to your system maps by adding these.

6.2.1 Causality Maps and Models

Adding causality to system maps establishes cause and effect between different

components. System Dynamics (Meadows, 2008) is one way to model causality, doing so by modeling stocks and flows, among other things. Stocks (or "buffers") are repositories that accumulate or are depleted, and flows are the addition or subtraction of things from stocks (Figure 6.11).

Figure 6.11 A generic stock with flows

For example, a refrigerator could be modeled as stocks and flows of food, where different compartments (e.g., the shelves, door, and freezer) are different stocks, and food flows in and out as the user buys groceries and eats them (Figure 6.12).

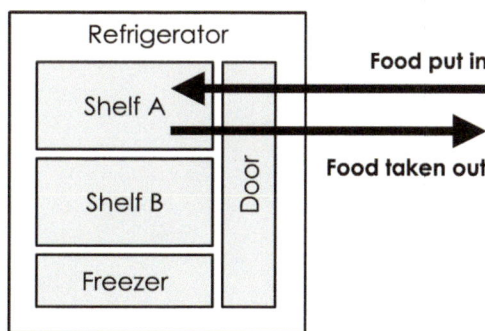

Figure 6.12 A refrigerator modeled as stocks and flows of food

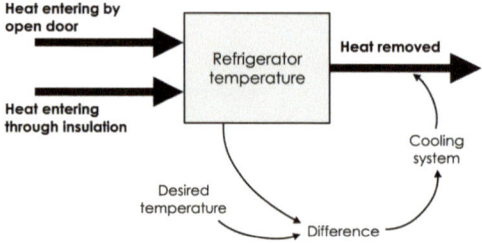

Figure 6.13 A refrigerator modeled as stocks and flows of heat, with a feedback loop (its thermostat)

The same refrigerator could also be modelled as stocks and flows of heat, to help engineer greater efficiency by identifying whether more coldness is lost through the insulation or to users opening and closing the door (Figure 6.13).

Causality models often include feedback. For example, a thermostat's stabilizing feedback is an important component in the refrigerator's temperature system. Such models are often qualitative, such as in Figure 6.13, but they can enable quantitative modeling.

6.2.2 Quantitative Models

System models can use differential equations to quantitatively predict the behavior of the system, its inputs, and its outputs. For example, SolidWorks CAD software can model a refrigerator's heat flows during design. During product operation, control algorithms use mathematical models. For instance, refrigerator thermostats usually use proportional–integral-derivative ("PID") controls to keep the temperature within an acceptable range (Figure 6.14). More sophisticated algorithms exist; controls are an entire engineering discipline. Systems engineering is the science of building and using these mathematical models, both for controls and for predicting things that can't be controlled.

6.2.3 Using Models

Modeling and changing systems can be done in many ways with varying degrees of effort and depth. Static qualitative system maps generally require the least time and effort; dynamic qualitative maps and models require slightly more; quantitative models usually require much larger investments of time and energy, plus detailed knowledge of the system.

All of these system models can involve analyses or interventions on multiple levels at once. Modeling a system on multiple levels can also be done in different ways,

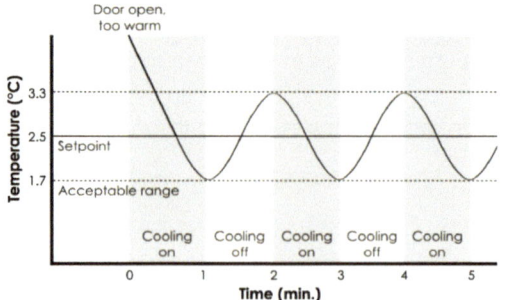

Figure 6.14 Mathematical prediction of refrigerator temperature based on a quantitative system model

as described above. Sustainable design and engineering often require multiple approaches, and critical thinking is required to decide when, where, and how much to think about your systems on different levels.

6.3 Twelve Leverage Points

A powerful tool for systems thinking in design is the "12 Leverage Points" by Donella Meadows (1999). It is simply a list of places to intervene in a system, listed in backward numerical order because it goes from easiest but lowest leverage to highest leverage but most difficult. It's simply a list to brainstorm from, but it helps push your ideation, first of all, to be more thorough by thinking at

many different levels, and, second, to be more radical by thinking at deeper levels (Figure 6.15). It's best used in Whole System Mapping's step 3 (ideation) when the system map includes stocks, flows, and feedback loops. It's also a central part of the Systemic Design Toolkit (Jones & Van Ael, 2022).

The 12 leverage points are:

12. Constants, parameters, numbers (e.g., subsidies, taxes, standards)

11. The sizes of buffers and other stabilizing stocks, relative to their flows

10. The structure of material stocks and flows (what connections and how many)

9. The lengths of delays, relative to the rate of system change

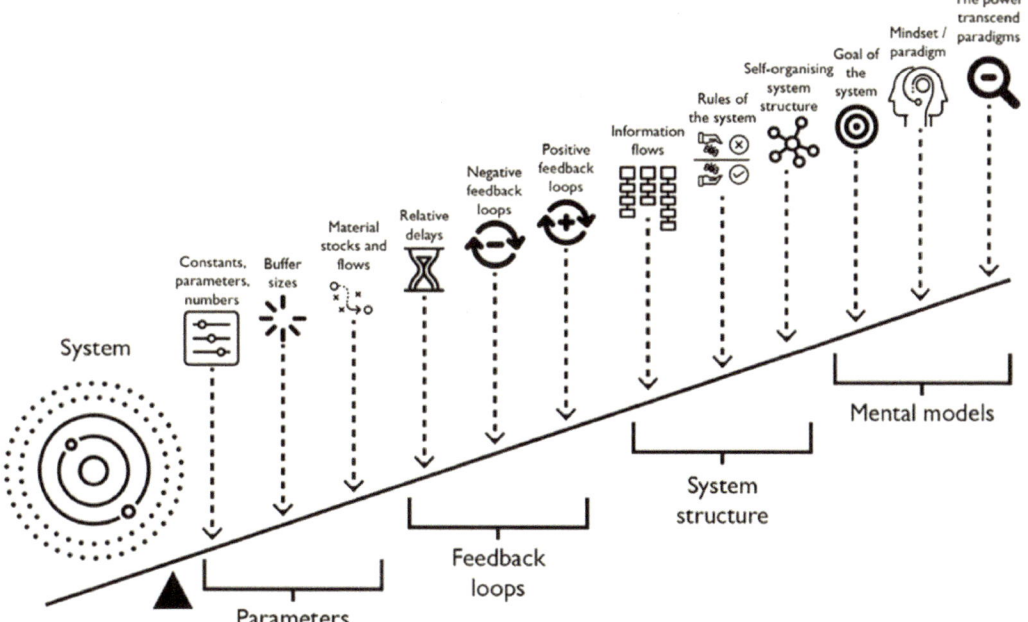

Figure 6.15 The 12 Leverage Points, illustrated

Source: Angheloiu & Tennant (2020).

8. The strength of stabilizing feedback loops (are they too weak or strong compared to the impacts they're working against?)

7. The strength of multiplying feedback loops

6. The structure of information flows (who does and does not have access to what kinds of information)

5. The rules of the system (such as incentives, punishments, constraints)

4. The power to add, change, evolve, or self-organize system structure

3. The goals of the system (what it does and why)

2. The mindset or paradigm out of which the system arises (switching to a different paradigm)

1. The power to transcend paradigms (new paradigm, or no paradigm)

For example, in designing a refrigerator, point 12 includes how many doors, how many shelves, thermostat temperature setpoint, etc.; also how much the product costs versus electricity costs. Point 11 includes the fridge volume versus how much food goes in and out every day, how often the door is open, how thick the insulation is. Point 10 includes where the freezer is, whether it has a separate door or is within the main door. Point 9 includes how often the thermostat samples temperature, to turn cooling on or off; also that electricity prices change yearly, sometimes hourly, while the product lasts 15 years. Point 8 includes how many watts of cooling energy the thermostat commands, versus the added heat from new food entering. Point 7 includes how the more energy refrigerators use, the more they drive global warming, requiring them to use more energy to cool inside, causing more global warming (though this is a weak effect). Point 6 includes how much detail the user's energy bills show for monthly or daily use; also government labeling of refrigerator energy efficiency and repairability. Point 5 includes taxes on inefficient appliances, or rebates on efficient ones. Point 4 includes the ability to rearrange shelves within the fridge, add an ice maker, or other features. Point 3 includes the fridge's purpose to cool food (to preserve it). Point 2 includes the paradigm that cooling food is the best way to preserve it, or if drying/salting, etc. might be better. Point 1 includes the question of whether food needs to be preserved at all, or if we could only eat fresh food or nourish ourselves other ways.

For more detail on all leverage points, read the free online PDF "Leverage Points: Places to Intervene in a System" (Meadows, 1999). Some of them apply to a refrigerator much more than others, and some are hard to even conceptualize for it, much less brainstorm. However, they will push your creativity to new places and higher levels. Remember to check which ideas actually improve impacts the most (Whole System Mapping step 4). For instance, canning all food might actually cause more impacts than a refrigerator over 15 years, even though it eliminates the product. But 12 Leverage Points can give you powerful insights and ideation.

Resources and References

Resources for Further Study

- Meadows, D. (1999). Leverage points: places to intervene in a system. The Sustainability Institute. Available at: https://donellameadows.org/archives/leverage-points-places-to-intervene-in-a-system/

- Senge, P. M. (1990). *The fifth discipline: The art and practice of the learning organization.* Doubleday.

- INCOSE, the International Council on Systems Engineering. Available at: www.incose.org

- Jones, P. H., & Van Ael, K. (2022). *Design journeys through complex systems: Practice tools for systemic design.* Bis Publishers.

References

Angheloiu, C., & Tennant, M. (2020). Urban futures: Systemic or system changing interventions? A literature review using Meadows' leverage points as analytical framework. *Cities*, 104, 102808.

Graessler, I., Hentze, J., & Bruckmann, T. (2018). V-models for interdisciplinary systems engineering. In *Proceedings of the DESIGN 2018 15th International Design Conference*, Dubrovnik, Croatia, pp. 747–756.

Meadows, D. H. (2008). *Thinking in systems: A primer.* Chelsea Green Publishing.

Nielsen, J. (1993). Iterative user interface design. *Computer*, 26(11), 32–41. (See also the resources listed above.)

Rittel, H. W. J., & Webber, M. M. (1973). Dilemmas in a general theory of planning. *Policy Sciences*, 4(2), 155–169.

Senge, P. M. (1990). *The fifth discipline: The art and practice of the learning organization.* Doubleday.

U.S. Department of Transportation. (2007). Systems engineering for intelligent transportation systems. Report FHWA-HOP-07–069. Department of Transportation, Office of Operations.

How to Apply #6: Apply Twelve Leverage Points to a Product Redesign
Time Estimate: 30 Minutes–2 Hours

STEP 1 (Optional): Add Dynamics to Your System Map
Time Estimate: 30–60 Minutes

If you have time, start with your Whole System Map of your product (or a similar product), and add system dynamics aspects. This means labeling/adding stocks and flows, and feedback loops. Mark whether feedback loops are stabilizing or multiplying, either by writing "+" and "-" signs or by arrow widths (both shown in Figures 6.2 and 6.3). If it's easier to start from scratch and draw a new system map, you can.

STEP 2: Brainstorm on 12 Leverage Points
Time Estimate: 20–40 Minutes

Brainstorm new design ideas for each of the 12 Leverage Points listed in the chapter, using the same priorities that you used for the Whole System Mapping exercise (e.g., reducing impacts from energy use, or creating positive impacts from materials). *Have at least 50 new ideas*; perhaps four per leverage point, or you can have more ideas for more relevant points and fewer ideas for others, but at least one per leverage point.

Follow the rules of brainstorming to generate ideas: don't say no to any ideas, but also stay focused on your goal. Often there can be misunderstandings or differences of opinion about whether an idea relates to one leverage point or another. Again, don't say no to ideas because they don't fit that point, simply move them to the more relevant point. Don't let this interrupt your flow too much—you might wait until the end to recategorize—but do discuss this if it helps everyone understand the leverage points or your system better.

Ideally brainstorm on your system map from Step 1, using a different color of sticky note than you did to make the system map, so you can tell your new ideas from the system components. When one idea relates to many parts of the system, you can draw lines connecting it to all of them.

STEP 3: Choose Winning Idea(s) Based on Your Priorities
Time Estimate: 10–40 Minutes

Use the same process as Whole System Mapping's step 4 to first narrow down your brainstormed list to 5–8 best options, then estimate their performance against your priorities for sustainability and business.

Finally, write down the winning idea (or combination of ideas). You could even illustrate it if you want.

Checklist for Self-Assessment

To score your success on this exercise, see if you…

☐   *Wrote down 50+ new ideas.*

☐   *Had 1+ idea for every leverage point.*

☐   *Correctly matched your ideas to the leverage points they fit best with.*

☐   *Decided on idea(s) to move forward with, and described it/them succinctly.*

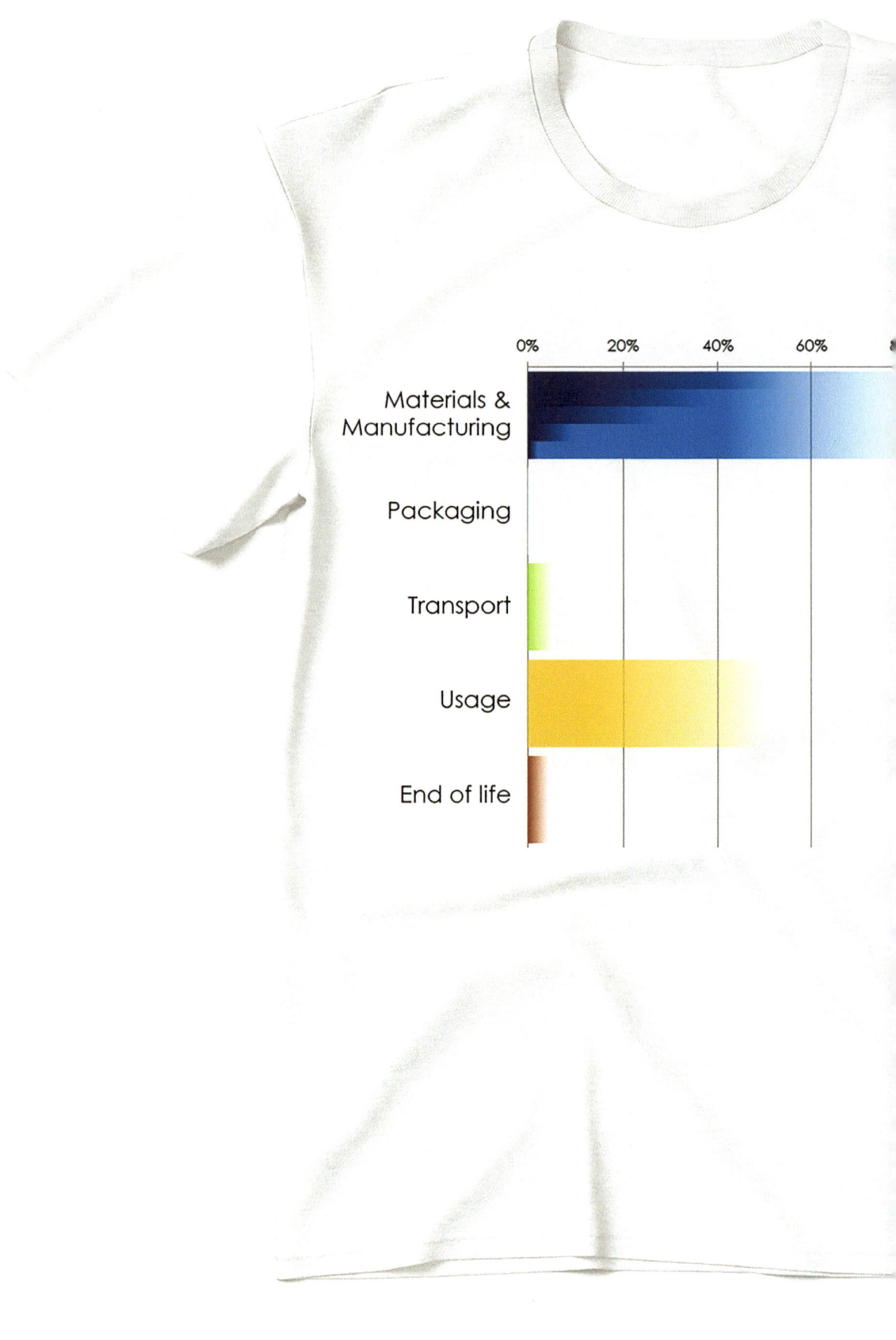

CHAPTER 7
Life Cycle Assessment (LCA)
Introduction

*Jeremy Faludi,
Ruud Balkenende,
and Conny Bakker*

Goals

- Recognize basic terms, methodologies, and tools of LCA

- Explain what life cycle assessment (LCA) measures and what it does not measure

- Read an existing LCA and describe what it means

- Set sustainability priorities using existing LCAs

- Compare design options using existing LCAs

DOI: 10.4324/9781003504672-8

Why It Matters

How do you tell greenwashing from real improvement? How do you find your top design priorities, compare design options, and decide when you've improved enough? Life cycle assessment helps you make evidence-based decisions by quantifying environmental impacts across the whole life cycle of your product or service. Every "carbon footprint" is an LCA that only measures global warming potential, but you can measure much more, from water to land use to radiation. LCA is far from perfect—it doesn't measure some important factors like social impacts, and it doesn't have perfect data on everything—but it's one of the best tools we have for science-based targets, certainly better than guesswork and superstition.

Summary

- LCA is a method to quantify environmental impacts of a product, service, or system.

- What an LCA measures depends on the methodology, boundaries, and other assumptions. "Carbon footprint" LCAs only measure greenhouse gas emissions, while others measure up to 18 impact categories, including almost all Planetary Boundary impacts.

- Multiple impact categories can be combined into a single score, such as "ReCiPe points."

- You don't need to do LCA yourself to use it in design—you can use existing analyses to make evidence-based decisions.

- Using LCA in design can mean setting priorities, comparing options, making tradeoffs, and setting science-based targets for improvement. LCA doesn't suggest designs—you still need creativity, critical thinking, and judgment to turn its scores into strategy.

- LCA can help avoid greenwashing or wasting your product development time and money on things that don't matter.

7.1 What Is LCA?

Life cycle assessment (LCA) is a way to quantify the environmental impacts of anything, from products to services to systems to nations. Quantifying the impacts lets you make evidence-based decisions when choosing between options or setting priorities for improvement, rather than deciding based on guesswork and prejudice. Thus, it also helps you avoid greenwashing.

LCAs measure many things, but not everything. They quantify the raw materials and energy extracted from the Earth and what solid waste or emissions were sent into the air, water, or land, for any process, from mining materials to driving a truck to lighting a lightbulb. See Figure 7.1. These could also be beneficial impacts, but they're almost always resource depletion and pollution, including land use, water use, fossil fuels and greenhouse gases, water pollution, toxic chemicals, and more (details later). However, LCA does not measure social or economic impacts, only environmental ones. Economic life cycle costing is its own field, and social LCA is an emerging field, both outside the scope of this book.

Where do LCAs quantify these impacts? They can be measured for any or all stages of a product's life cycle, including material production, manufacturing, distribution, customer use, and end of life (again, see Figure 7.1). Different LCAs are organized differently, depending on the goals, but most are organized by life cycle stages.

For example, Figure 7.2 shows the impacts of a T-shirt by simplified life cycle stage, in two common formats. Both graphs group all raw material extraction and processing, parts production, and assembly into one bar in each graph. they also group all transport from

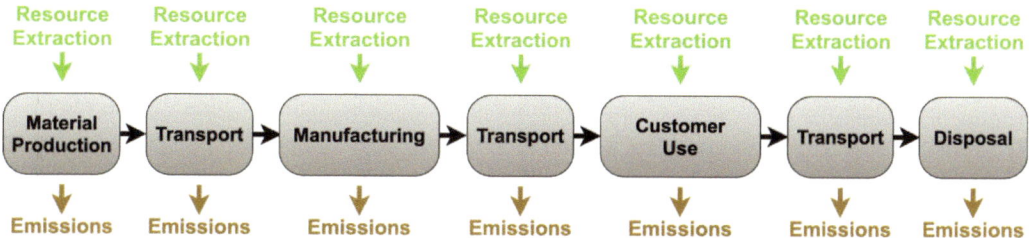

Figure 7.1 LCA quantifies extractions and emissions for any process throughout a life cycle

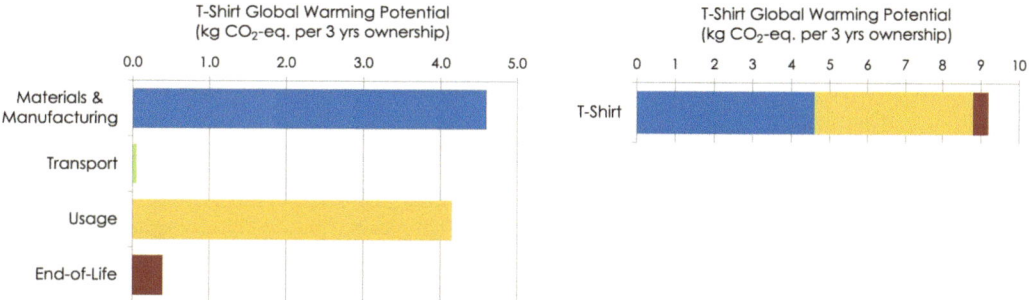

Figure 7.2 Carbon footprint of a T-shirt, assuming an average three-year lifetime. The left and right graphs show the same data, in two different common formats.

the supply chain to distribution to retail to home into one bar. They group all electricity and hot water and soap to wash the shirt into one bar. Even if different materials had different fates at end of life (e.g., some recycled and some incinerated), those are all grouped into one bar.

The depth and breadth of LCAs can vary greatly, depending on the goal and scope of analysis. They're commonly used for reporting environmental impacts after a product, service, or system has been developed. However, as mentioned above, designers and engineers use LCAs during product development:

- to set priorities (target big impacts)
- to choose between options
- to make tradeoffs
- to benchmark and set targets to achieve
- to avoid greenwashing

7.1.1 Setting Priorities

Setting priorities means identifying the major problems (or "hotspots") that need solving. Don't waste product development time and money on things that don't matter. This also helps avoid greenwashing. For example, Figure 7.2 shows a carbon footprint of a T-shirt, with the same data graphed in two different ways because both formats are common in published LCAs. The abbreviation "kg CO_2-eq." is kilograms of carbon dioxide equivalent emissions, which will be explained later. In both graphs, materials and manufacturing are the biggest hotspot, but the impacts of usage are also high, so both could be priorities to redesign. Transport and end of life are not. If a company marketed a new "eco-shirt" based on nothing but local production, that would be greenwashing. (Though it might have social or economic benefits.)

Note that Figure 7.2 doesn't tell you specifically what caused the impacts, for example, is the usage impact mostly due to energy of washing, or its water use, or soap? To answer that, you'd need to read the report that produced the graph.

When setting your design priorities, if you can't fix the worst impact, go to the next one down, or get creative. You may surprise yourself with clever solutions. If you can address multiple impacts with one design intervention, even better.

Once you've set design priorities, you can ideate new designs with the Ecodesign Strategy Wheel, Whole System Mapping, or other methods from later chapters. You can check how good your new design options are by modeling their impacts in LCA (see Chapters 8 and 9 on LCA). This is usually an iterative process, going back and forth between LCA and the other methods.

7.1.2 Choosing Between Options

Besides setting priorities, LCA's other main use in design is helping you choose between design options. Let's say your ideations

from Ecodesign, Whole System Mapping, or other methods resulted in the three design options shown in Figure 7.3 next to the original design. "Green materials" uses sustainably-grown cotton and better dyes to reduce material and manufacturing impacts by one quarter, but everything else is the same. "Nanosilver" coats the cotton threads in nanosilver particles to be anti-bacterial and thus cuts the frequency of washing in half, but it causes much bigger material and manufacturing impacts, with transport and end of life the same. You can see it actually has larger total environmental impacts than the original shirt, so even if well-intended, it would be greenwashing if this analysis is right. "Double lifetime" uses better construction methods with negligible extra impacts that double the shirt's lifetime, with everything else the same—the graph shows material and manufacturing, transport, and end of life all cut in half because it's graphing kg of CO_2 equivalents per three years of shirt ownership. You can see that this "double lifetime" option is the best.

LCA can compare options at any level. Graphs like Figure 7.3 could compare individual material choices, or the impacts

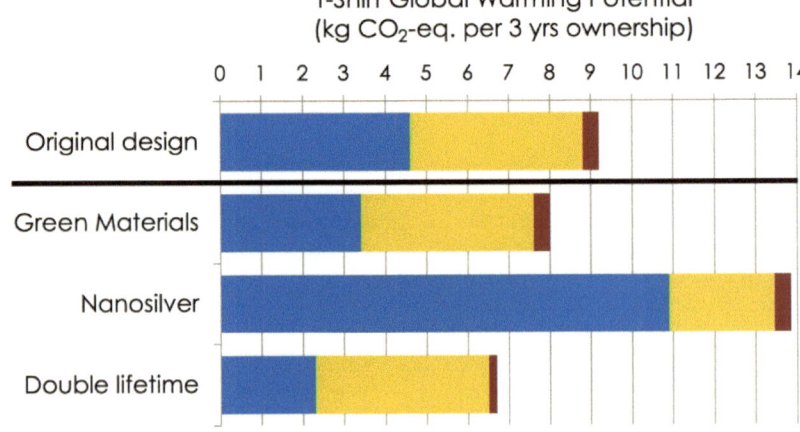

Figure 7.3 Comparing design options with LCA

of the whole system life cycle of the t-shirt in different design scenarios, or anything in between.

7.1.3 Tradeoffs and Synergies

Part of choosing between options is handling tradeoffs or synergies. When choosing between complex options, different options have both pluses and minuses compared to others. For example, if you redesigned Figure 7.2's shirt to reduce the need for washing by making it with antibacterial silver-coated threads, that would lower the usage impacts but raise the materials and manufacturing impacts. Would the payoff be worth it? It depends on what the total impacts for that shirt's whole life cycle are, compared to the original design, as shown in Figure 7.3.

The opposite of tradeoffs are synergies, when different benefits multiply each other. LCA can also help you find these. For example, using less material in a t-shirt reduces the material impacts; it also reduces transport impacts and end-of-life impacts because there is less material to ship and dispose of. If reducing material doesn't reduce lifetime too much (a tradeoff), this could be a good design strategy.

7.1.4 Benchmarking and Targets

Benchmarking is about determining what your impacts are today, to set a baseline for improvement. You can then set achievement targets, for example, improve total impacts by 20% by next year, or 80% in 5 years. Be bold—failing an extremely ambitious goal can often get you further than succeeding at a small goal. You can estimate the feasibility of goals by making quick estimated LCAs of different design options, as discussed in the following chapters.

Ideally, you want to shoot for a "negative" LCA score. That means the system not only causes less environmental damage than other options, its impacts actually improve the environment. This might mean sequestering more carbon than is emitted, restoring wild lands, refilling groundwater reservoirs, etc. Negative LCA scores are very rare, but not impossible—they're the ultimate goal we want to achieve. It's Cradle to Cradle's idea of regenerative design, not doing less bad but doing good.

7.1.5 Avoiding Greenwashing

"Greenwashing" is when a company makes false claims about their product's sustainability. More commonly, what they claim is true but doesn't matter (a minor improvement with major marketing). While it sounds malicious, most greenwashing is unintentional: well-intentioned people making design decisions based on guesswork rather than data. Our intuition about which impacts will be large or small is often wrong. For example, some companies spend large amounts of money to make products locally, and loudly market the sustainability of their efforts; however, Figure 7.2 shows that in the case of a t-shirt, this would probably be greenwashing and a waste of company money. LCA and other quantitative assessments let you avoid such problems with real data.

Use critical thinking when evaluating greenwashing—numbers can lie as well as text, if the assumptions behind the numbers are flawed, or if they're presented in misleading ways.

7.1.6 Uncertainty

To decide between options, make tradeoffs, and benchmark, you have to get comfortable

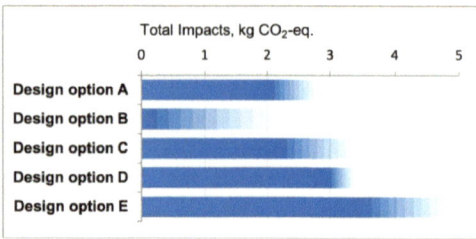

Figure 7.4 Comparing design options with large uncertainties

with uncertainty. No data are perfect, and environmental impact data are often terrible because the market doesn't incentivize tracking or publicizing the data. But even with uncertain data, you can still make certain decisions.

When deciding between options with uncertainties, one option is clearly better or worse than another if their uncertainties don't overlap. Figure 7.4 shows an LCA of several design options, where the blurred bars show uncertainty for a 95% confidence interval—they are the same as error bars, but harder to ignore because you cannot see exactly where they begin and end. People often subconsciously ignore error bars when deciding between options.

In Figure 7.4, option B is clearly the best choice, despite its high uncertainty. Even B's worst-case (highest) value is lower than the best-case (lowest) value of the other options. Despite option D's high precision, it's not the best choice. We can't definitively say whether option A is better or worse than option C because their uncertainty ranges overlap; also the same for C and D. However, D is clearly better than A because they don't overlap. Design option E is worst, and we don't need any improved precision to know

that. To decide between options where error bars overlap, you'd need more precise data, or judge based on other factors. For more on uncertainty, see Chapter 9.

7.2 What Does LCA Measure?

The most common form of LCA is a "carbon footprint," measuring the global warming potential of a product or system in units of kilograms of CO_2-equivalent emissions ("kg CO_2 eq"). However, LCAs can measure much more. Table 7.1 shows that most planetary boundaries are directly measured by LCA, and LCA measures several additional things, too. The boundary "biosphere integrity" is not directly measured because it is an effect of several of the other impacts, such as land use, toxicity, eutrophication, climate change, and more. The boundary of ocean acidification (not to be confused with LCA's terrestrial acidification) is also not directly measured, but it is caused by CO_2 emissions, which are included in LCA's global warming potential. "Novel entities" is the only planetary boundary not measured by LCA at all.

Each of Table 7.1's impact categories also counts many sub-categories, measured in the same units. For example, climate change's "kg CO_2" includes methane, nitrous oxide, and other greenhouse gas emissions, all translated into kg CO_2 equivalents. Methane causes 28–36 times as much global warming per kilo as CO_2 over a 100-year timeframe, so 1 kg of methane is counted as 28–36 kg CO_2 eq. (Gillenwater, 2010). Many

Table 7.1 Units for various LCA impact categories

Impact category	Planetary boundary?	Unit of measurement
Climate change ("Global Warming Potential," "carbon footprint")	Yes	kg of CO_2, the most common greenhouse gas
Stratospheric ozone depletion	Yes	kg of CFC-11, a commonly used and well-documented chlorofluorocarbon
Terrestrial acidification	No	kg of sulfur dioxide, the primary ingredient in "acid rain"
Eutrophication ("biogeochemical flows")	Yes	kg phosphorus in fresh water or kg nitrogen in the sea (the rate-limiting factor for chemical reactions in each environment)
Human toxicity and ecotoxicity (terrestrial, freshwater, and marine)	No	kg of 1,4 dichlorobenzene, a well-documented toxin
Photochemical oxidants ("smog")	No	kg non-methane volatile organic compounds (VOCs)
Particulate matter ("atmospheric aerosol loading")	Yes	kg of particulates with 10 micron diameter ("PM10")
Ionizing radiation ("radioactivity")	No	kilo-becquerels of uranium 235, a standard measure from the nuclear industry
Land use (agricultural, urban, and natural)	Yes	square meters of land area
Water depletion	Yes	cubic meters of aquifer depletion
Mineral depletion	No	kg of iron depletion from the Earth's crust ("critical" materials have high multipliers)
Fossil fuel depletion	No	kg of oil depletion from the Earth's crust

such equivalencies exist for all other impact categories, so there's no need to count thousands of different impact categories separately.

Measuring so many kinds of impact means LCA publications have many graphs. For example, Figure 7.5 shows an analysis of a metal 3D printer measuring 18 different environmental impact categories, using Table 7.1's units but subdividing different kinds of land use and ecotoxicity. The units are listed on the right axis. The top bar shows 9 kg of total CO_2 eq. emissions, of which roughly 75% is caused by electricity use, 12% by material use, 10% by machines, etc.

7.3 Limitations to LCA

There are many things the LCA doesn't measure. It measures no social or economic impacts, it's purely about the environment and human health. Even there, not all environmental or health impacts are

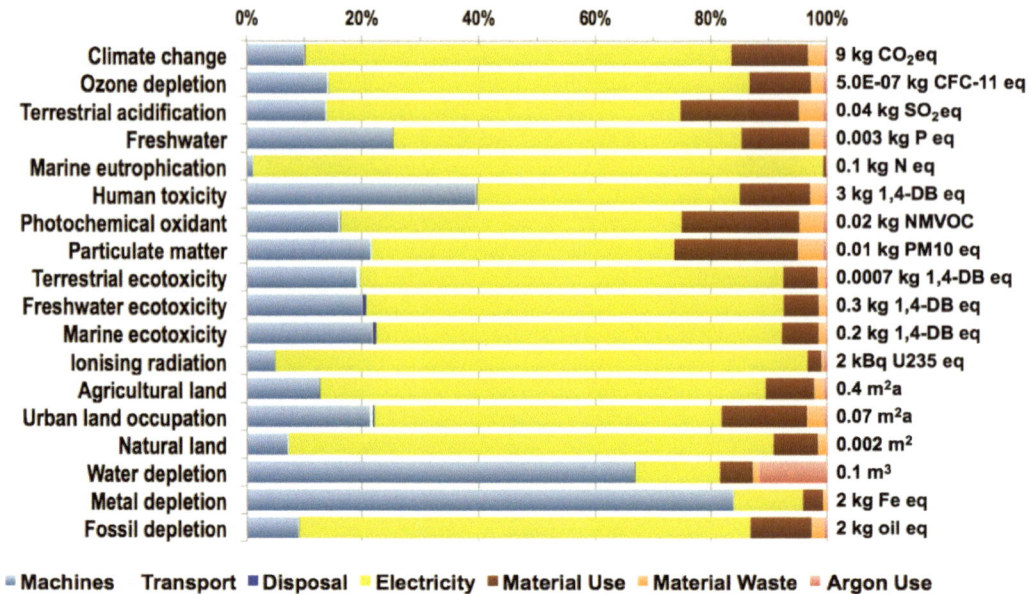

Figure 7.5 An LCA measuring 18 different impact categories. Each bar shows the percent of that impact category caused by machines, transport, electricity, etc.

Source: Faludi et al. (2016).

captured: for example, microplastics aren't measured, and the toxicity measurements are conservative, not including factors like endocrine disruption, due to low data availability. Many LCAs only measure climate change, while all impact categories above can be calculated and reported with similar time and effort to just reporting carbon. Try to find more comprehensively reported LCAs whenever you can.

Still, LCA can provide the most thorough and comprehensive picture of a product or system's environmental impacts, when assessing all the categories above. Even the simplest carbon footprint analysis enables much better decision-making than guesswork and rules of thumb. Just beware analyses with hidden agendas. Just as with any quantitative analyses, LCAs can be distorted by skewed assumptions or a scope that intentionally excludes important parts of the system.

7.4 Merging Impact Categories

As described previously, LCA can help decide between tradeoffs, quantifying whether worse impacts in one area outweigh improvements in another area. There are not only tradeoffs between life cycle stages, but also tradeoffs between impact categories. For example, the silver-infused t-shirt causes less washing energy use but increases mining rare materials. A simple carbon footprint can quantify whether the decreased climate change from less washing outweighs the increased climate change from extra mining. But what about comparing the CO_2 impacts to mineral depletion impacts? For this tradeoff, you can't just add up the total impacts of each bar like before, because they're measured in different units. But there is a solution.

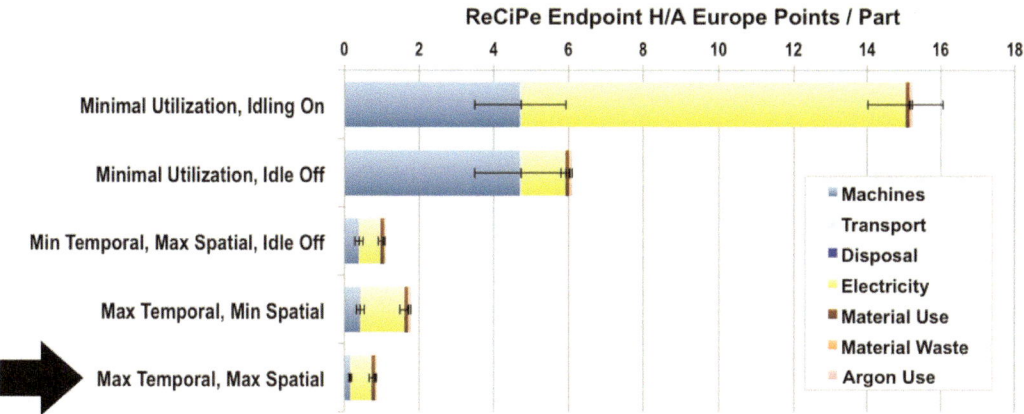

Figure 7.6 The same 3D printer impacts from Figure 7.5 shown as a single score in ReCiPe points (bottom bar), compared with other usage scenarios (other bars)

Source: Faludi et al. (2016).

Single-score LCA metrics exist to combine many different environmental impacts into one number. The "ReCiPe" methodology (Huijbregts et al., 2016), developed by a coalition of EU governments and universities, is one of the most comprehensive available today. Another comprehensive method is Environmental Footprint (Bassi et al., 2023). ReCiPe combines all 18 impacts shown in Figure 7.5 into a single score, measured in units of "ReCiPe points".

Figure 7.6 lets you easily compare different design scenarios, not only including tradeoffs between impacts in different life cycle stages (like materials versus energy use), but also tradeoffs between different impact categories (like climate change versus mineral depletion). But how were those 18 impact categories combined?

Figure 7.7 shows how different impact categories ("midpoints") from Figure 7.5 become a single score in Figure 7.6: by first "normalizing" them into the same units of damage caused to people and the planet ("endpoints"), then weighing those

endpoints. All impacts damaging human health are measured in disability-adjusted life years (DALYs). This metric comes from the insurance industry; "life years" calculates the odds of a person dying a year earlier because of, for example, extreme storms or drought from climate change. Or if they don't die and are merely injured, there's a "disability adjustment" counted in the same units of life years. All impacts damaging ecosystems are measured in the percentage of species likely to go extinct every year. Finally, all impacts causing resource depletion are measured in cost increase of commodity prices. This is because it's hard to predict when a mineral in the Earth's crust will be completely gone—more extreme mining can extract ever lower grades of ore—but such mining is more expensive, so predicting the cost increase of mining is easier. Some impact categories, like global warming, are counted both in human health and ecosystem damage.

Once the kilos of CO_2 and cubic meters of water depletion, etc. are normalized into the three endpoints of actual damage to

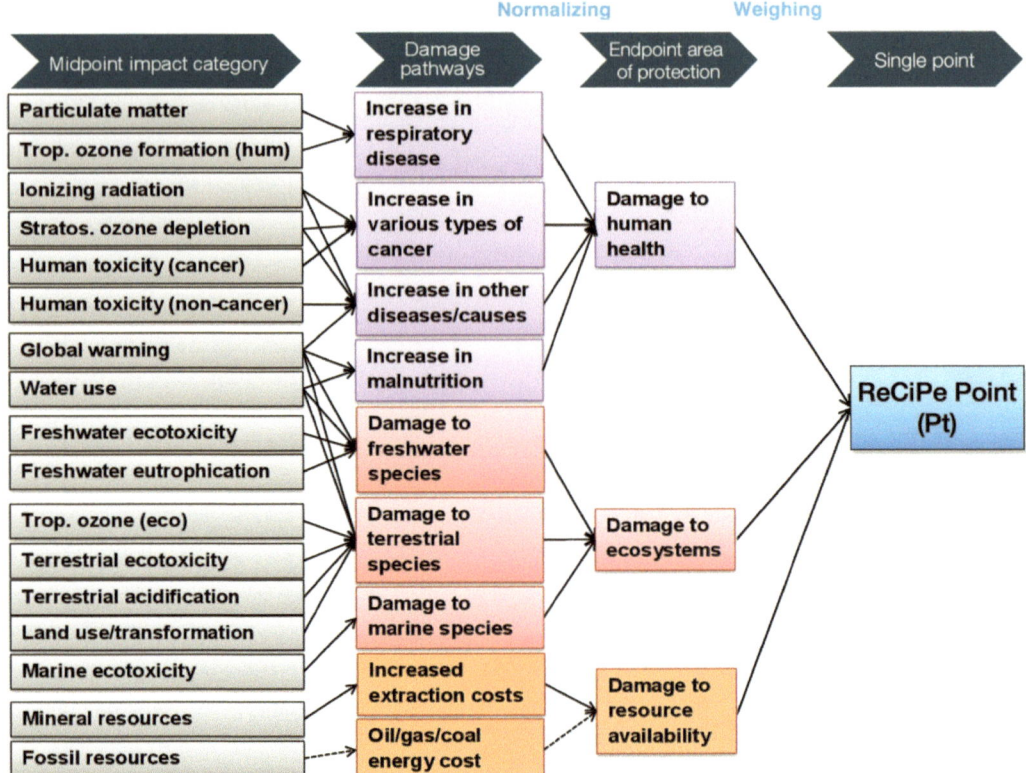

Figure 7.7 How 18 impact categories (midpoints) are normalized into three endpoints and weighed into a single score in the ReCiPe method

Source: Reproduced with modification from Huijbregts et al. (2016).

human health, ecosystems, and resource depletion, the last step is to weigh these three endpoints against each other. This is a subjective value judgment, so the ReCiPe method provides different options. Such subjective value judgments already exist in all LCAs; determining how many kg of methane is equivalent to 1kg of CO_2 depends on the time scale, which is a subjective judgment. The details of these choices are beyond our scope here, but most people use a moderate scheme called "ReCiPe endpoint H/A."

Single-score LCA metrics like this are helpful because most designers, product managers, and business executives are not environmental scientists, and thus are not qualified to weigh different environmental impact categories, much less make such weighing decisions consistently repeatable over time and across different products or services. The ReCiPe metric is the most widely accepted metric, but other commonly-used metrics include Environmental Footprint, EcoIndicator and the US government's "TRACI" methodology.

Some LCA professionals don't use single-score LCA methods because they dislike subjective value judgments, or want more subjectivity to accommodate the company's priorities. However, subjective weighing is not eliminated by only graphing one impact category, or by graphing all 18 impact categories separately. Graphing only carbon footprint is weighing climate change as worth 100% and all other impact categories as worth zero. Graphing all impact categories separately, as in Figure 7.7, visually suggests that all impact categories are weighed equally.

When reading an existing LCA, how do you make decisions from papers where many different impact categories are graphed separately, like Figure 7.5? First, do not assume all impact categories are equally important. They are not, and even if they were, the total amount of impacts in one category (e.g., water depletion) might be much smaller than the amount in another category (e.g., climate change). If you don't have the time or environmental science expertise to normalize and weigh the different impact categories yourself (which most of us do not), look for the similarities between bars in the graph. Luckily, the causes of climate change impacts are usually also the causes of acidification, eutrophication, etc., so the biggest causes are often the same across almost all impact categories. If impacts are very different and you need a tie-breaker, climate change is often considered a top priority, followed by acidification and eutrophication. But this is subjective, it's best to focus on where the data agree.

For example, in Figure 7.5, the yellow bars for energy dominate every impact category except water depletion and metal depletion, so you can be confident the energy is the biggest impact. Similarly, the blue "machines" bars and brown "material use" bars are almost always bigger than the orange

"material waste" bars, giving you another confident decision. However, there's less consistency about whether machines cause more impact than material use, so you should not consider this a clear conclusion.

7.5 Using Existing LCAS to Guide Design

You, the designer or engineer, do not have to perform LCAs yourself. You can have LCA experts perform them, or use existing published LCAs to set priorities or choose options, as long as they're close enough to your situation. To find your priorities for green redesign, you can find an LCA of a similar product or service and see what its "hotspots" (worst impacts) are. Then you can choose sustainable design strategies to address those problems.

Find published LCAs from credible sources like government reports, academic journals, or product manufacturers that you trust. For academic journal articles, search Google Scholar, Web of Knowledge, or other academic search engines. For company or government reports, use any search engine with keywords targeting your product category. Adding company names or other details may help, and you might search for "carbon footprint" or "life cycle analysis" as well as "life cycle assessment" or "LCA." You might limit your search to only Environmental Product Declarations or ISO 14040 certified reports, discussed below, but most product categories have so few published LCAs that you won't have a choice.

The trick here is finding LCAs of similar enough products or services or systems. Judge this based on the materials, energy, and other physical aspects of the system, not the product functionality. For example, if you're designing a cereal box, an existing LCA of a cracker box will be better than an existing LCA of a cereal bag, because the materials and manufacturing drive environmental impacts much more than whether it's used for cereal or crackers.

If you can find existing LCAs that explore various design choices, this helps you even more by suggesting good design options (or eliminating bad ones). For example, if the existing LCA of a cracker box modeled scenarios where the box was made from 100% recycled paper, or used soy-based inks, or other options, you could see the likely ramifications for your own design choices.

Of course, use critical thinking about whether the existing LCA is credible and relevant to your situation. Try to find several LCAs and compare their results, especially when they make different assumptions.

You can be more confident in the results they agree on. When they disagree, try to see why. If you can't find an LCA of a product like yours, find multiple LCAs of related products. For example, when the iPhone was first invented, it was a new product category, but its impacts could have been predicted by comparing LCAs of existing mobile phones, MP3 players, and GPS navigators.

7.5.1 Finding High-Quality, Relevant LCAs

LCAs vary widely in quality and relevance. LCAs are more relevant to you if they're similar to your situation in time (recent), space (similar region or transport distances), and usage conditions (product lifetime, functionality, user behavior, etc.). LCAs showing uncertainty are more trustworthy than those that don't. LCAs measuring more of the product life cycle or larger system show you better context than those that don't. The most trustworthy LCAs are usually peer-reviewed academic journal articles, government reports, or an "Environmental Product Declaration (EPD)." An EPD uses a standard methodology agreed to by the product manufacturer and other industry stakeholders, so you can compare LCAs of competing products. Another measure of credibility is certified as meeting the International Standards Organization (ISO) 14040 standard, though it has more flexibility than an EPD. Finally, check the LCA's assumptions—the devil is in the details. If you think its assumptions are unreasonable or not relevant to your context, don't use it. See Chapter 8 to learn which assumptions go into an analysis.

7.5.2 Examples of Common Product LCAs

Figures 7.8–7.14 are LCAs of some common products, also listing sustainable design strategies relevant to their largest impacts. The impact uncertainties are high because these are each averaged from several publications and/or direct measurements (see citations), where each study concerned a different product with different materials, energy use, locations, etc., and some studies used different methods or assumptions.

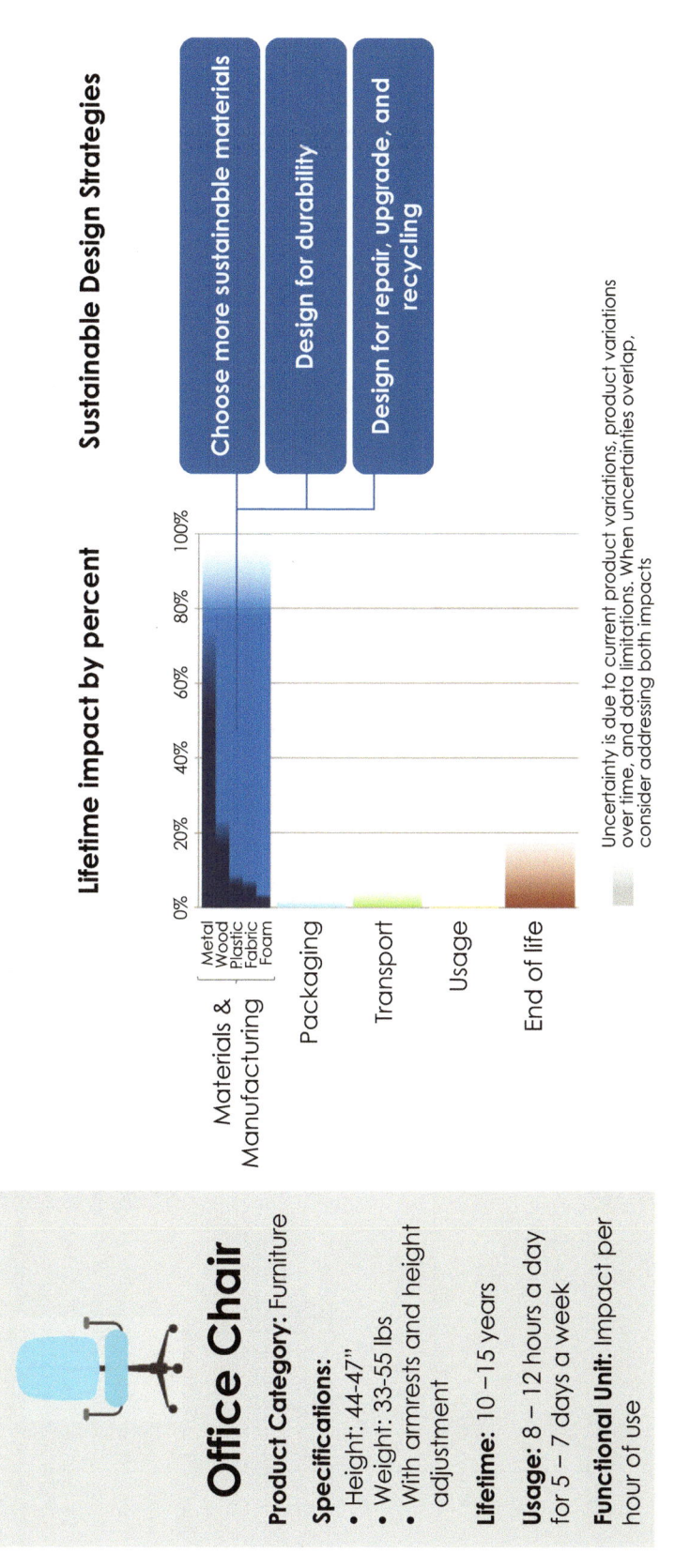

Figure 7.8 LCA of an office chair

Source:

- Martinez E. (2018). Empirical Analysis of product teardown of an office chair [unpublished].
- The Norwegian EPD Foundation (December 2014). Environmental Product Declaration of the Håg H03 – 330 office chair for Flokk.
- The Norwegian EPD Foundation (August 2019). Environmental Product Declaration of the Håg RH New Logic office chair for Flokk.
- Herman Miller (September 2016). Environmental Product Declaration of the New Aeron chair.
- Steelcase (June 2004). Environmental Product Declaration of the Think task chair.
- Wiesner-Hager (July 2019). Environmental Product Declaration of the poi swivel chair with multifunctional arms.
- Missing data on packaging in some publications supplemented by calculations using the Idemat 2020 and Ecoinvent 3–5 database.

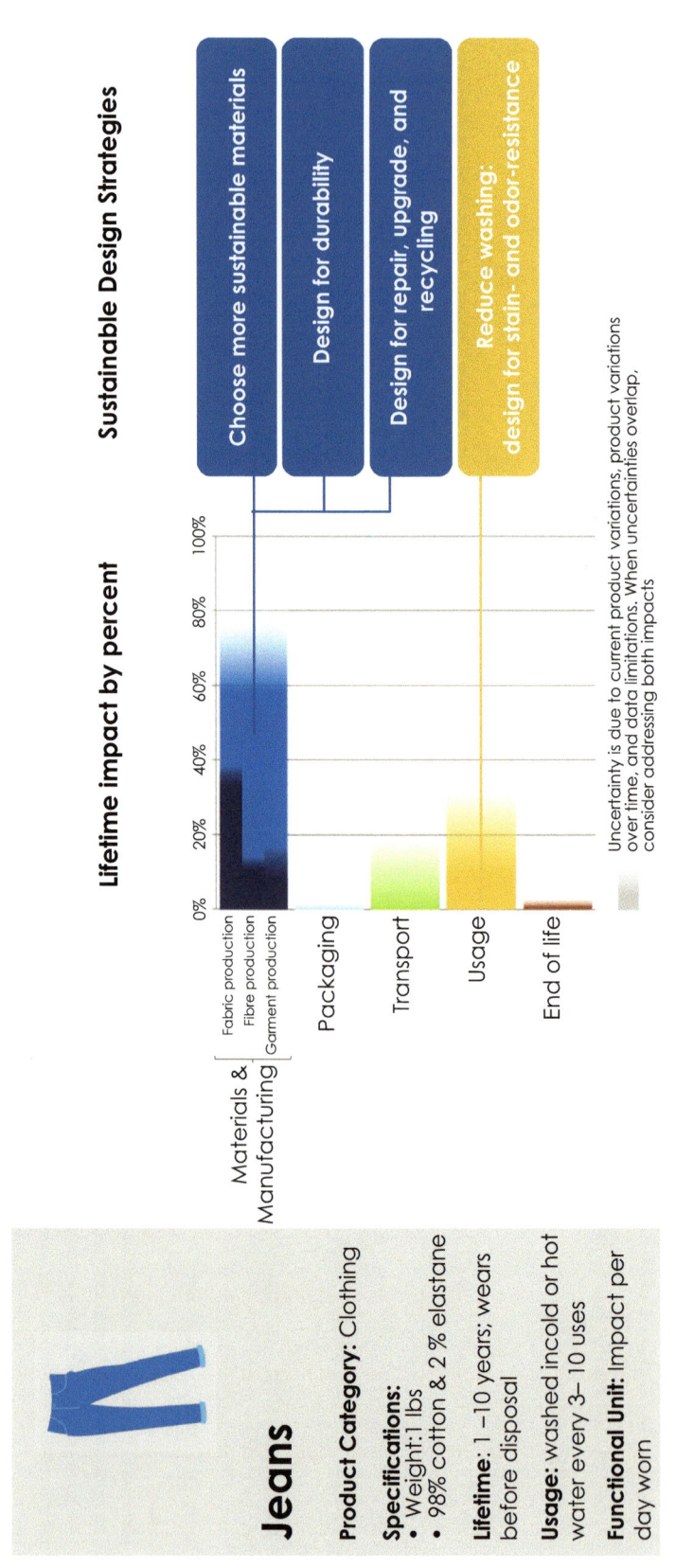

Figure 7.9 LCA of jeans

Source:

• Levi Strauss & Co. (2015). The Life Cycle of a Jean – Understanding the environmental impact of a pair of Levi's 501 Jeans.
• Roos, S., Sandin, G., Zamani, B., & Peters, G. (2015). Environmental assessment of Swedish fashion consumption. Five garments–sustainable futures. Mistra Future Fashion.
• De Haan, B. R., & Liefferink, J. D. (2017). Exploring life cycle sustainability in the fashion industry. A case study on the impacts in the life cycle garments and the application of Life Cycle Assessment in a company. [master thesis]
• Åslund Hedman, E. (2018). Comparative Life Cycle Assessment of Jeans: A case study performed at Nudie Jeans.
• Missing data on packaging and transportation in some publications supplemented by calculations using the Idemat 2020 and EcoInvent 3–5 database.

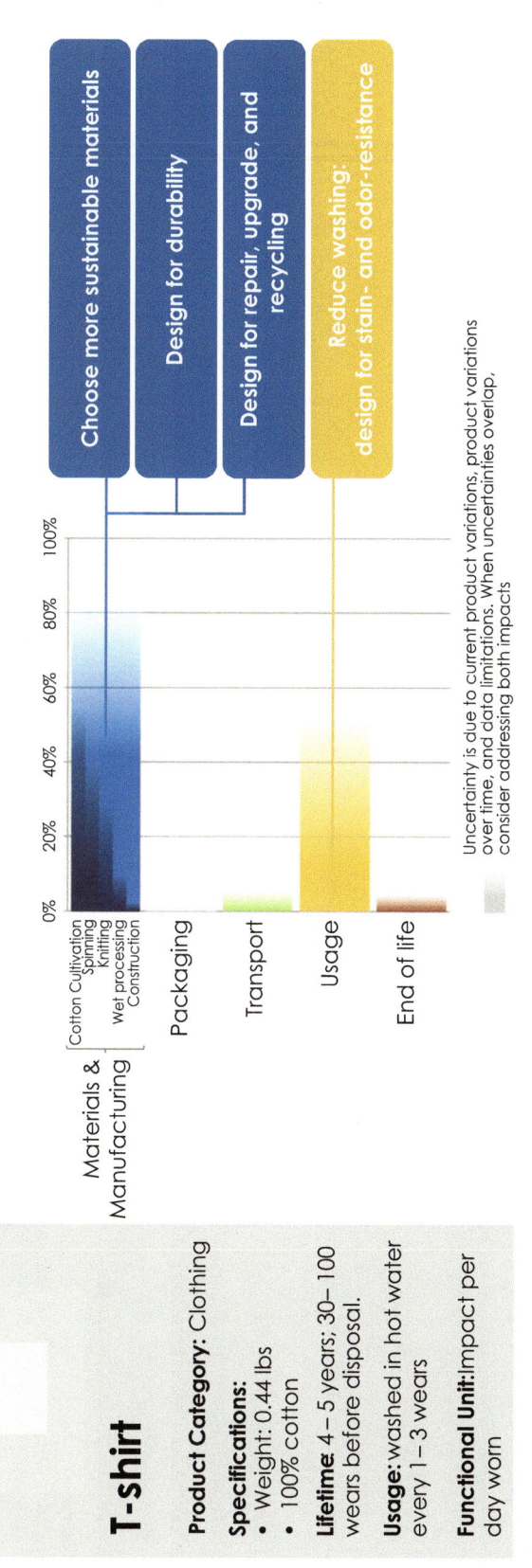

Figure 7.10 LCA of t-shirt

Source:

• Zhang, Y., Liu, X., Xiao, R., & Yuan, Z. (2015). Life cycle assessment of cotton T-shirts in China. The International Journal of Life Cycle Assessment, 20(7), 994–1004. https://doi.org/10.1007/s11367-015-0889-4

• Baydar, G., Ciliz, N., & Mammadov, A. (2015). Life cycle assessment of cotton textile products in Turkey. Resources, Conservation and Recycling, 104, 213–223. https://doi.org/10.1016/j.resconrec.2015.08.007

• Wang, C., Wang, L., Liu, X., Du, C., Ding, D., Jia, J., ... & Wu, G. (2015). Carbon footprint of textile throughout its life cycle: a case study of Chinese cotton shirts. Journal of Cleaner Production, 108, 464–475. https://doi.org/10.1016/j.jclepro.2015.05.127

• Steinberger, J. K., Friot, D., Jolliet, O., & Erkman, S. (2009). A spatially explicit life cycle inventory of the global textile chain. The International Journal of Life Cycle Assessment, 14(5), 443–455. https://doi.org/10.1007/s11367-009-0078-4

• Missing data on packaging in some publications supplemented by calculations using the Idemat 2020 and EcoInvent 3–5 database.

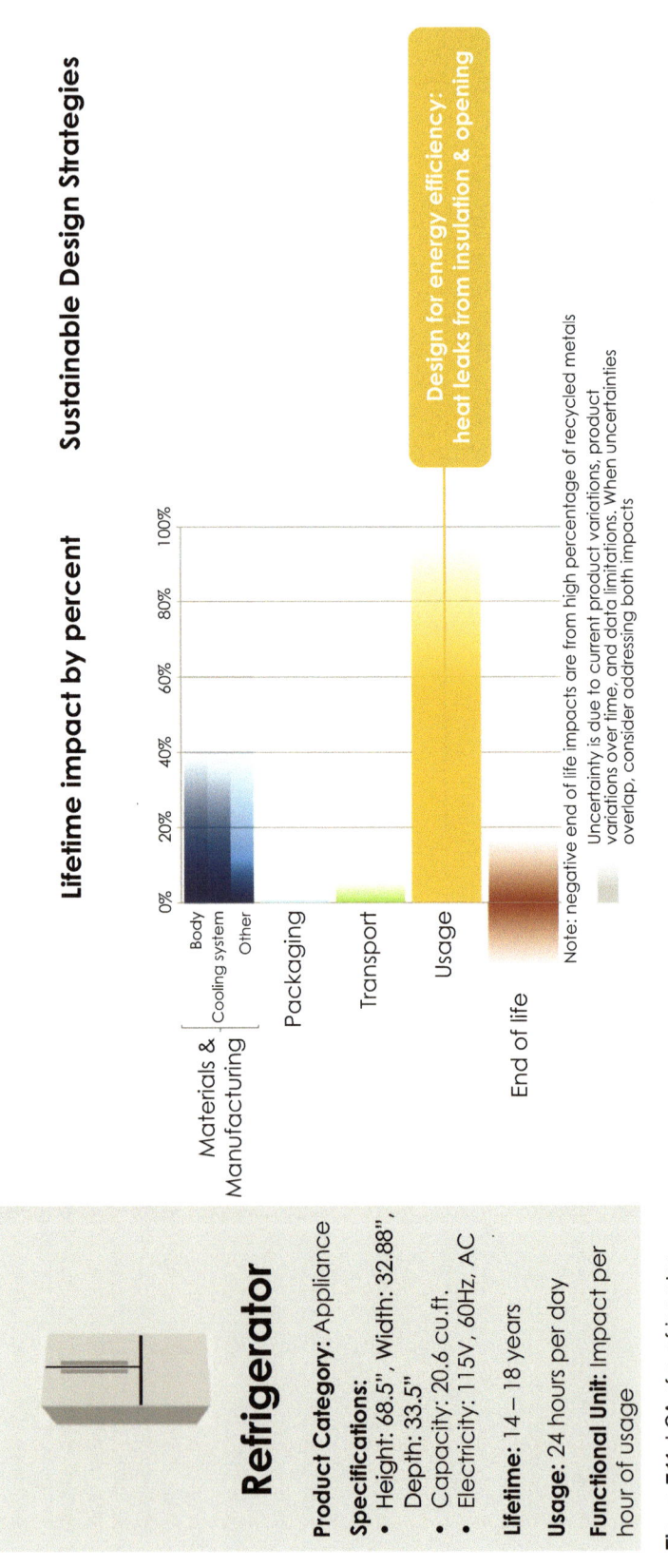

Refrigerator

Product Category: Appliance

Specifications:
- Height: 68.5", Width: 32.88", Depth: 33.5"
- Capacity: 20.6 cu.ft.
- Electricity: 115V, 60Hz, AC

Lifetime: 14 – 18 years

Usage: 24 hours per day

Functional Unit: Impact per hour of usage

Lifetime impact by percent   **Sustainable Design Strategies**

Design for energy efficiency: heat leaks from insulation & opening

Materials & Manufacturing — Body, Cooling system, Other
Packaging
Transport
Usage
End of life

0%   20%   40%   60%   80%   100%

Note: negative end of life impacts are from high percentage of recycled metals

Uncertainty is due to current product variations, product variations over time, and data limitations. When uncertainties overlap, consider addressing both impacts

Figure 7.11 LCA of a refrigerator

Source:

- Xiao, R., Zhang, Y., Liu, X., & Yuan, Z. (2015). A life-cycle assessment of household refrigerators in China. Journal of Cleaner Production, 95, 301–310. https://doi.org/10.1016/j.jclepro.2015.02.031
- Gehin, A., Zwolinski, P., & Brissaud, D. (2009). Integrated design of product lifecycles—The fridge case study. CIRP Journal of Manufacturing Science and Technology, 1(4), 214–220. https://doi.org/10.1016/j.cirpj.2009.05.002
- Ma, J., Yin, F., Liu, Z., & Zhou, X. (2012). The eco-design and green manufacturing of a refrigerator. Procedia Environmental Sciences, 16, 522–529. https://doi.org/10.1016/j.proenv.2012.10.072
- Yang, Q., Yu, S., & Sekhari, A. (2011). A modular eco-design method for life cycle engineering based on redesign risk control. The International Journal of Advanced Manufacturing Technology, 56(9–12), 1215.
- Faludi, J. (2010). Autodesk Sustainability Workshop [Measuring Sustainability Life Cycle Assessment: Example 2]

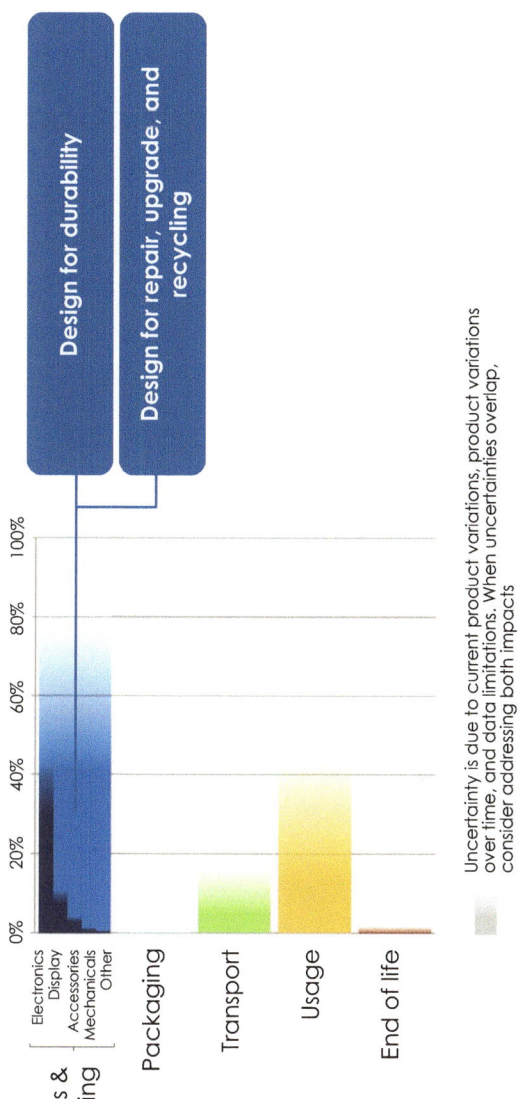

Smartphone

Product Category: Electronics

Specifications:
- Height: 5.44", Width: 2.64", Depth: 0.27"
- Weight: 4.55oz
- 32 GB, 4.7" LED-Backlit widescreen Multi-Touch display with IPS technology

Lifetime: 2 – 4 years

Usage: 2 – 4 hrs use, 20-22 hrs idle per day

Functional Unit: Impact per hour of usage

Figure 7.12 LCA of a smartphone

Source:
- Apple (2017, September). iPhone X Environmental Report. Retrieved 4 February 2020, from https://www.apple.com/environment/pdf/products/iphone/iPhone_X_PER_sept2017.pdf
- Güvendik, M. (2014). From smartphone to futurephone: assessing the environmental impacts of different circular economy scenarios of a smartphone using LCA.
- Ercan, E. F. (2013). Global Warming Potential of a Smartphone.
- Yu, J., Williams, E., & Ju, M. (2010). Analysis of material and energy consumption of mobile phones in China. *Energy Policy*, 38(8), 4135–4141.
- Andrae, A. S., & Vaija, M. S. (2014). To which degree does sector specific standardization make life cycle assessments comparable?—the case of global warming potential of smartphones. *Challenges*, 5(2), 409–429.
- Unpublished empirical analysis by Eric Munsing for Thinkstep (2014). Munsing, E. (2014). Unpublished empirical analysis of a cell phone.
- Missing data on packaging in some publications supplemented by calculations using the Idemat 2020 and Ecoinvent 3–5 database.

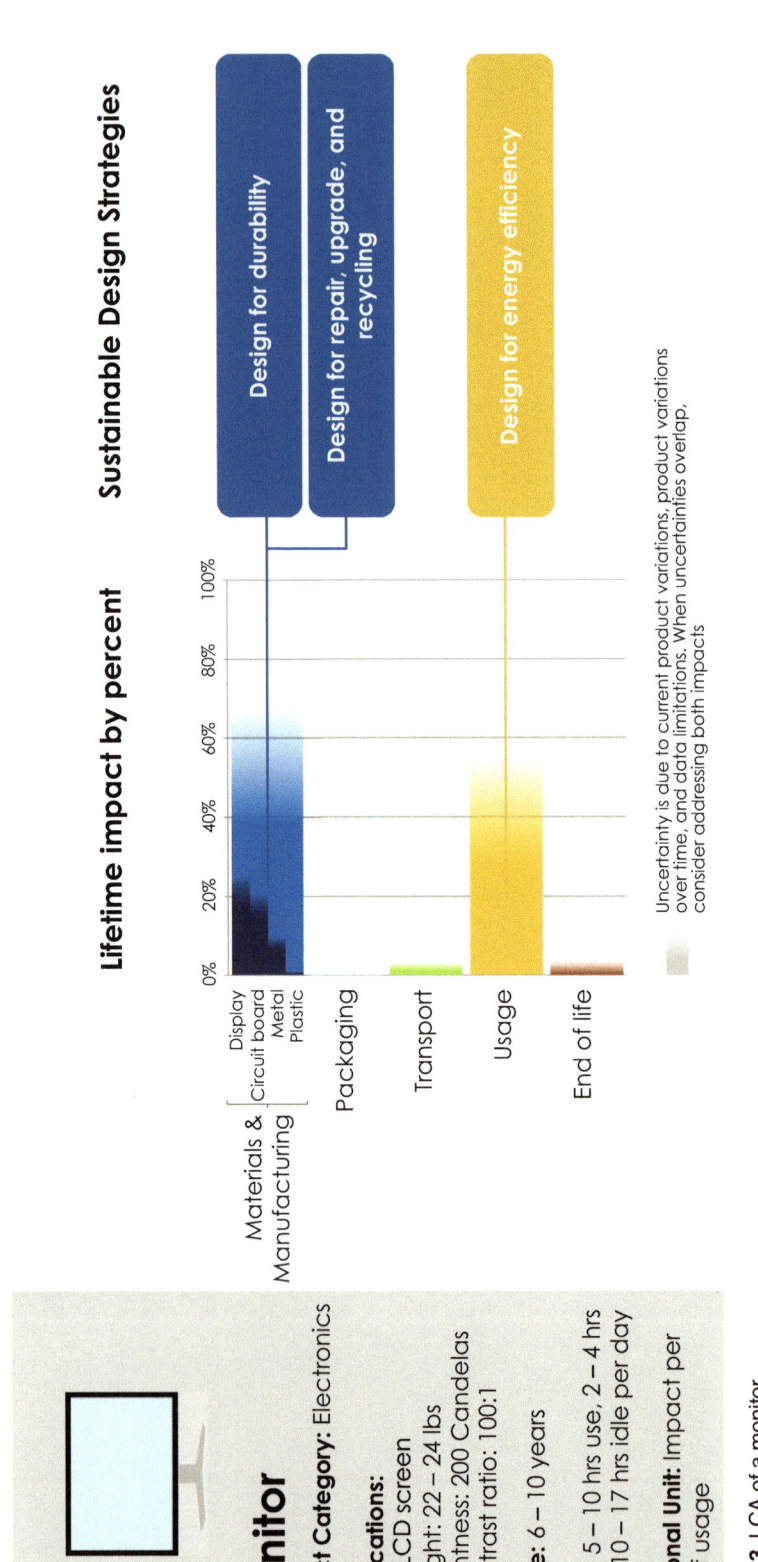

Figure 7.13 LCA of a monitor

Source:

• Bhakar, V., Agur, A., Digalwar, A. K., & Sangwan, K. S. (2015). Life cycle assessment of CRT, LCD and LED monitors. Procedia CIRP, 29, 432–437.
• Socolof, M. L., Overly, J. G., & Geibig, J. R. (2005). Environmental life-cycle impacts of CRT and LCD desktop computer displays. Journal of Cleaner production, 13(13–14), 1281–1294. https://doi.org/10.1016/j.jclepro.2005.05.014
• Apple (2011, July). Thunderbolt Display Environmental Report. Retrieved February 6, 2020 from https://www.apple.com/environment/pdf/products/archive/2011/ThunderboltDisplay_PER_july2011.pdf
• Zhang, L. (2014). Empirical Analysis of product teardown of a Dell monitor [unpublished].
• Missing data on packaging and transportation in some publications supplemented by calculations using the Idemat 2020 and EcoInvent 3–5 database.

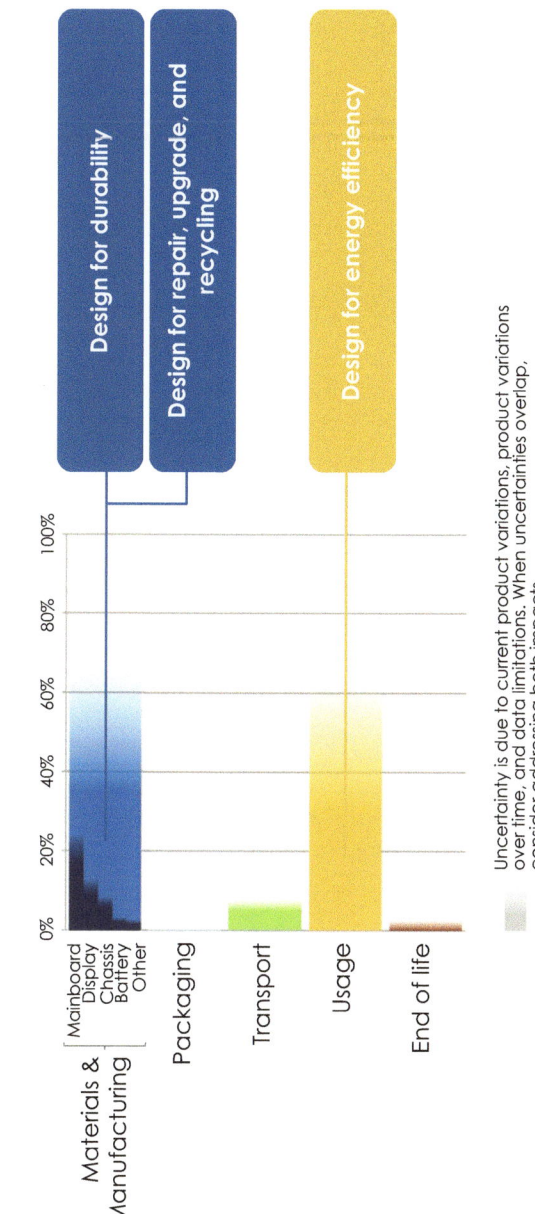

Laptop

Product Category: Electronics

Specifications:
- 15" LCD Screen
- 1.5 – 2.5 GHz CPU
- 500 MB – 1.5 TB Hard drive
- Weight: 5 – 8 lbs

Lifetime: 2 – 4 years

Usage: 4 – 6 hrs use, 18 – 20 hrs idle per day

Functional Unit: Impact per hour of usage

Figure 7.14 LCA of a laptop

Source:

- Tekwawa, M., Miyamoto, S., & Inaba, A. (1997, May). Life cycle assessment; an approach to environmentally friendly PCs. *Proceedings of the 1997 IEEE International Symposium on Electronics and the Environment. ISEE-1997* (pp. 125–130). IEEE. https://doi.org/10.1109/ISEE.1997.605287
- IVF (2007) European Commission DG TREN Preparatory studies for eco-design requirements of EuPs. IVF Report 07004. Laptop used in office.
- O'Connell, S., & Stutz, M. (2010, May). Product carbon footprint (PCF) assessment of Dell laptop-Results and recommendations. *Proceedings of the 2010 IEEE international symposium on sustainable systems and technology* (pp. 1–6). IEEE.
- Deng, L., Babbitt, C. W., & Williams, E. D. (2011). Economic-balance hybrid LCA extended with uncertainty analysis: case study of a laptop computer. *Journal of Cleaner Production, 19*(11), 1198–1206.
- Apple (2019, May). MacBook Pro 15-inch Environmental Report. Retrieved February 6, 2020 from https://www.apple.com/environment/pdf/products/notebooks/15-inch_MacBookPro_PER_may2019.pdf
- Missing data on packaging in some publications supplemented by calculations using the Idemat 2020 and EcoInvent 3–5 database.

Resources and References

Resources for Further Study

- Vogtlander, J. G. (2014). *LCA: A practical guide for students, designers and business managers* (2nd ed.). Delft Academic Press.

- OVAM. (2011). Ecolizer 2.0 Ecodesign Tool. Danny Wille Publishers, Openbare Vlaamse Afvalstoffenmaatschappij.

References

Bassi, A., et al. (2023). Updated characterisation and normalisation factors for the environmental footprint 3.1 method. Publications Office of the European Union.

Faludi, J., Baumers, M., Maskery, I., & Hague, R. (2016). Environmental impacts of selective laser melting: Do printer, powder, or power dominate? *Journal of Industrial Ecology*, 21(S1).

Gillenwater, M. (2010, June 28). What is a Global Warming Potential? And which one do I use? Greenhouse Gas Management Institute.

Huijbregts, M. A. J., et al. (2016). ReCiPe 2016: A harmonized life cycle impact assessment method at midpoint and endpoint level. RIVM Rapport 2016–0104.

How to Apply #7: Using Existing LCA(s) to Set Design Priorities
Time Estimate: 2–10 Hours

STEP 1: Find Relevant LCA(s)
Time Estimate: 5 Minutes–5 Hours

If someone already did an LCA of your exact product, use it. If not, find an LCA of a product that's close to yours—not in functionality or marketing, but in terms of energy use, materials, and lifespan. Use the LCAs on the preceding pages, or search the web or scientific literature. Ideally find at least three or four LCAs of the product category from different companies and/or academics to explore varying assumptions with different biases. Cite your sources.

When judging existing LCAs, also consider whether they use reasonable and relevant assumptions about what is and isn't counted, product lifetime, and product usage.

STEP 2: Identify the Biggest Impacts You Can Change
Time Estimate: 5–30 Minutes

This is easy—find the largest bar in the graph and write down what causes it. If you have multiple LCAs and they agree about the causes of the largest impacts, that's ideal. If they disagree, you may need to dig into their assumptions a bit more, or decide to prioritize both big impacts. Note that uncertainties may be high—if the LCAs aren't for your exact product, the impacts in the existing LCAs might differ from your product by ±50%, maybe even over ±100%. Use your best judgment and write down your reasoning.

STEP 3: Prioritize Design Strategies for the Whole System
Time Estimate: 5–15 Minutes

Choose two or three design strategies to maximize improvement in the total overall LCA score (even have a positive environmental impact if you can), and write them down. For example, long life, remanufacturing, recycling, sourcing greener materials; you might use the Ecodesign Strategy Wheel as a guide. Usually this means targeting the largest impact(s) you just wrote down, but remember to consider the whole system and use critical thinking about what causes those impacts, and where you might find synergies.

If the largest impacts come from things you can't change for functionality or legal reasons, you might choose strategies to improve lower-priority improvements. Or you can be more creative about new solutions to the biggest impacts.

STEP 4: Brainstorm Improvements
Time Estimate: 20–40 Minutes

Brainstorm to generate at least 20 new ideas (ideally 50+ new ideas) for each design strategy in the last step. Be specific and concrete—this also helps you come up with many ideas, as small variations. But have bold ideas, too. To be thorough, you can brainstorm on a system map (see Chapter 5).

STEP 5: Estimate Improvement of Your Best Ideas
Time Estimate: 15 Minutes–1 Hour

Take your list of brainstormed ideas and narrow it down to your three to six favorites, then estimate how much of an improvement they will make. Will they use half the material? Twenty percent less energy? Will there be any synergies that multiply improvements? Or will there be tradeoffs, for example, having to use materials with higher environmental impact in order to extend lifespan? (If using 20% more material would triple the product's lifespan, overall impacts per unit of customer use are greatly improved; see "functional units" in Chapter 8.) Or will there be "rebound" effects, where improving product energy efficiency means that it's cheaper to run, and thus users use it more, partially negating the efficiency improvement?

Once you've estimated improvements of these favorite ideas, sketch out graphs to compare to the original existing LCA graph you started with. Compare the graphs of the new design options to decide what will likely cause the most improvement. This is your winning idea (or set of ideas). If you want to be thorough about the calculations, do your own LCA as described in Chapter 8. This can also happen later in your design process, once prototypes become more clearly defined.

STEP 6: Illustrate Your Best Idea
Time Estimate: 15 Minutes–2 Hours

Once you've determined the best-performing idea(s), illustrate it—either with a simple text description, visual sketch, CAD rendering, or other mockup to make the idea attractive.

Checklist for Self-Assessment

To score your success on this exercise, see if you…

☐ *Cited and displayed a graph from at least one published LCA relevant to your product or service (ideally four LCAs from different sources).*

☐ *Listed what causes the biggest impacts in the whole system, and your design strategies to improve them.*

☐ *Brainstormed 20+ new ideas (ideally 50+).*

☐ *Calculated estimated improvements of your favorite ideas to identify the best one(s).*

☐ *Illustrated the winning idea(s).*

CHAPTER 8
Doing Fast-Track LCA

*Jeremy Faludi,
Ruud Balkenende, and
Conny Bakker*

Goals

- Compare environmental impacts of materials, manufacturing methods, transport, energy, and end-of-life options using life cycle assessment (LCA)

- Calculate the environmental impacts of a product, service, or system using LCA, including:

  - set scope
  - set boundaries
  - choose functional unit
  - perform inventory (with proxies)
  - analyze results
  - communicate results

DOI: 10.4324/9781003504672-9

Why It Matters

You often can't find a pre-existing LCA of your product or system, and even if you can, it's often better to make your own. LCA is complicated but manageable, just break it down into steps. Even if you'll never do one, it's still good to learn how so that when you read someone else's LCA, you know where to look for questionable assumptions or missing details. It also helps you talk to LCA practitioners to know what to ask for and ensure their models accurately reflect your system and its usage.

Summary

To do an LCA:

1. Set the goal and scope (which questions your LCA will answer).

2. Set boundaries (what your LCA model will and won't include).

3. Choose the functional unit (to make fair comparisons of environmental impacts per unit of product functionality).

4. Inventory everything within your boundaries (count how many per functional unit, and look up impacts).

5. Analyze results (calculate and compare, consider uncertainty, maybe different scenarios).

6. Communicate findings to your design team (to support design decisions).

Here it's written as linear steps, but in practice it's an iterative process.

8.1 Set the Goal and Scope

Why are you doing this LCA—what's your goal? Is it a quick and dirty aid to make a simple design decision, is it for a published report facing legal scrutiny, or something else? As discussed in Chapter 7, during product development you're most likely setting design priorities by finding the biggest impacts, or choosing between design options (including tradeoffs). You might also be benchmarking and setting achievement targets, or reporting after product development, or other things. LCA modeling is an infinitely deep rabbit hole, and your scope (how deep you dive) is determined by your goal and your available time, see Figure 8.1.

As Figure 8.1 shows, checking simple options like plastic versus wood or grid electricity versus a solar panel can be modeled in a few minutes. That's called "streamlined LCA." We focus on "fast-track" or "screening" LCA to discover the biggest impacts in a system, or to compare one design option against another. It can take a few hours to a day or two. Fast track LCA also lets you establish a baseline: what the impacts of your product are today, and how much better they could theoretically be.

"Full" LCAs are for published reports that might face academic or legal scrutiny, like an

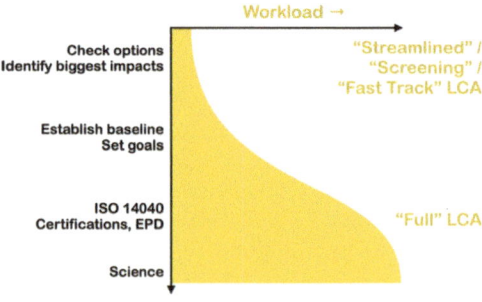

Figure 8.1 Different scopes of LCA

Environmental Product Declaration (EPD), getting an eco-label certification, or for ISO 14040 certification. These LCAs can take months or even a year for complex systems, and thus are expensive. Also, you can only do them after the design is finished, when you know the exact grams of every material, transportation distances, etc. Obviously that means you can't use them to make design decisions until the next product revision. To do a full LCA, you need more depth than we cover here.

Thus, here we discuss fast track LCA for two kinds of scope: (1) where are the biggest impacts in the system, so you can focus your design time and money? Or (2) which of your design options causes the best environmental impacts, so you can choose it over other design options?

8.2 Set Boundaries

Boundaries are the specific list of what's in scope and out of scope—what your LCA

includes and leaves out. The more you include, the more complete a picture you'll get of the environmental impacts of your system, but, the more you include, the less certain you can be about all the details. Figure 8.2 gives an idea of the different boundaries you may want to set. It's often standard to count the whole life cycle of the product—a "cradle to grave" boundary. But in some circumstances, you might only consider early stages ("cradle to gate"), late stages ("gate to grave"), or one specific stage ("gate to gate").

Boundaries are both in time and space. Figure 8.2 shows that a "Scope 1" boundary is roughly what happens in your factory, a "Scope 2" boundary includes energy used in your factory but purchased from offsite (usually the grid), and a "Scope 3" includes use phase, energy, materials and end of life, wherever they occur. Many product brands who contract out their manufacturing have over 90% of their impacts in Scope 3.

When setting boundaries, don't just choose parts of the system where data are easily

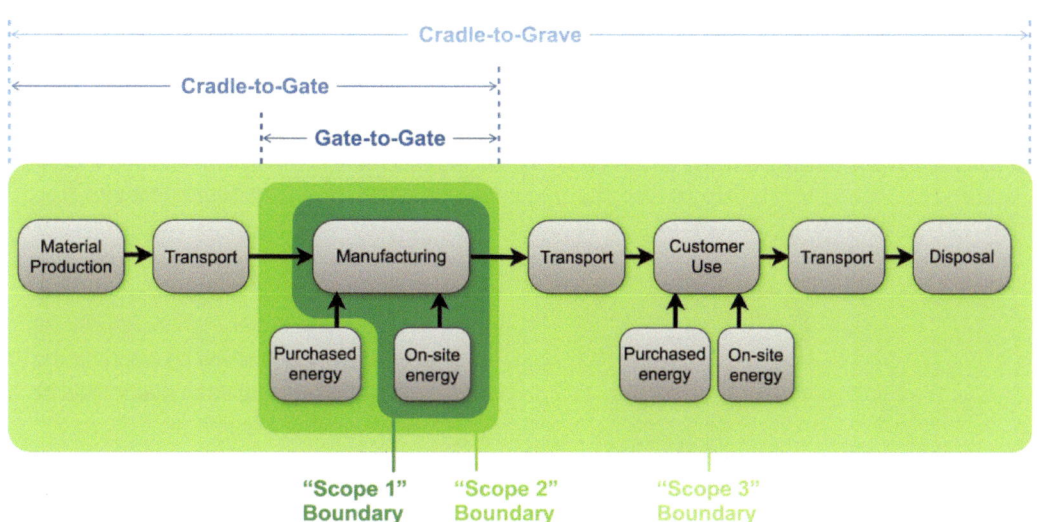

Figure 8.2 Different system boundaries in time and space

available—include everything important, and estimate values where data aren't available. For example, if comparing a disposable coffee cup to a ceramic mug that's used for years, you can't know precisely how much soap and hot water your user will use to wash the mug, but it may have a big impact, so you should at least estimate it. You might model several scenarios, as discussed later.

Do choose to exclude parts of the system that won't matter but would take a lot of time to gather data on (like factory worker commutes). How do you tell the difference? Guess and check. For example, the impacts of a factory worker commute would be divided by the number of products manufactured that day; if 10,000 coffee cups are produced each day, the impacts of commuting would be divided by 10,000. This would turn the impacts to nearly zero for even extreme commutes.

8.3 Choose Functional Unit

How many of everything should you count in your LCA inventory? LCAs calculate the environmental impacts per unit of functionality, for example, kg of CO_2 equivalent emissions per cup of coffee drunk, as in Figure 8.3. To set your functional unit, consider what function your product or service provides, and what units you would measure the function.

If your LCA comparing paper cups to ceramic mugs simply counted all the materials and manufacturing processes of each cup, the impacts of the paper cup would look much smaller. There's much less mass of material, and much less manufacturing energy, as ceramics are fired in high-temperature kilns. But this would be calculating the environmental impacts per cup produced. Thinking critically about this, it's clearly wrong—it doesn't match how the products are used by the user. The paper cup is only used once, while the ceramic mug lasts for years, perhaps being used 2000 times.

Instead of calculating the impacts per cup produced, you want to calculate environmental impacts per cup of coffee drunk, because that's the cup's function for the user. Thus, when you inventory how much mass of materials and how many kWh of energy are used for each product, you divide the mug's mass by 2000 (multiply by 0.0005) to account for the fact that only 1/2000th of a ceramic mug is produced per cup of coffee drunk. Calculating the number of items per functional unit is common sense, but check your units to verify. Will your final LCA list environmental impacts per coffee cup produced, environmental impacts per cup of coffee drunk, or something else?

Calculating environmental impacts per cup of coffee drunk will likely show the ceramic mug scores much better than the paper cup, unless usage impacts such as washing the mug with hot water and soap every time it's used outweigh this advantage.

No functional unit will capture all details of a system, so keep it simple if you can,

Figure 8.3 Comparing coffee cups

matching it to your LCA's goal and scope. A good functional unit is critical for fair comparisons, whether comparing one design option to another or comparing one life cycle stage to another in the same product to set priorities. Chapter 9 discusses functional unit choice in more depth.

8.4 Inventory and Model

Now the scope, boundaries, functional unit have been decided, you can build the LCA model (a "digital twin" of your product or service), inventorying everything in the system that causes impacts. This is a system Bill of Materials table, including the amounts and kinds of materials, manufacturing methods, transportation, energy and other resources used during product lifetime, and what happens at end of life. With this model, the impact can be calculated, either with LCA software or a spreadsheet and a lookup table from a reputable source like Idemat (Vogtlander, 2023) or Ecolizer (OVAM, 2011). Technically, this is three steps (inventory, finding proxies, and modeling), although they're so intertwined, they mostly happen together.

8.4.1 Inventory

The life cycle inventory is exactly what it sounds like: a list counting up how many kg of this material and that material, how many MJ of this energy or that energy, etc. It would be nice if the product's bill of materials contained all this data, but it never does; still, it's a good place to start if you have it. You can make this inventory on its own first, then enter it into your LCA software or spreadsheet to make your LCA model, or you can enter your list directly in the software/

spreadsheet, building the model as you inventory.

The inventory includes allocations: how much of a material or process is allocated to your system versus outside systems? For example, if you're doing an LCA of an app that runs on a phone, do you include all the impacts of manufacturing the phone, or do you assume the user already had the phone and not count any hardware, or somewhere in between? You can allocate impacts partially crossing your boundaries; for example, you might assume 10% of the reason your user bought the phone was to use your app, and 90% for other uses, thus allocating 10% of the phone manufacturing to your system. You might also allocate things within your inventory, for example, allocating a certain percent of energy use to one function versus another, based on customer usage data.

8.4.2 Modeling

The LCA model is how you calculate impacts and test different design ideas. To make the model, you look up everything in your inventory in the LCA software or lookup table, such as TU Delft's free open access spreadsheet shown in Figure 8.4, which uses data from the Idemat database (see download link in the resources list at the end of this chapter). If you have some recycled aluminum, you find "Aluminum (secondary)." (Secondary means recycled.) Sometimes the material you have doesn't appear in the list, and you need to find a proxy. More on that later. Sometimes the list has several kinds of the material, so you need to be more specific to choose the right one. For example, Idemat lists over 100 kinds of steel, and Figure 8.4 shows three kinds of aluminum—100% primary ("virgin"), 100% secondary, and the average world market mix of the two. As you choose what database items correspond to your product or service inventory, you also enter the quantities used.

The software or spreadsheet then lets you calculate the impact of each item. In Figure 8.4, recycled aluminum's carbon footprint impact is 0.8527 kg CO_2eq. per kg of material, which is shown by the "carbon footprint" column heading listing kg CO_2eq. as the unit, and the "Aluminum (secondary)" row listing kg as the unit. By contrast, "Anodizing" lists units of m^2 (area), "Hydro-electric power

"(Norway)" lists units of MJ (energy), and "Train, freight, diesel USA" lists units of tkm (ton-kilometers of weight transported across a distance). Be careful about entering your data in the right units, and entering the right amount per functional unit! When you enter your inventory data, the software or spreadsheet will calculate and graph impacts for everything, creating your LCA model.

For example, 4 kg of recycled aluminum for a chair would have a carbon footprint of (4kg aluminum) • (0.8527 kg CO_2eq./ kg aluminum) = 3.4 kg CO_2eq per chair

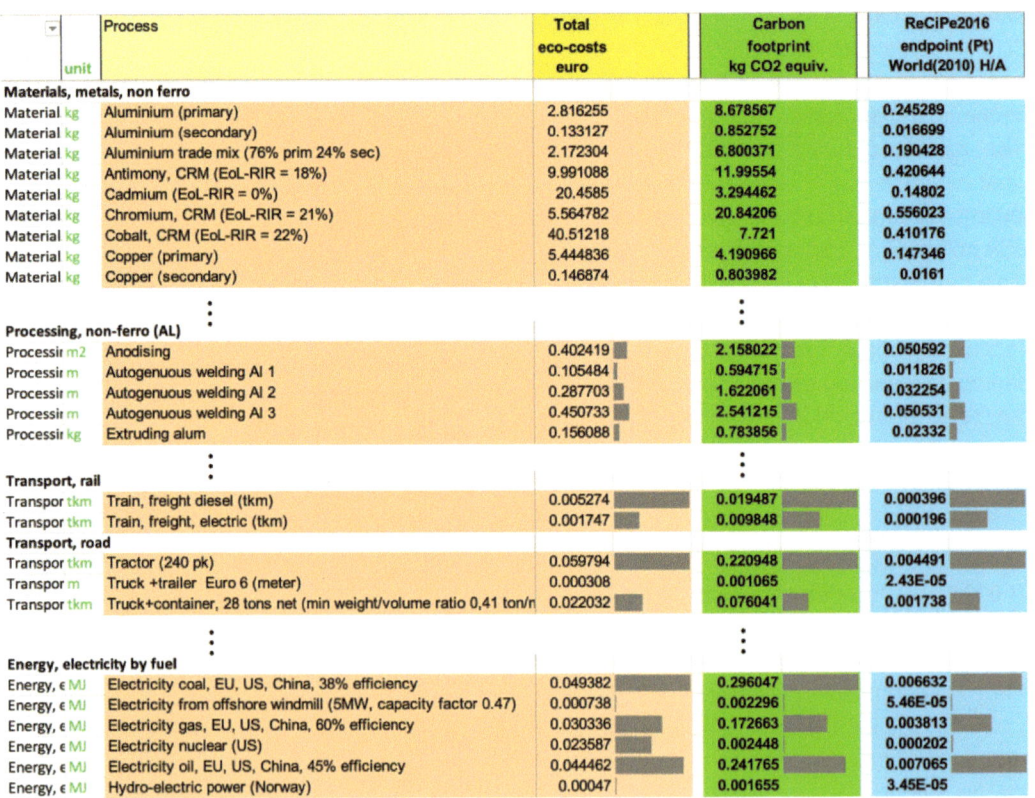

	Process	Total eco-costs euro	Carbon footprint kg CO2 equiv.	ReCiPe2016 endpoint (Pt) World(2010) H/A
unit				
Materials, metals, non ferro				
Material kg	Aluminium (primary)	2.816255	8.678567	0.245289
Material kg	Aluminium (secondary)	0.133127	0.852752	0.016699
Material kg	Aluminium trade mix (76% prim 24% sec)	2.172304	6.800371	0.190428
Material kg	Antimony, CRM (EoL-RIR = 18%)	9.991088	11.99554	0.420644
Material kg	Cadmium (EoL-RIR = 0%)	20.4585	3.294462	0.14802
Material kg	Chromium, CRM (EoL-RIR = 21%)	5.564782	20.84206	0.556023
Material kg	Cobalt, CRM (EoL-RIR = 22%)	40.51218	7.721	0.410176
Material kg	Copper (primary)	5.444836	4.190966	0.147346
Material kg	Copper (secondary)	0.146874	0.803982	0.0161
Processing, non-ferro (AL)				
Processir m2	Anodising	0.402419	2.158022	0.050592
Processir m	Autogenuous welding Al 1	0.105484	0.594715	0.011826
Processir m	Autogenuous welding Al 2	0.287703	1.622061	0.032254
Processir m	Autogenuous welding Al 3	0.450733	2.541215	0.050531
Processir kg	Extruding alum	0.156088	0.783856	0.02332
Transport, rail				
Transpor tkm	Train, freight diesel (tkm)	0.005274	0.019487	0.000396
Transpor tkm	Train, freight, electric (tkm)	0.001747	0.009848	0.000196
Transport, road				
Transpor tkm	Tractor (240 pk)	0.059794	0.220948	0.004491
Transpor m	Truck +trailer Euro 6 (meter)	0.000308	0.001065	2.43E-05
Transpor tkm	Truck+container, 28 tons net (min weight/volume ratio 0,41 ton/m	0.022032	0.076041	0.001738
Energy, electricity by fuel				
Energy, € MJ	Electricity coal, EU, US, China, 38% efficiency	0.049382	0.296047	0.006632
Energy, € MJ	Electricity from offshore windmill (5MW, capacity factor 0.47)	0.000738	0.002296	5.46E-05
Energy, € MJ	Electricity gas, EU, US, China, 60% efficiency	0.030336	0.172663	0.003813
Energy, € MJ	Electricity nuclear (US)	0.023587	0.002448	0.000202
Energy, € MJ	Electricity oil, EU, US, China, 45% efficiency	0.044462	0.241765	0.007065
Energy, € MJ	Hydro-electric power (Norway)	0.00047	0.001655	3.45E-05

Figure 8.4 Excerpt from TU Delft Idemat LCA lookup tables, showing impacts in carbon footprint, EcoCost, and ReCiPe points

Source: Vogtlander (2023).

produced, plus impacts of manufacturing processes, etc. If your functional unit was one year of sitting and the chair lasted ten years, the impact of the aluminum material would be divided by ten: 0.34 kg CO_2eq per year of sitting.

Continue modeling every part of your whole system, adding everything up to find the total impacts of your system, or any subsystem.

Sophisticated LCA professionals usually use software like SimaPro or GaBi to do full LCAs for Environmental Product Declarations, ISO certification, and academic papers. There are also easier to use, inexpensive web-based programs like SustainableMinds. However, you might find it easier or more productive to work in the Delft Idemat spreadsheet because of its fast and easy visualization of uncertainty. Figure 8.5 shows the interfaces of two programs. Some CAD programs like SolidWorks have an LCA plug-in where, as you build your CAD model and assign the materials, it calculates impacts. Ansys' Granta Selector and Granta Edupack also have LCA data. Try different programs to see which you prefer. Since the learning curve is often steep, many people prefer simple spreadsheets. The free Delft/ Idemat spreadsheet tool also automatically graphs uncertainty, so we recommend it for early-stage fast track LCAs.

8.4.3 Uncertainties

You likely won't know the exact mass of every part down to the gram, or the exact steel alloy you'll use, or the exact energy use or transport distances, especially given wide variations in customer use or shipping to different markets. So your model

should include uncertainty. This can involve sophisticated statistics, but at least use a best-case/worst-case. Uncertainty can also save you huge amounts of time: rather than spending days or weeks tracking down exact masses, transport distances, alloys, etc., you can make an estimate with a large uncertainty (even ±100% or ±10x). When you graph your results, if the estimate with its large uncertainty is still insignificant compared to other things, don't waste time on more precision. Only spend time on precision where it actually affects your decision-making. This is especially important in fast-track LCAs.

For example, if you don't know which material or transport mode to choose from the LCA database, you can guess by looking at impacts of the various choices, seeing how different they are, and including that range in your uncertainty. This is particularly easy with the Delft Idemat spreadsheet, where Excel can calculate an average and two standard deviations of the options to give you a guess with a 95% confidence interval above and below it. You can also just try switching between worst-case and best-case options and see if they make much difference in your results. (More on this in the next chapter.)

8.4.4 Proxies

Inventory and modeling sound like simple data entry, but LCA databases never include every material or process you want. Professional LCA software generally has more data, but you still usually have to use proxies for some things. This can be tricky. You can model proxies two ways: either find a direct substitute in the database, or combine several components.

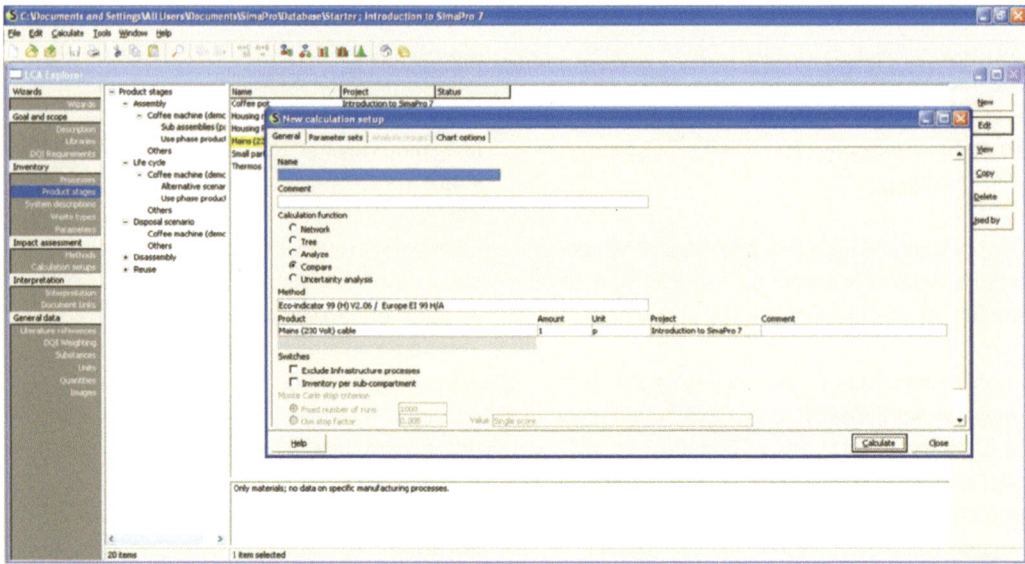

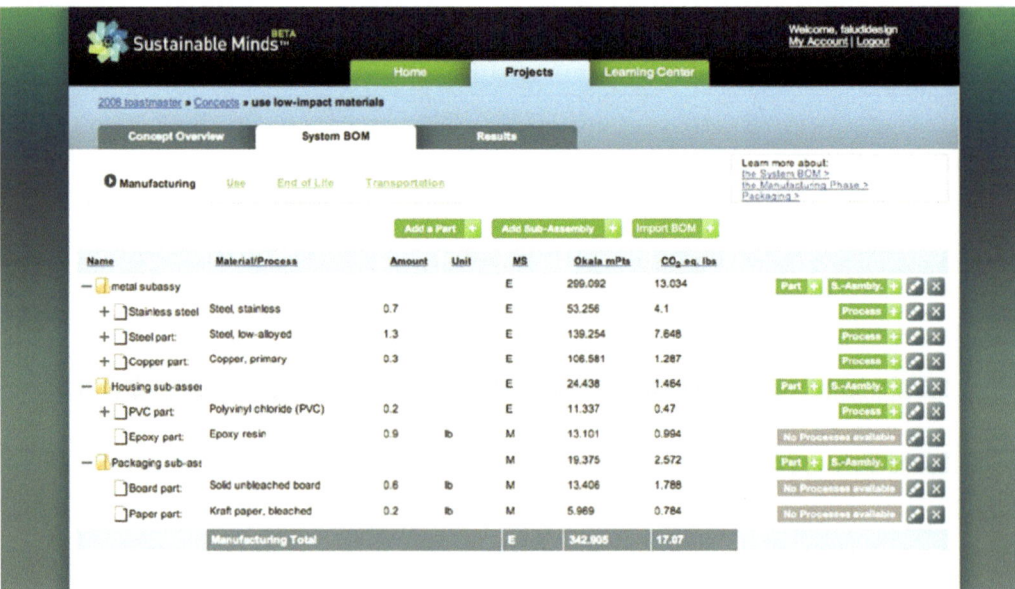

Figure 8.5 LCA software interfaces (2023): SimaPro (top) and Sustainable Minds (bottom)

For example, Figure 8.6a shows mica, used as a fireproof material in a heated hair straightener product. Mica is not in the database, so you need to find a similar material. That does not mean functionally similar, it means similar environmentally— similar abundance in the Earth's crust, similar mining energy, similarly non-toxic, etc. This is where proxies get tricky. Gypsum is a decent substitute for mica environmentally.

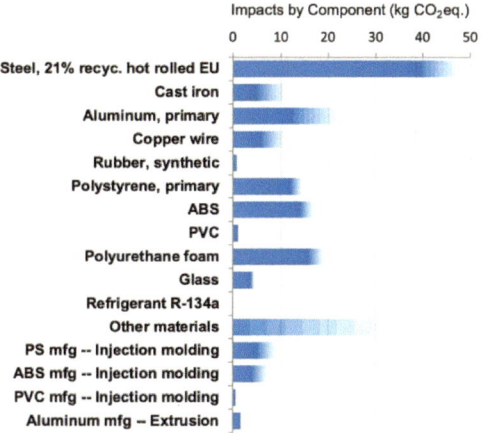

Figure 8.6 (a) gypsum is a proxy for mica; (b) copper and plastic are combined to model a cable

If there is no single database item that's a good proxy, add up components to model the missing item. In Figure 8.6b, the item missing from the database is a cable. You can model the cable as the mass of copper with a manufacturing method of wire drawing, plus the mass of plastic with a manufacturing method of extrusion. You can even do this for chemical processes or additional manufacturing heat or electricity, or anything else used to create your component.

8.5 Interpreting and Communicating Results

Remember the goal of a fast-track LCA in design is usually to decide on design priorities or decide between design options. How do you interpret the impact assessment data for your decisions, or communicate your data to others who decide?

For example, Figure 8.7 shows the material and manufacturing impacts of a refrigerator. The tallest bar (the biggest impact) is the steel material. But perhaps adding up all the plastic materials (polyurethane foam insulation, polystyrene drawers, and ABS and PVC) together would be a higher impact? Likely not, considering steel's high impacts

for cold rolling also. But there are cases where the breakdown by component or by material could change your priorities for sustainability. Think critically, and maybe group the data to reveal key factors if needed. In this case, even if the plastic parts were a worse impact, it doesn't matter, because of Figure 8.8.

Figure 8.7 LCA of refrigerator materials and manufacturing, with blurs showing uncertainties

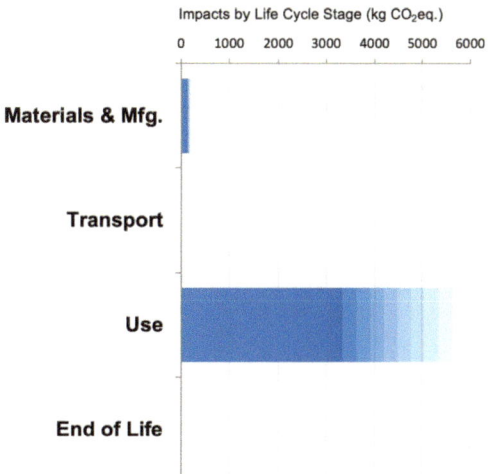

Figure 8.8 LCA of the refrigerator by life cycle stage, with blurs showing uncertainties

Figure 8.8 shows the same refrigerator, but all material and manufacturing impacts are combined in the first bar. The second bar is transport from China to Europe, but its impacts are so small they're invisible. The third bar is energy use during the refrigerator's life, which obviously dominates. The end-of-life bar appears to be zero because landfill of inert materials like steel and plastic do not emit significant amounts of greenhouse gases. This graph suggests that energy impacts are so large that no material or manufacturing choices are very important. Of course, this depends greatly on the source of electricity—this graph assumes average residential grid electricity in the Netherlands in 2023. You could check this assumption by trying other electricity sources—such tests are called "sensitivity analysis," covered in Chapter 9. But for this specific situation, a redesign that doubled material impacts in order to cut energy usage by even 15% or more would improve the carbon footprint.

Figure 8.9 shows another way of visualizing the same data as Figure 8.8, adding up all impacts of the refrigerator's materials and manufacturing in one bar, with each color representing the impact of a component or process (see the legend), and without the uncertainties. Stacking everything into one bar lets you easily compare this design option to another, while still seeing which components cause the biggest impacts in each refrigerator design. Another data visualization is a "Sankey diagram," which is like a quantified flow chart.

Finally, interpretation of results can include any other subjective aspects of your LCA, such as your boundary choices, functional units, data sourcing, allocation, etc. For

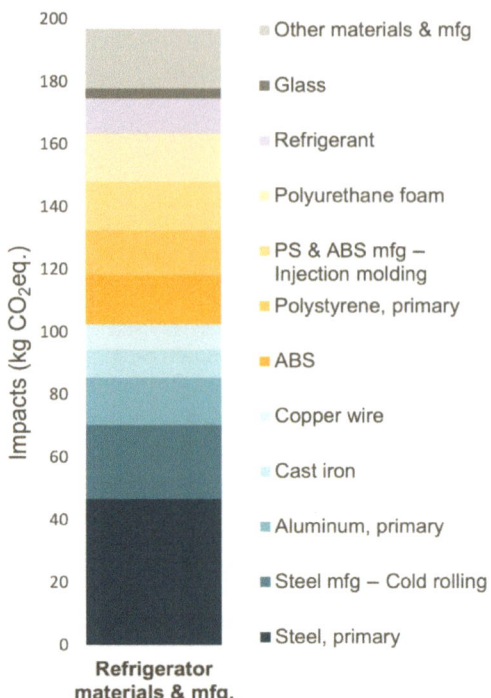

Figure 8.9 Impacts of the refrigerator's material and manufacturing, stacked into one bar

example, comparing two refrigerators that are different sizes might make you redo your functional units. But if you decide the user wouldn't use the extra capacity, you'd leave the comparison as it is. Again, sensitivity analysis can test assumptions or explore possibilities.

8.6 Iteration

LCA is often presented as a linear process, but in reality, it's iterative. The ISO14040 method for LCA (Figure 8.10) shows the iteration among all steps listed above.

In the coffee cup example, you might have started your inventory and impact assessment counting environmental impacts per cup produced. But once you realized that did not account for the many paper cups needed to perform the same function as one ceramic mug, you realized you needed to change the functional unit and redo the inventory. Such realizations can happen at any step. While it would be nice to do the entire analysis perfectly the first time, it's an expected part of the process to go back and redo things after your critical thinking makes you understand the problem better (Figure 8.10).

To sum up, doing LCA can be complicated, but breaking it down into steps and expecting iteration make it manageable. Running the numbers to quantify impacts helps set design priorities and choose options based on

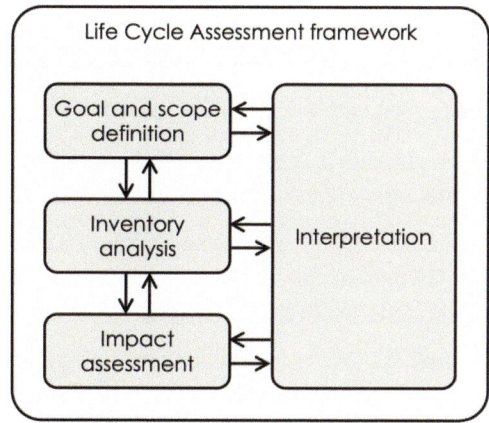

Figure 8.10 ISI 14040 LCA framework Iteration

Source: used with the permission of NEN, Delft, www.nen.nl

evidence rather than guesswork and prevents unintentional greenwashing as you make design decisions.

Resources and References

Resources for Further Study

- Vogtlander, J. G. (2014). *LCA: A practical guide for students, designers and business managers* (2nd ed.). Delft Academic Press.

- Faludi, J. (2018). Life cycle inventory data for a refrigerator. Downloadable from link on LCA tutorial page: https://venturewell.org/tools_for_design/measuring-sustainability/life-cycle-assessment-content/

- Idemat LCA Calculator template spreadsheet. (2021). Data by Vogtlander, interface by Faludi. Downloadable from link on LCA tutorial page: https:// venturewell.org/tools_for_design/measuring-sustainability/life-cycle-assessment-content/ or https://www.ecocostsvalue.com/data-tools-books/

References

OVAM. (2011). Ecolizer 2.0 Ecodesign Tool. Danny Wille Publishers, Openbare Vlaamse Afvalstoffenmaatschappij.

Vogtlander, J. G. (2023). Idemat LCA Database. Sustainable Impact Metrics Foundation. Available at: https://www.ecocostsvalue.com/data-tools-books/ (accessed July 1, 2023).

How to Apply #8: Fast-Track LCA of Your Product, or a Refrigerator
Time Estimate: 2–4 Hours

Perform a fast-track LCA. This exercise uses a refrigerator as an example, so you can see a correct answer, but ideally do the exercise once for the refrigerator and then do it for your own product or service or system.

STEP 1: Decide on Goal, Scope (Boundaries), and Functional Units
Time Estimate: 10–30 Minutes

The goal for this LCA is to quickly estimate priorities for eco-design. That is, identify the biggest eco-impacts of your product. You can also do this exercise for a service or software, but you'll make modifications as mentioned below.

- **Choose your boundaries**. What you will and won't consider in your analysis. Ideally your boundaries would choose everything in your Whole System Map, "scope 3" cradle to grave, but you might exclude things out of your control, or too peripheral to the system. For example, with the refrigerator, you'll likely exclude the food stored in it.

- **Choose the functional unit**. For example, a good functional unit for a refrigerator would be "impacts per 15-year fridge lifespan," or "impacts per year of fridge use

STEP 2: List Your Inventory
Time Estimate: 1–5 Hours

Make a table of your product's system Bill of Materials (SBOM) as best you can: how many kg of each material, with what manufacturing method(s), an estimate of energy and other resource use during product lifetime, transportation from the manufacturing site to end consumer, and disposal scenario—everything that was within the boundaries you set above. If you don't know all the details (e.g., which materials or what transport distances), make guesses and include large uncertainties (e.g., ±50% or ±100%).

If you're not modeling a product but modeling a service or software, make a table of the energy use, transportation, and other things that will be caused by people using your system and/or performing its services. You might ignore materials and manufacturing of things the users already have (like a phone they run your app on), but don't ignore materials and manufacturing your service causes to be used (e.g., printouts of art that users designed in your app).

Table 8.1 is an SBOM for a refrigerator (a large family-sized one). Note that in the Idemat LCA database, some items contain both the material and manufacturing method (e.g., hot rolled steel, cast iron); others do not (e.g., aluminum and plastics do not include extrusion or wire drawing).

STEP 3: Model and Calculate Impacts
Time Estimate: 1–3 Hours

Enter all of the data above into the LCA calculator template spreadsheet (see "Resources" section for download link), or use LCA software like SustainableMinds, SimaPro, or GaBi. If you don't know which database items to use for a material, component, or process, you can try to look them up online with any search engine, Wikipedia, or material database, or you can just guess and include a large uncertainty (e.g., ±50% or 100%), For multiple best-case/worst-case scenarios, see Chapter 9.

Table 8.1 SBOM for a refrigerator

Material	Manufacturing method	kg	lbs	Where used
Steel, 21% secondary	Hot rolled	47.6	104.8	Exterior paneling, structural
Iron	Cast	4.5	10	Compressor housing
Aluminum, primary	Extruded	2.1	4.7	Equipment for refrigeration cycle
Copper, primary	Wire drawing	2.7	6	Equipment for refrigeration cycle
Rubber, synthetic	Cast	0.2	0.4	Seals and gaskets
Polystyrene, primary	Injection molding	6.3	13.8	Shelving, drawers, interior surfaces
ABS	Injection molding	5.1	11.2	Shelving, drawers, interior surfaces
PVC	Injection molding	0.5	1.2	Shelving, drawers, interior surfaces
Polyurethane foam	N/A	5.6	12.3	Insulation
Glass	Sheet	2.9	6.3	Shelving
Refrigerant "R-134A"	N/A	0.1	0.2	Cooling/radiator tubing
Other materials	Other	7.0	15.3	Misc.
Use and end of life				
Lifetime	15 years			
End of life	Landfill			
Utilization	24 hrs/day, 7 days/week			
Power (avg.)	70.0 Watts			
Yearly energy	613.2 kWh/year			
Transport				
Ocean freighter	10,000 km (6,000 mi)			
Rail	800 km (500 mi)			
Truck	80 km (50 mi)			

You likely won't find all the right items in the database. Make your best guess about what an equivalent item would be, or assemble a proxy out of several items (e.g., you might estimate a motor with a certain mass of forged steel, copper drawn into wire, and magnets). Your estimates may be wrong, but leaving out components means you're modeling them as zero impact. If you include your large uncertainties in the analysis (again, ±50% or 100%), you can see whether you need more precision, or if the impacts are small enough that it wouldn't change the final overall result.

When you've entered all the data, you can graph the results.

STEP 4: Interpret Results
Time Estimate: 5–15 Minutes

Look at the graph(s) of your results. What do you see? What are the biggest impacts per functional unit? Where do you need to know more before moving ahead (too much uncertainty)? Use your final interpretations to estimate your priorities for eco-design. Write your interpretation briefly (20–100 words).

Note: remember, if interpreting your graphs causes you to redo earlier steps, that's okay, don't expect it to be a linear process.

Checklist for Self-Assessment

To score your success on this exercise, see if you…

☐   *Listed boundaries: what you did and didn't consider in your analyses.*

☐   *Listed functional units, to fairly compare each part of the system.*

☐   *Built system bill of materials table, including material masses and manufacturing methods, energy/resource use during product life, transportation, and disposal scenario (whatever was within your scope).*

☐   *Graphed results from LCA software or your own hand-calculations, and labeled graphs with their impact units (e.g., kg CO_2 equivalents, ReCiPe Endpoint H points, etc.).*

☐   *Wrote brief (20–100 word) interpretation: what are the biggest impacts and what does that mean for your design priorities?*

☐   *If you did the LCA of the refrigerator using the BOM listed above, the graphs showing correct answers are in this chapter, Figure 8.7 and Figure 8.8; and a diagram of the system boundaries is Figure 5.5 in Chapter 5. The SBOM's "other" item is up for interpretation. What did you choose and why?*

CHAPTER 9
LCA

Scenarios, Functional Units, Uncertainty

Jeremy Faludi,
Ruud Balkenende, and
Conny Bakker

Goals

- Decide between life cycle assessment (LCA) options considering uncertainty
- Incorporate appropriate amounts of uncertainty in your LCA models
- Choose functional units more precisely
- Perform sensitivity analysis in LCA to test your assumptions

DOI: 10.4324/9781003504672-10

Why It Matters

As the old saying goes, "there are lies, damned lies, and statistics." LCAs can be misleading when done poorly, and the most common flaws are mistakes with functional units, failing to acknowledge uncertainty, and failing to test assumptions with sensitivity analysis. Handling these three factors well in your own LCA, or judging them in someone else's LCA, helps ensure you make the right decisions.

Summary

- Functional units enable fair comparisons between different design options.

- Functional units divide environmental impacts by how much service is performed for users. Be clear how many items per functional unit are involved, or your LCA will be wrong.

- All LCAs have uncertainty. You should display and acknowledge it.

- When uncertainty ranges overlap, don't assume one impact is larger or smaller than another—improve your precision so they don't overlap, or make decisions including the uncertainty.

- Uncertainty can save you time—when LCA modeling, start with guesses that have huge uncertainties; anything whose impacts are too small to matter even in the worst case does not require more precision.

- Sensitivity analysis is making multiple LCA models to test whether your conclusions change when your assumptions change, to see if different situations change your design decisions.

What does your LCA measure— environmental impacts per product produced? Environmental impacts per person served? Environmental impacts per year of system operation? To make fair comparisons between design choices, or between different stages of one product's life cycle, you need a good functional unit. It's not about the masses of materials or the emissions produced, it's about the functionality to the user. There are no set formulas for this, you make up the math through common sense and checking the units work out.

Some product categories have standards agreed upon by academics and/or industry. For example, passenger vehicle LCAs usually use the functional unit of "passenger kilometers," which means the LCA measures impacts per person per unit distance they travel. So while a hybrid-electric car might burn much less fuel per km traveled than a diesel bus, shown in Figure 9.1, the bus carrying an average of 50 people will have its impacts per km divided by the 50 passengers. Despite using a dirtier fuel and using more energy than one car, it will likely be the more environmentally responsible solution, because it replaces many cars. Smart choice of functional units shows this, and poorly chosen units can bias results.

In deciding your unit, think about how you'd measure its function. What's the function of a pair of jeans? One year of ownership? What if one pair of jeans is worn much more than another because it's more stylish or more comfortable? Should it be one day of wearing? What's the functionality of a refrigerator—one year of cooling? What about refrigerators that are larger or smaller

Figure 9.1 Comparing a hybrid car versus a diesel bus

than others, should it be one year of cooling one cubic meter of food, or a similar metric? Consider the goal of your LCA and what would serve it best.

There comes a point where no functional unit will capture all details, so match your functional unit to the scope of what question you're answering with the LCA. For example, the bus often has just one or two riders, thus its impacts per passenger kilometer become higher. But the bus always runs according to schedule whether anyone rides it or not, so any private car trip will increase pollution and fossil fuel depletion compared to riding a bus. Such details cannot be handled by the functional unit, they must go into the scenarios modeled for comparison (sensitivity analysis), and the range of usage should be displayed as uncertainty in the results.

9.1.1 Items Per Functional Unit

When making your life cycle inventory, how many of each inventory item are there per

functional unit? You need to make up an equation so the units work out right. For example, Chapter 8 compared a ceramic mug to a disposable cup, measuring environmental impacts per cup of coffee drunk, not per cup produced, to make a fair comparison. One disposable cup per drink is easy to calculate: there's one item per functional unit. The mug requires making up an equation. If the user drinks one cup of coffee per day from the ceramic mug, every day of the year (365 days/year), and an average mug's lifetime is 6 years before being broken or lost, then the equation is in Figure 9.2.

Figure 9.2 shows how units of days cancel out, as do units of years. The final unit "drinks/mug" verifies you're not calculating drinks/year or mugs/year or others. But your end goal is to calculate environmental impacts per cup of coffee drunk, so you don't want units of drinks/mug, you want mugs/drink. Thus, the number of items per functional unit that you enter into your lifecycle inventory will be 1/(2190 drinks/

$$1 \, \frac{\text{drink}}{\text{day}} \cdot 365 \, \frac{\text{days}}{\text{year}} \cdot 6 \, \frac{\text{years}}{\text{mug}} = 2190 \, \frac{\text{drinks}}{\text{mug}}$$

Figure 9.2 Calculating items per functional unit by simple unit analysis

mug) = 0.000457 mugs/drink. That number of items per functional unit will give answers in environmental impacts per cup of coffee drunk.

If the mug lasted two years or ten years or was shared among multiple people, the numbers would change accordingly--the 0.0005 in Chapter 8 assumed the mug lasts 5.48 years. Check your units to check the calculations.

You can also adjust your functional unit for convenient calculation. For example, after finding that 0.000457 is an awkward number, you might decide to calculate environmental impacts per 2190 cups of coffee drunk. Then your inventory would contain one mug and 2190 disposable paper cups. Similarly, you might decide a refrigerator's functional unit is 10 years of cooling, or the unit comparing a bus and car is a million passenger kilometers. These multipliers make no difference functionally, as long as they're consistent. They're just more convenient to calculate.

9.2 Uncertainty

No data are perfect. Get comfortable with uncertainty—acknowledging the possible range of error in your data. Even the most rigorous LCA is never better than ±10% precision (Ashby, 2020), and early-stage estimates can be off by orders of magnitude. When uncertainty ranges overlap, don't assume one impact is larger or smaller than another.

There are many sources of possible error; some major ones are these: time variability is when the data in LCA databases were several years old, sometimes 20 years old or more. Some processes stay similar over time, but others, like electronics manufacturing, have changed greatly over time. Spatial variability is when the database data were gathered in a different location from your operations—maybe on a different continent. Some processes cause fairly similar impacts wherever they happen, like coal-fired steel production; others, like farm irrigation, have very different impacts in different locations. Some database data were not empirically measured, just modeled/estimated. Other kinds of variability also exist.

Even if the data in the LCA database are perfect, there is still uncertainty in your LCA model. "Parameter uncertainty" is when you lack exact data on how much material mass or energy use or travel distance, etc. This you can fix yourself, by gathering more precise data and/or modeling multiple scenarios of different possibilities, but it's usually your largest source of error, particularly in early stage design.

Uncertainty is not always bad—you can use uncertainty to save time. When building an LCA model, the most time-consuming part is gathering precise data. By taking rough guesses with huge uncertainties, you can build a fast track LCA model quickly. Things whose impacts are too small to matter even in the worst-case estimate can be left as vague guesses. Only things whose impacts matter (and especially whose uncertainty ranges overlap with other things you're comparing them to) need the precision of exact data.

For example, Figure 9.3 shows the impacts of 3D printing a plastic part compared to CNC milling or injection molding the same

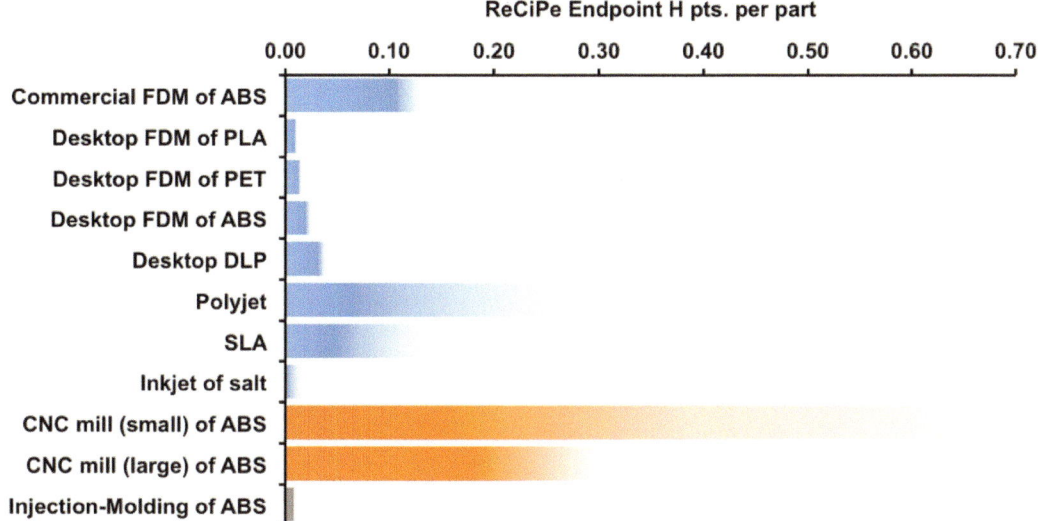

Figure 9.3 Comparing impacts per part produced for various 3D printers (blue), two CNC mills (orange), and injection molding at scale (gray). Blurs represent uncertainty (95% confidence interval).

Source: Faludi et al. (2014).

part at large scale (millions of parts). The blurred bar ends are uncertainty, showing 95% confidence intervals (i.e., there is a 95% chance the real number lies somewhere within the blur). You don't need to understand the different kinds of 3D printers to see which have better or worse impacts. You can see that almost everything is worse than injection molding, although "desktop FDM of PLA plastic" and "inkjet of salt" may be within uncertainty of it. The 3D printer "desktop FDM of PLA plastic" is clearly better than Polyjet, but it's not clear whether Polyjet is better or worse than "CNC mill (large) of ABS," because their uncertainty ranges overlap significantly. Comparing Polyjet to SLA, the latter seems likely better, but you can't be completely positive because the uncertainties overlap somewhat. To be safe, gather more precise data to shrink the error bars and determine which is clearly best. SLA is clearly better than either CNC mill, because their uncertainties don't overlap.

In early design stages, there will be lots of uncertainty, but you can still make many decisions confidently. As described above, Figure 9.3 has many large uncertainties but enables many clear judgments. More precision is required to decide between Polyjet and CNC or between inkjet of salt and injection molding of ABS, but you don't need more precision on everything. Thus, you can manage uncertainty versus time and effort.

Some LCA graphs don't show uncertainty, but that doesn't mean it isn't there. Again, no LCA data are ever better than ±10% precision, so when you report LCA results, you should generally not list more than two significant figures, and you should not decide between items with less than 10% difference between them, unless you have other ways of determining precision.

Uncertainty is often displayed as error bars in graphs, as in Figure 9.4. These

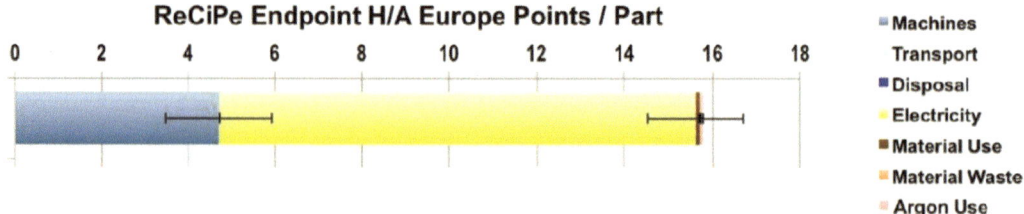

Figure 9.4 Uncertainty (95% confidence interval) shown as error bars in a stacked graph

are easy to ignore, which is why many of the figures here show uncertainty as blurred bars—the blur both communicates uncertainty to non-experts, and forces experts to acknowledge that the exact value is not known. While uncertainty often feels uncomfortable, you need to acknowledge it to make better decisions. In Figure 9.4, error bars are required because the different bars are stacked on top of each other. As noted in Chapter 8, this allows you to compare impacts within a single design while also comparing that design to other designs.

How much uncertainty should you use? Error bars, blurs, and other visualizations can represent different levels of uncertainty. The most common are one standard deviation (68% likelihood of any value falling within the uncertainty range) or two standard deviations (95% likelihood of any value falling within the uncertainty range). Sometimes people show three standard deviations (99.7% likelihood), though this is extreme. Figure 9.3 and Figure 9.4 both show 95% confidence intervals (two standard deviations); it's a good standard providing near certainty without extremely large error bars. As mentioned in Chapter 8, one advantage of the Delft Idemat spreadsheet is that Excel lets you quickly calculate averages and standard deviations of selected cells, If you're choosing between several kinds of steel or several transport modes, for instance.

What if you don't know the statistics for your data, then how much uncertainty should you use? A non-scientific but helpful rule of thumb is: assume ±10% for precise data and a perfect match between your inventory item and the database item (e.g., the exact alloy of steel or the exact truck size for shipping), because of the uncertainty of LCA database data. Assume ±30% for a very plausible proxy (e.g., drawn copper wire and extruded PVC for a cable), and assume this includes both LCA database uncertainty and parameter uncertainty. Assume ±50% for reasonable but less certain proxies, and assume ±100% or more for wild guesses. You can check such guesses with sensitivity analysis, and refine your percentages.

9.3 Sensitivity Analysis

Sensitivity analysis is trying out different assumptions to make better-informed decisions, seeing how sensitive your final results are to changes in input assumptions. Try a best case and worst case, or multiple cases. You might test your assumptions with different materials, different product usage scenarios, different energy sources, different transportation distances, or anything else you're uncertain about. If the different input changes the LCA result so much that you'd make different design decisions, then you

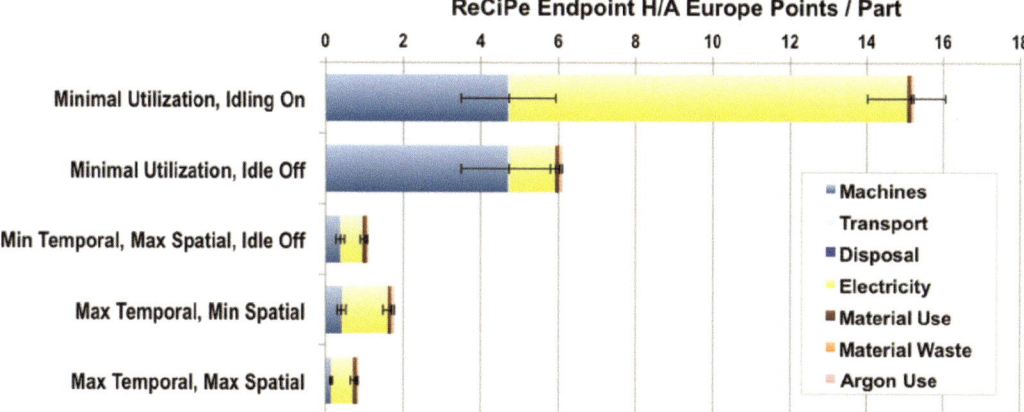

Figure 9.5 An LCA showing uncertainty (95% confidence interval) in stacked columns using error bars

should clarify what the real-world data are. If several cases might happen in reality (such as different customer usage patterns changing whether you focus design on better energy efficiency or better materials), you might make different product lines for different scenarios. Such sensitivity analysis may even cause you to rethink your boundaries or functional unit; that's okay, even encouraged.

For example, imagine you are planning a sustainable redesign of the metal 3D printer in Figure 9.5. Figure 9.5 shows its impacts per part printed in several different scenarios. "Minimal utilization, idling on" is printing one part per week while sitting turned on while idle the rest of the week. "Minimal utilization, idling off" is printing one part per week but turned off when not in use. "Max temporal, max spatial" is printing parts 24 hours per day 7 days per week and printing as many parts at once as possible. For "minimal utilization, idling on," the largest impact is energy use, suggesting you prioritize energy efficiency in design. However, for "minimal utilization, idling off," the largest impact is the printer hardware ("machines"), suggesting your design should prioritize material

reduction or greener materials. But both of these scenarios cause vastly worse impacts per part printed then "max temporal, max spatial," so perhaps the highest priority is to maximize printer utilization. You could analyze other scenarios also, such as renewable electricity instead of normal grid electricity.

As this example shows, you may not have one single answer for top priority in sustainable redesign—critical thinking and judgment are required! However, the LCA helped narrow down priorities. All scenarios agreed that impacts of material use, material waste, transport, and end of life were not high priorities. Even when an LCA gives one clear answer, your actual design changes depend not only on the LCA, but what's feasible to build technologically and economically, etc.

When modeling scenarios, choose ones that answer your main questions, for example, where you lack data or where different circumstances will cause different impacts. You can have many different scenarios and even have different uncertainties in different scenarios. Whatever helps you make informed design decisions is good.

References

Ashby, M. F. (2020). *Materials and the environment: Eco-informed material choice* (3rd ed.). Butterworth-Heinemann.

Faludi, J., et al. (2014). Sustainability of 3D printing vs. machining: Do machine type & size matter? In *Proceedings of EcoBalance Conference*, Japan.

Faludi, J., Baumers, M., Maskery, I., & Hague, R. (2016). Environmental impacts of selective laser melting: Do printer, powder, or power dominate? *Journal of Industrial Ecology*, 21(S1).

How to Apply #9: Refined Fast-Track Refrigerator LCA
Time Estimate: 2–4 Hours

Refine your LCA from Chapter 8 by considering uncertainty and revisiting functional units, as well as modeling different design scenarios and/or performing sensitivity analysis.

STEP 1: Refine Goal and Functional Units
Time Estimate: 10–30 Minutes

Start with your LCA from Chapter 8, but edit the goal to be about comparing design options and/or performing sensitivity analysis to test assumptions you made in the original LCA model.

Keep the same boundaries of what you will and won't consider, for fair comparison between all design options/scenarios.

Reconsider the functional units for your analysis, to ensure fair comparisons between all design options/scenarios. Is there something your original units missed? For example, if you compare a mini-fridge to a full-sized refrigerator, the unit should include the volume of food cooled. Write down the new units and say why. (Or if you keep the old units, say why.)

STEP 2: Estimate Uncertainties
Time Estimate: 10–30 Minutes

In your LCA model in the spreadsheet or LCA software, write down how much uncertainty you have for every item, and briefly write why. These can be copy-pasted or listed by category for similar uncertainty levels (For example, maybe all materials and manufacturing are ±50% due to early-stage design vagueness, but energy use is ±10% because some design specification tightly dictates it.)

Not all LCA software has places to enter uncertainty data, so you might have to export data from the software to graph with uncertainties in Excel or other software. The free Delft Idemat LCA calculator spreadsheet graphs uncertainty for you. To determine uncertainty amounts, you can use the rough rules of thumb in the chapter, or proper statistical methods if you know them.

Remember, no LCA results anywhere are ever less than ±10% uncertainty, due to variations between database data and your product or service.

STEP 3: Model Design Scenarios and/or Sensitivity Analysis
Time Estimate: 1–3 Hours

Make variations of your LCA model, either to compare different design options or for sensitivity analysis (testing the assumptions you made in the original model). Make at least 3–5 model variations, listing the different assumptions and graphing them all to compare results. Technically, sensitivity analysis and testing different design scenarios are different things, but they follow the same process, so you might do them all together.

For sensitivity analysis, consider where your LCA model 's uncertainties could make a big difference in your final interpretations, and model best-case and worst-case (or several) scenarios to test those assumptions. These could be different materials, usage scenarios, end- of-life options, etc., especially where you used proxies or guesses. For example, changing a refrigerator's source of electricity from US grid average to 100% solar would radically reduce impacts and change your priorities for green redesign.

For design alternatives, simply model any variations you're curious to test (e.g., a mini-fridge, or a fridge with a glass door so users don't need to open it to see inside). For each one, consider not only your desired design changes, but all the consequences of those changes (e.g., a mini-fridge would use less material, but more material per unit volume of food; a fridge with a glass door would save energy from less door opening, but would lose more energy the rest of the time, due to poorer insulation).

To make these alternate models, make copies of your LCA model so you only need to change the things that are different. In the Delft Idemat spreadsheet, you can copy the whole sheet or copy and paste within the sheet. The sheet is built to compare two scenarios, you could edit it to compare more. In other LCA software, the process can be more involved. Some software doesn't let you copy models, so you'll have to export results as you change the model and change it back.

STEP 4: Sensitivity/Scenario Interpretation
Time Estimate: 5–20 Minutes

Compare the graphs for the different model scenarios and interpret them. If you modeled different design options, which one is the winner and why? For sensitivity analysis, did it change your priorities for improvement? For either, do you need to gather more data or model more variations? Write down your conclusions briefly (20–100 words).

If this causes you to rethink your boundaries or functional unit, that's okay. Redo them and make new graphs to interpret, just be consistent about the same boundaries and functional unit for all scenarios being compared.

Checklist for Self-Assessment

To score your success on this exercise, see if you…

☐   *Listed the new functional units, and why you did or didn't change them from last time.*

☐   *Listed uncertainties for every item in your inventory, and listed why.*

☐   *Modeled 3–5 or more variations, sensitivity analysis and/or design scenarios.*

☐   *Wrote a brief (20–100-word) interpretation of the results: Which design option was best? Did sensitivity analysis change your priorities for improvement?*

PART II

Action

Design Strategies

Visions with concrete goals and metrics are a great start, but how do you get there? As mentioned in Part I, we have had good enough theories of sustainability for decades. We mostly lack implementation at scale. You can use standard design tools and methods to apply Part I's ideas. Some of the tools and methods in Part I provide not just goals but roadmaps. But more specific design tools and methods multiply your effectiveness. Different products or services have different needs and different contexts, so finding the right tool for the job (or the right combination of tools) gets you much farther much faster. Part II's chapters provide more of these specific design tools, methods, and strategies.

Many of these chapters describe what to think about in different life cycle stages, corresponding directly to the eight strategies of the Ecodesign Strategy Wheel. However, not all are obvious, and some have tools or methods that address several life cycle stages or ecodesign strategies. This list may help (see Figure P2.1):

- Materials chapters (introduction, metals, plastics, and bioplastics) help with low-impact materials, clean manufacturing, efficient distribution and packaging, extended use (durability), and end of life recovery.

- Energy chapters (energy literacy, efficiency, and generation and storage) help with use efficiency.

- Behavior change helps with use efficiency, extended use, recovery for reuse, end of life recovery, and system level (e.g., buying less stuff).

- Product lifetime extension helps with extended use.

- Product service systems help with extended use and recovery for reuse.

DOI: 10.4324/9781003504672-11

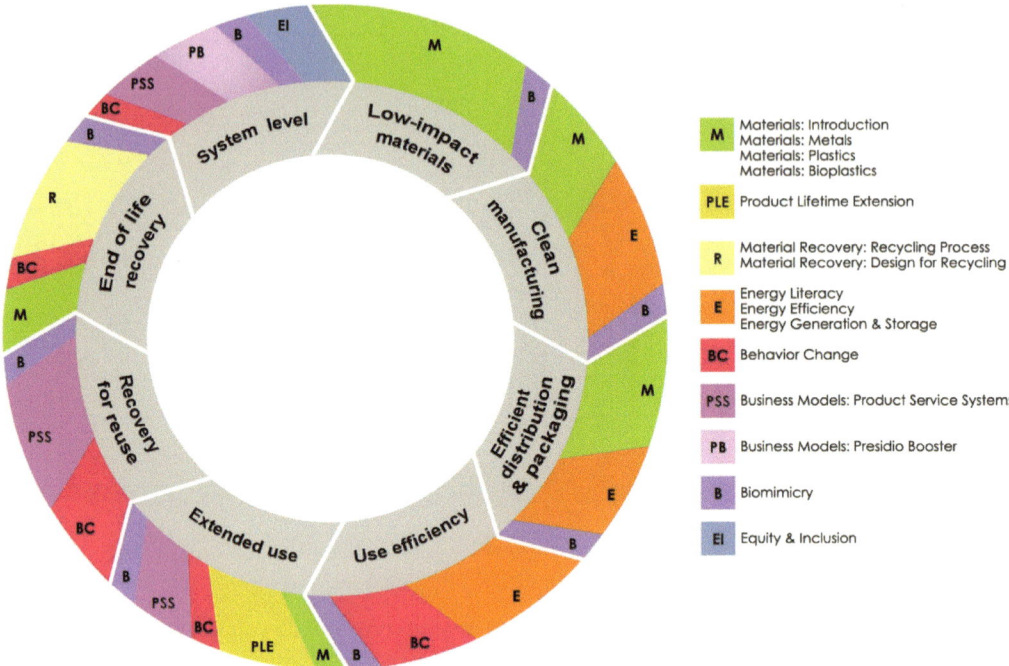

Figure P2.1 Part II's chapters in the Ecodesign Strategy Wheel

- Recycling chapters (process and design for recycling) help with end of life recovery.

- Business models: Presidio Booster helps with the system level (and thus potentially also any other life cycle phase).

Integrating into Your Design Process

As with Part I, these strategies don't replace standard design practice, they add to it. Some of these strategies are most useful early in the design process, while others are more useful later, in detailed design. Figure P2.2 shows the chapter topics as lines across the standard "double diamond" process. The lines are blurred to show there aren't strict limits, but generally speaking,

business model tools are useful in the earliest Discovery stage. Product service systems span from Discovery to Define, with tweaking in Development. Behavior change can span all stages, from inspiring a new product or service to fine-tuning detailed "nudge" strategies in Develop and Deliver stages. Material choices mostly span Define and Develop, while energy considerations mostly span Develop and Deliver, though inspiration or changes may come at any stage. Product lifetime extension and recycling usually apply to Develop and Deliver, and often must be further refined in future product generations.

Considering what stage of product or service development you're working at, and what life cycle stages you need to focus on to meet your sustainability goals, will help you choose the best tools for the job. This, in

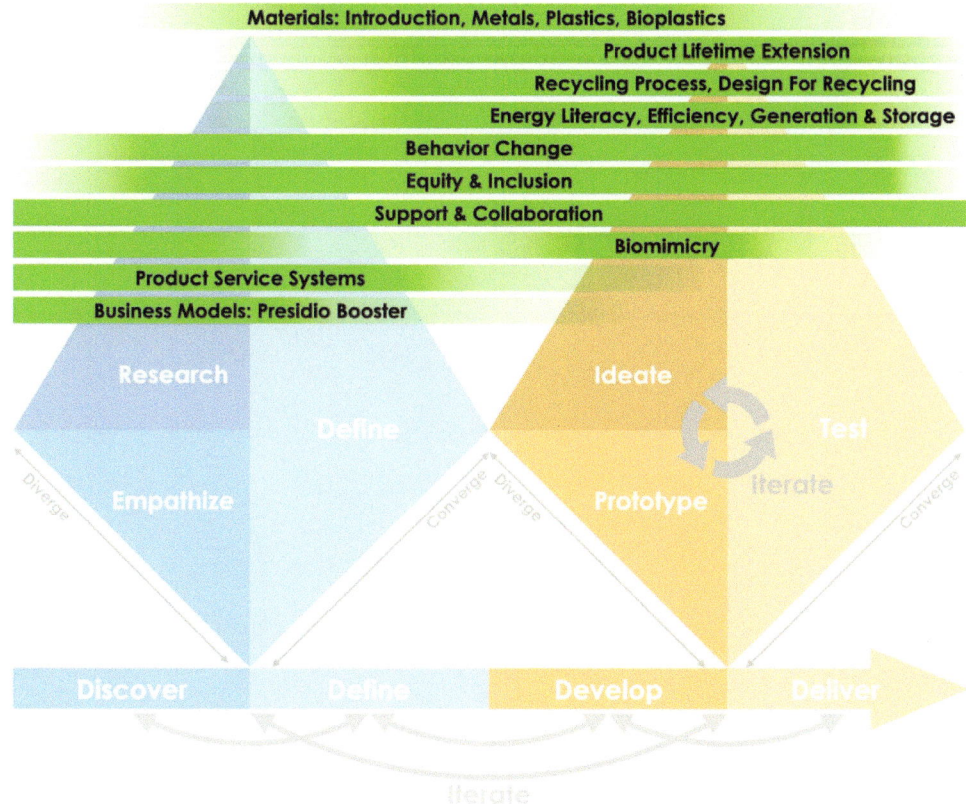

Figure P2.2 Part II's chapters overlaid on the product development "double diamond"

turn, helps you save time, money, and effort without compromising your vision of a more sustainable product or system. This is where your vision of healthy abundance, justice, and beauty in the world becomes action. Where you create a better world one nut and bolt at a time, one sewing stitch at a time, or one user interface click at a time. And as you scale up to millions of products, millions of systems, and millions of designers, it changes the world.

Innovating without Products

As with Part I, some of these tools and strategies can also be useful if you're not developing a product, but trying to reduce overconsumption by reducing the number of products developed. Governments, nonprofits, and communities can especially use behavior change, product service systems, and the Presidio business model canvas sustainability booster to design programs reducing consumption, whether designing regulations or repair cafés or events for connecting communities. Again, while the economics are very different, the design process is similar. educing the amount of stuff in our lives can be a joy, not a hardship, if we design for gratitude and the concept of having enough.

CHAPTER 10
Materials

Introduction

Ruud Balkenende,
Jeremy Faludi,
and Conny Bakker

Goals

- Describe the major effects of materials
  with respect to sustainability

- Compare sustainability aspects of different
  material classes

- Select appropriate material properties,
  including environmental aspects

DOI: 10.4324/9781003504672-12

Why It Matters

Materials play a dominant role in the design of physical products. The choice of materials not only determines product performance and cost, but also to a considerable extent the environmental impact of a product over its lifetime, from the effects of sourcing and processing to opportunities for recovery and reuse at the end of life of a product. Mining and processing often do not comply with basic human rights standards. The increasing global use of materials is, on the one hand, a major source of environmental impact, while, on the other hand, increasing amounts of critical materials are required for the renewable energy systems to achieve sustainability goals regarding the energy transition, so tradeoffs must be considered.

Summary

- The number of different materials, as well the amount of materials used, keep increasing and already largely exceed the carrying capacity of the planet.

- In addition to performance-determining product functionality and economic aspects, environmental and social impacts of materials should be considered from the start when designing a product.

- A major distinction can be made between renewable materials and non-renewable materials.

- Many strategies in the Ecodesign Strategy Wheel directly relate to materials and the way they are applied in products.

10.1 Uses of Materials

Materials are crucial to all products and are used in enormous quantities. Even seemingly dematerialized digital services are enabled by materials, for example, in smartphones and data centers, powered by fossil fuels, solar cells, or wind turbines. And not all materials that are used during manufacturing end up in the final products, for example, manufacturing waste, processing chemicals, machinery that wears out and needs repairs, disposables used in machines, all add to the material bill of products. Our demand for resources still rises every year, from 9.34 tons per capita in 2000 to 12.33 tons per capita in 2019, a 32% rise (Our World in Data, 2020). Until the 1970s, global consumption remained within the natural capacity to replenish these resources. Since then, we have been using more resources than one planet Earth can provide. Currently, we live as if we had 1.7 planet Earths available to sustain our needs.

For millennia, humans have used a limited number of readily available materials like wood, leather, grasses, stone, sand, copper, silver, tin, and iron. The past two centuries have seen a huge rise in the use of many more materials that are harder to obtain and require more elaborate processing. Where electronics in the 1950s were still mainly based on nickel, copper, tin, zinc, lead, and tungsten (for electrodes in vacuum tubes), a current smartphone contains over half of the elements in the periodic table, see Figure 10.1 (Torrubia et al., 2022).

Many of these materials are in the meantime referred to as "critical." The concept of criticality usually addresses two criteria:

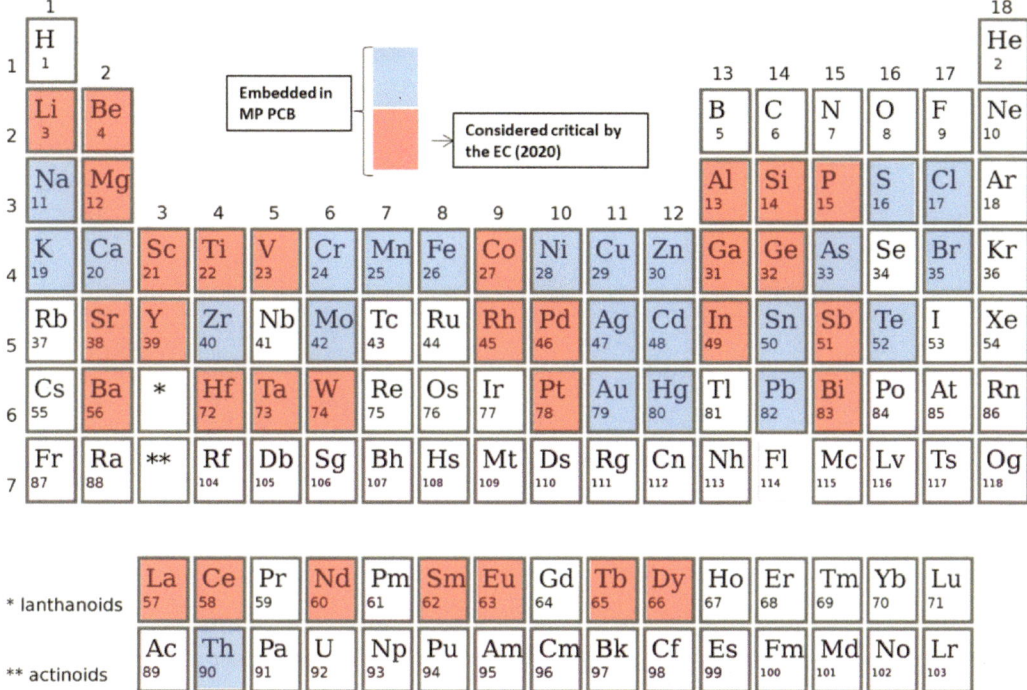

Figure 10.1 The periodic table, indicating elements used in smartphone PCBs
Source: Torrubia et al. (2022).

supply risk and vulnerability to supply restriction. See Figure 10.2, which also considers a third criterion, environmental implications. This implies that the materials are considered essential economically, while they bear a risk that access could be limited (whether due to physical scarcity or geopolitics). Note that an evaluation of these criteria will lead to different results for different geographical regions and that at a company level the results can again be different, for example, dependent on their supply chain. Recently a third criterion has been added: environmental implications. For this criterion, LCA data can be used. This addition provides useful information to designers, for example, on the potential impacts of using a specific material.

Materials are not used separately but are combined in all kinds of ways in products. On a molecular scale, alloys contain multiple metals, plastics have many additives to tune their properties. On a microscopic scale, composite materials derive their unique and often tailored structural properties from the combination of particles or fibers that are embedded in a matrix material. At the macroscopic level, different materials are connected to each other, often with joints like screws or glues that add yet another material. And coatings are applied to many surfaces for protective or aesthetic purposes. This combination of materials at various scales improves functional performance of products, but simultaneously poses problems when it comes to reusing and recycling products and materials.

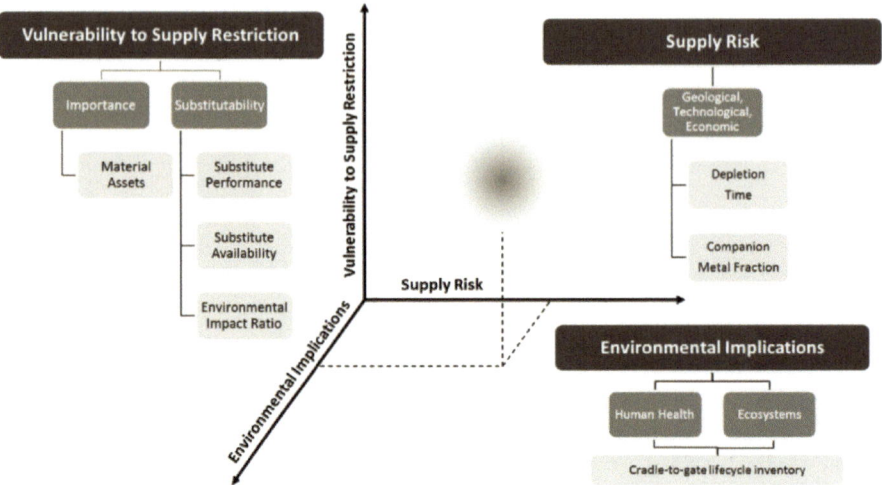

Figure 10.2 The Yale analytical framework for determining metal criticality at the global level
Source: Graedel et al. (2013).

10.2 Properties of Materials

Sustainable products need materials that are sustainable and that enable sustainable use of the product. The sustainability qualities needed vary by application and can contradict each other. For example, to improve fuel efficiency, Boeing's 787 Dreamliner replaced aluminum with stronger and lighter carbon fiber for most of its body and claimed a 20% improvement in fuel efficiency. But the 787's carbon fiber is not easily recyclable at the end of life, like aluminum. Nevertheless, for an airplane body it is an excellent choice, because most of the airplane's environmental impacts are due to burning fuel during the use phase. Determining which materials are more or less sustainable for your application requires quantifying the impacts of the whole system, to set priorities and balance tradeoffs. Sometimes the functionality of a material can be its greatest sustainability leverage.

In product design, materials are usually primarily selected for their function during use: mechanical/structural, electrical, thermal, magnetic, optical, or combinations of these properties. But products also need to be manufactured, and the manufacturing process brings about additional requirements regarding, for example, ductility and workability. Further, products need to be affordable, so the costs of materials and manufacturing come into play. The latter, unfortunately, is mostly limited to direct financial costs, like material prices and equipment investments. Social and environmental costs are too often neglected, as are the potential side effects of use and at the end of life of a product.

When selecting materials, it is important to directly consider environmental and social indicators. These are as important as the physical and chemical properties that determine product functionality and the cost that determines competitiveness of a product in the market. Figure 10.3 provides an overview of properties that

Environmental

- Impacts
  - Global warming
  - Human toxicity
  - Ecotoxicity
  - Durability
  - Chemical pollution
  - Ozone depletion
  - Acidification
- Efficiency
  - Materials and energy use
  - Recyclability
  - Product durability
  - Assessment of suppliers

Social/Economic

- Standards of conduct
  - Business
  - Child labor
  - Fair prices
- Work
  - Income distribution
  - Work satisfaction
  - Staff turnover
  - Staff development
- Economic
  - Value added
  - Environmental liabilities
  - Ethical investments

Functional

- Mechanical
- Thermal
- Optical
- Electrical
- Magnetic
- Weight (density)

Figure 10.3 Material properties that are easily and commonly quantified

for many materials are readily accessible, distinguishing between environmental, social/economic, and functional indicators. These indicators should be dealt with in an integral way: considering environmental and societal indicators in a late stage of product design will lead to sub-optimal considerations regarding for example, global warming effect or recyclability.

10.3 Classification by Physical and Chemical Properties

Traditionally, materials science distinguishes between four classes based on physical and chemical properties. These classes and some examples are shown in Figure 10.4 and some will be discussed in more detail in subsequent chapters.

From a sustainability perspective, these classes are less distinct, as all classes contain low and high impacts, dependent on the specific material and its processing. For example, see the global warming potential of materials in Figure 10.5. In general, high-performance metal alloys (like titanium alloy), specialty ceramics (like barium titanate), and composites (like carbon fiber reinforced plastic), have a relatively high carbon footprint, engineering materials (e.g., common steel, aluminum, and plastics) are intermediate, and materials based on natural

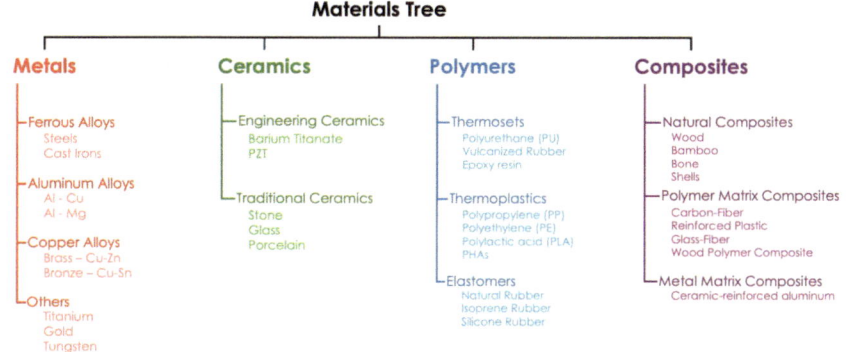

Figure 10.4 Material physical/chemical types
Source: Idemat (2024) database.

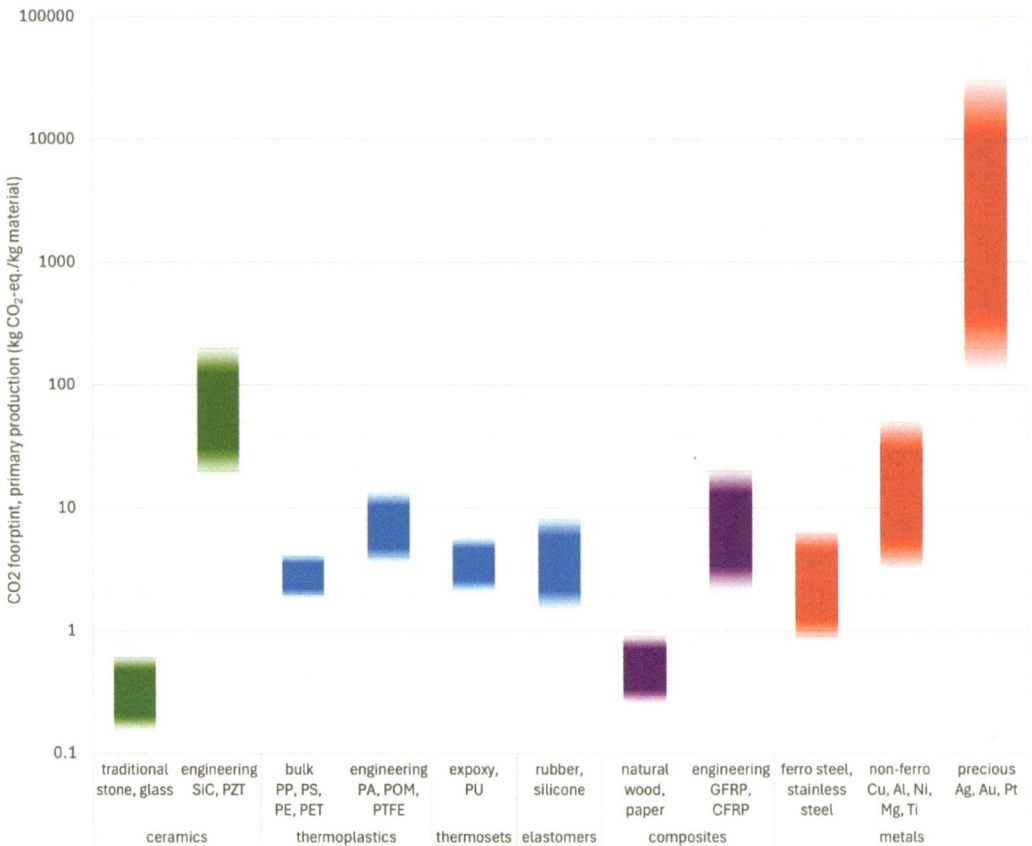

Figure 10.5 Carbon footprint per kilo of material for various material types

resources that need little processing (wood, leather, stone, glass) have a relatively low impact. Materials also have very different levels of health or toxicity hazard. A material can be toxic to the end user, but also to workers during harvesting/extraction, manufacturing, and disposal. Toxicity also includes ingredients/substances the material emits or decomposes into. For example, plywoods are safe to touch, but many outgas formaldehyde fumes that are harmful for people to breathe. Gold is harmless to the end user, but usually requires poisonous cyanide and mercury to extract it during mining.

Social impacts are not properties of the materials themselves, but are related to the companies and governments involved. Particularly bad examples are "conflict minerals" (tin, tungsten, tantalum, and gold), named for their mining profits supporting wars between various armies and rebel groups, such as in the eastern provinces of the Democratic Republic of the Congo (DRC). These elements are used in many electronic components, undoubtedly contained in your mobile phone and computer. Diamond is another example. However, not all these materials are conflict minerals; it depends on the mining location.

Other social impacts are less dramatic, but still important—many material extraction jobs, such as mining and farming, pay poorly and don't offer benefits such as health care or education assistance. You can avoid these problems by sourcing from reputable companies and governments, so workers do get fair pay, benefits, and community support. Some relevant certifications include Fair Trade, SA8000, and Global Reporting Initiative. But this is at best a partial solution, that should be accompanied by developing new projects that can sustain the poorest.

10.4 Renewable and Non-Renewable Materials

Instead of characterization by their properties, materials can also be characterized by their origin as renewable or non-renewable. Renewable resources are resources that are replenished naturally in a human timeframe (Figure 10.6). In contrast, non-renewable resources (also called finite resources) are those that are available in limited quantities, or those that are renewed so slowly that the rate at which they are

consumed is too fast. This means that their stocks are getting depleted before they can replenish naturally.

Not all natural materials are renewable. Sand and stone, for instance, are not, as they are replenished on geologic timescales. Sand is the second most used natural resource after water, with 40–50 billion tons used annually (18 kg per person). Despite the apparent abundance of sand, this even leads to scarcity for high quality sands used for silicon semiconductors, glass, and cement. Living materials that grow, like wood or cotton, are technically renewable but may not renew fast enough for the rate used in industry. For example, mahogany wood often comes from endangered species. Sustainably harvested materials, such as Forest Stewardship Council (FSC) certified wood products, are grown in places managed so they re-grow to full size as rapidly as they are harvested, and with good working conditions for laborers. "Rapidly renewable" materials make this sustainable harvesting easier by growing back quickly. For example, bamboo and cork can be harvested without killing the plants, so they grow back much faster than pine or oak wood.

Not all renewable materials are sustainable—in fact, most are grown and harvested unsustainably, even though they have the potential to be sustainable. Actual sustainability depends on growth, processing, and production conditions. For example, cotton (Figure 10.7) is usually grown at large scale in a monoculture, needing large amounts of water and fertilizer, occupying about 2.4% of the world's arable land, while consuming 11% of the world's pesticides. Also, subsequent processing of cotton has considerable environmental impacts. Further, basic social standards are often not met. In

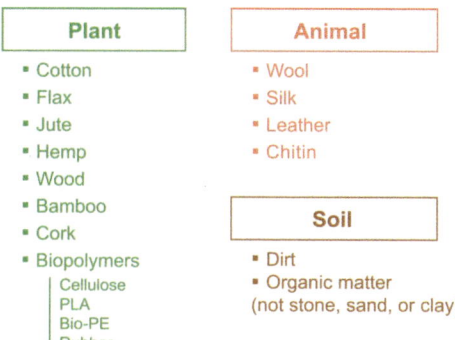

Figure 10.6 Examples of some common renewable materials

comparison, "organic" or "bio" cotton scores much better; especially on pesticide use and labor conditions.

Thus, sustainable choices cannot be made based on material type alone (although this is often a useful starting point), you should have specific information on the actual source. Certifications, like the FSC or "organic" eco-labels, can save you large amounts of time and provide due diligence you would not have the resources to pursue. However, some eco-labels are weak or even greenwashing, so do enough research to find credible certifications for your industry.

10.5 Life Cycles and System Thinking

Above, we zoomed in on specific indicators. It makes sense to do so, as this allows for data-based decisions. However, it also might provide too much focus. Sustainability is not just about using less of a material for an incremental reduction of the carbon footprint of a product. Looking at the bigger picture, at the level of the product life cycle, the lower weight might reduce the product's lifetime, or it might limit recyclability. Zooming out further to the societal level, the number of products

Figure 10.7 The sustainability of materials, such as cotton, often depends greatly on how they are grown and processed

Source: Unsplash.

that are put on the market can be disputed; is everything necessary, is there stuff that we can share?

Zooming out to the global level, huge amounts of materials will be needed for the energy transition. Oil and coal have the benefit of a high energy density and can be obtained with relatively small installations, mainly consisting of commodity materials (especially steel). In contrast, solar radiation and wind are very dilute low-density energy sources, implying that a huge area needs to be filled with solar panels or wind turbines. This equipment needs expensive critical materials with high environmental impact to effectively convert the captured energy to electricity and store it. For designers focusing on materials and their impact, this creates a hard-to-answer paradox.

Considering the Ecodesign Strategy Wheel, designers developing or improving a particular product will be tempted to focus on the use of clean materials and manufacturing processes and to enable recovery for reuse and proper end-of-life treatment. However, zooming out from the product level emphasizes the importance of also looking into solutions that create a net-positive impact at the level of the product ecosystem and society. This will hardly be addressed in the following chapters on materials but should be kept in mind (see Chapters 4 and 5 on C2C and Whole System Mapping).

Resources and References

Resources for Further Study

Books

- Ashby, M. (2020). *Materials and the environment* (3rd edn). Butterworth-Heinemann.
- Allwood, J. M., et al. (2012). *Sustainable materials: With both eyes open*. UIT Cambridge Limited.

Material Catalogs

- Cradle to Cradle Certified Products Registry (free access; lists all C2C certified products): http://www. c2ccertified.org/products/registry/
- Fair Trade certified products database by Flocert (free access; lists Fair Trade certified companies): https://www.flocert.net/about-flocert/customer-search/
- FSC Certificate Database of wood and paper products (free access; lists all FSC certified products): https:// info.fsc.org/certificate.php
- Cleangredients (free access; lists chemicals meeting US EPA Safer Choice standard; largely cleaning products): https://cleangredients.org/
- MaterialDistrict.com (free access; only some materials are green): https://materialdistrict.com/
- Material Connexion (only some materials are green; has physical libraries in several cities): https://www. materialconnexion.com/
- Transmaterial (only some materials are green): http://transmaterial.net/
- Materiom (free access; do-it-yourself local renewable biomaterial recipes): https://materiom. org/

Databases and Analysis Software

- Granta CES Selector (paid desktop software; mostly physical material properties, but some LCA data)
- Paid LCA software (e.g., SimaPro desktop, GaBi desktop, SustainableMinds online, etc.)
- Idemat LCA phone app (free app by TU Delft, extensive data)
- SolidWorks Sustainability plugin (free and pro LCA-light plugins for SolidWorks CAD software)
- Higg Index online calculator (non-LCA sustainability scores for fabric/soft goods): https:// apparelcoalition.org/the-higg-index/

- MatWeb online database (free access; no sustainability data, but detailed mechanical and other data): http://www.matweb.com/

Chemical Toxicity Libraries

- Pharos Chemical & Material Library (easiest for non-experts to read; toxicity data for both good and bad materials): https://pharos.habitablefuture.org

- NIOSH pocket guide to chemical hazards (free access; toxicity data for both good and bad materials): https://www.cdc.gov/niosh/npg/default.html

- ChemSec "Sin List" (free access; hazardous chemicals with toxicity data): http://sinlist. chemsec.org/

- European Chemical Agency (ECHA)'s Registered Substances Database (free access; difficult for non-toxicologists to read): https://echa.europa.eu/information-on-chemicals/ registered-substances

- US National Library of Medicine's Hazardous Substances Data Bank, "HSDB" (free access): https://toxnet. nlm.nih.gov/newtoxnet/hsdb.htm

- International Programme on Chemical Safety (IPCS)'s INCHEM (free access; difficult for non-toxicologists to read): http://www.inchem.org/

- Cradle to Cradle Restricted Substances List (list of chemicals that will immediately disqualify a product from Cradle to Cradle certification): http://www.c2ccertified.org/

References

Graedel, T. E., Harper, E. M., Nassar, N. T., & Reck, B. K. (2013). On the materials basis of modern society. *Proceedings of the National Academy of Sciences*, 112(20), 6295–6300.
Our World in Data. (2020). Domestic material consumption per capita. Available at: https:// ourworldindata. org/grapher/domestic-material-consumption-per-capita?tab=table (accessed October 20, 2022),
Torrubia, J., Valero, A., & Valero, A. (2022). Thermodynamic rarity assessment of mobile phone PCBs: A physical criticality indicator in times of shortage. *Entropy*, 24(1), 100. https://doi. org/10.3390/e24010100

How to Apply #10: Find Greener Materials
Time Estimate: 1–5 Hours

Find an alternative material that could replace a high-environmental-impact material in your product, and get a cost estimate for it.

STEP 1: Decide on a Material in Your Product to Replace
Time Estimate: 5–20 Minutes

Identify a high-environmental-impact material in the product you are working on. This doesn't have to be a homogeneous material, it can be an amalgamation of materials, like a composite. State the reasons why this material may be an environmental concern and/or an opportunity to reduce the product's impact. Optional: ideally use an LCA or other quantitative analysis you've done of your product.

STEP 2: Explore Material Libraries or Other Resources to Find Green Materials
Time Estimate: 30 Minutes–3 Hours

Go material hunting in online catalogs and databases from the Resources for Further Study list, and/or other search engines, blogs, physical material libraries, or whatever means you see fit. Don't limit yourself to practical things, feel free to find wild, avant-garde materials from exotic suppliers, or waste materials from industry or agriculture. Anything that could be acquired at a large production scale.

- List at least five interesting materials.

- List in just a few words why each one could be a sustainability improvement, based on LCA score, material health (non-toxicity), circular life cycle, or other data from the provider.

- List in just a few words why each one might have sustainability drawbacks, or functionality drawbacks.

STEP 3: Get Price Quotes
Time Estimate: 20–60 Minutes

Find a price quote for each material by calling the manufacturer or looking on their website. They may refer you to third-party distributors; if so, it's often best to ask them for two vendors, in case one doesn't know the material you're asking about. Make sure to get the quote for a realistic production-level amount of material (e.g., for a popular consumer product selling 100,000 units/year with 0.1 kg of the material per product, you might ask for a price on 10,000 kg; for a low-to-medium sales volume product, you might ask for 1/10th or 1/100th as much).

STEP 4: Choose a Final Material and Document Your Findings
Time Estimate: 20–60 Minutes

Choose one final material, using whatever criteria you like. Get an image of what the chosen material looks like.

Make a PDF showing:

- The product you're working on.

- What material you're working to replace.

- A list of all five materials you found, with URLs or other means for someone to find them.

- A short (5–20 words) description of each new material's sustainability benefits and drawbacks. This may simply be a caption to an illustration, but must be clear to a newcomer.

- Each material's cost per kg (or pound/square meter/whatever units it is sold in). Include sources of where you got the price quotes from (company name and website, or address and phone number).

- Which material is your #1 recommendation for use, including an image of it (remember to attribute image sources!)

- A short 20–50-word description of the criteria or reasoning you used to choose the winning idea. Include cost, functionality, and environmental factors.

Checklist for Self-Assessment

To score your success on this exercise, see if you…

- ☐ *Listed what material you're replacing.*

- ☐ *Listed five interesting alternative materials.*

- ☐ *Listed sustainability benefits / drawbacks of all materials.*

- ☐ *Listed price quotes for all materials.*

- ☐ *Chose a final recommendation.*

- ☐ *Included an image and description of the final recommendation's benefits.*

CHAPTER 11
Materials

Metals

Ruud Balkenende,
Jeremy Faludi,
and Conny Bakker

Goals

- Apply sustainability considerations regarding use and recovery of metals

- Identify which combinations of metals can and cannot be recycled simultaneously

DOI: 10.4324/9781003504672-13

Why It Matters

Metals exhibit structural, electrical, magnetic, and thermal properties that make them indispensable in many products. However, mining and processing metals are energy-intensive and lead to considerable environmental impacts, sometimes also social impacts, so metals should be recycled as much as possible. Although most metals can be recycled very well, the actual recovery of most metals is low. This is largely due to the use of multiple metals in close connection to each other, such as mixed in alloys or micro-structured in electronics. Product design must enable improved separation and recycling of metals at end-of-life stage.

Summary

- Metals are obtained from ores through intensive processing: numerous steps from mining to pure metal that all require large amounts of energy and cause emissions of often toxic substances.

- Metals are generally used as alloys, meaning other elements are added to the major metal to modify the properties.

- Alloying means that in recycling, where alloys are usually mixed, a "new" alloy is formed. This means the properties of the original alloys are downgraded.

- Electronics contain several elements that are compatible with each other in pyrometallurgical recycling, implying that printed circuits boards can be recycled as an entity.

11.1 Use of Metals

Metallic materials have been used since the Bronze Age. Metallic materials exhibit useful mechanical properties like strength, ductility, toughness, and workability, combined with functional properties such as thermal and electric conductivity; some also provide magnetism. While some corrode (rust or oxidize), others resist corrosion. The electric kettle shown in Figure 11.1 demonstrates a variety of metal properties.

The kettle housing is made of stainless steel, which is an alloy of iron, chromium and other elements like nickel, molybdenum and silicon, with relatively low heat conductivity and high corrosion resistance. The heating element consists of a nickel-chromium alloy (with some silicon and calcium) that has relatively high electrical resistivity, good

Figure 11.1 Electric kettle using several combinations of metals

Source: Unsplash.

corrosion resistance, and is easy to bend due to high ductility. Heating is controlled by a bimetallic iron-brass element (the latter a copper-zinc alloy) that is copper-plated to protect it against corrosion. Copper is used for the electrical wires for its high conductivity. Further, a small printed circuit board (PCB) is present with an electrical switch and an LED light indicator. This PCB has copper wire traces, solder (e.g., a tin-silver-copper alloy with a low melting point) and additionally metals like aluminum, silicon, and indium in its circuit components. Finally, mild steel screws are used to join several components. Thus, in a single product, a variety of metals may be encountered, often closely connected to each other—either mixed as alloys, assembled on a PCB, or joined as parts. This brings clear advantages when it comes to the performance of the product (including energy efficiency during use), but has considerable disadvantages at end of life, as these metals, which are almost inseparable, cannot be jointly recycled. This is further explained later in this chapter.

11.2 Metal Life Cycle

Metals are not readily available and not renewable resources, but need to be mined from ores. The ores are subsequently crushed, concentrated, and purified, using energy-intensive and chemically polluting processes, leaving large amounts of waste rock ("tailings"). The material that is obtained (usually the oxide or carbonate of the desired metal) is subsequently refined, which requires high temperature and a "reducing atmosphere" (without oxygen) to obtain the metal. The metal obtained directly from ore is referred to as "primary" or "virgin" metal. Subsequently, alloying is done by mixing

with other metals. The semi-finished product may then be pretreated to give it a desired microstructure co-determining its properties. Parts can then be made using a variety of techniques, for example, shape casting, sheet metal rolling, extrusion, forging, and machining. At the end-of-life stage, metals can be recycled, although for many metals this is only achieved to a limited extent.

For aluminum, this process is shown in Figure 11.2. It starts as bauxite, a mineral that contains a relatively high fraction of oxidic aluminum compounds. Bauxite is strip-mined because it is usually present near the surface (major producers are Australia, Republic of Guinea, and China). Grinding and chemical processing result in pure aluminum hydroxide. The waste material, highly caustic red mud, is often disposed of in badly managed conditions, with considerable risks for local ecosystems. The aluminum hydroxide is subsequently turned into aluminum oxide at 1000°C, which is dissolved in highly corrosive Na_3AlF_6 at 960°C for

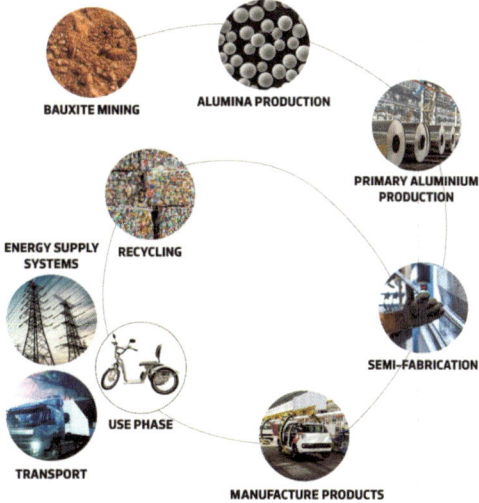

Figure 11.2 The life cycle of aluminum
Source: Ferrá et al. (2024).

electrochemical reduction to pure aluminum metal. The tapped metal is then alloyed and cast into ingots or rolled into sheets to be sent to product manufacturers. The entire process, but especially the electrochemical processing, is extremely energy-intensive; greenhouse gas emissions are therefore high. Also, as an undesired result of this process, highly stable perfluorocarbon gases are formed, compounds with only carbon and fluorine atoms. These last for over 10,000 years, and if their emission is not carefully prevented, they cause 5,000x–10,000x more global warming per kg than CO_2 does, and also cause stratospheric ozone depletion. Other notable effects of aluminum production include acidification (due to sulfur oxide emissions), and eutrophication (mainly due to NO_x emissions).

Light weight, high strength, and corrosion resistance make aluminum attractive for applications varying from packaging to household products, transportation, and structural elements in buildings. The relatively low weight of aluminum parts especially brings advantages for transportation (e.g., when using in automotive parts or as packaging material). For application in products, two main classes of alloys are used: wrought alloys and cast alloys. For example, foils, deep-drawn cans, and extruded window frames are typical wrought products, while an engine block is a typical cast product. Both classes have similar alloying elements, but at lower concentration for wrought alloys, to maintain the ductility that is important in rolling and extrusion processes. Further, for both wrought and cast alloys, tens of different grades exist, all tuned to optimize specific properties.

At the end-of-life stage, aluminum is attractive to recycle because the energy

associated with processing "secondary" (recycled) metal is only 10–20% of the energy needed to produce primary aluminum. However, scrap material from different products has a different elemental composition, due to the specific alloys used for each original product. Although advanced recycling facilities might be able to collect and sort the various alloys, processing them together is more common. This implies that a new specification is obtained that needs to be tuned to desired properties by mixing with primary material and adding other elements. But in contrast to polymers that are often downcycled, implying less-defined properties, the final result of the metal recycling process is an alloy with well-defined performance (although high-performance alloys are usually not obtained through recycling).

11.3 End-of-Life Recovery

All metals can be relatively easily recycled when compared to primary production, and generally have much lower environmental impacts. However, this does require that the metals are easy to access and economically interesting to process. The actual status of metal recycling is quite disappointing for most metals. You should use metals that are actually recycled, not merely recyclable, such as steel and aluminum.

Figure 11.3 shows the periodic table of elements colored by post-use End-of-Life Recycling Rate (EoL-RR). This represents the percent recycled at end-of-life of products in the EU, relative to the amount that is put on the market in the EU in the same year. (Recycling rates are often much lower elsewhere in the world, including North America.) The EU's recycling rate is over

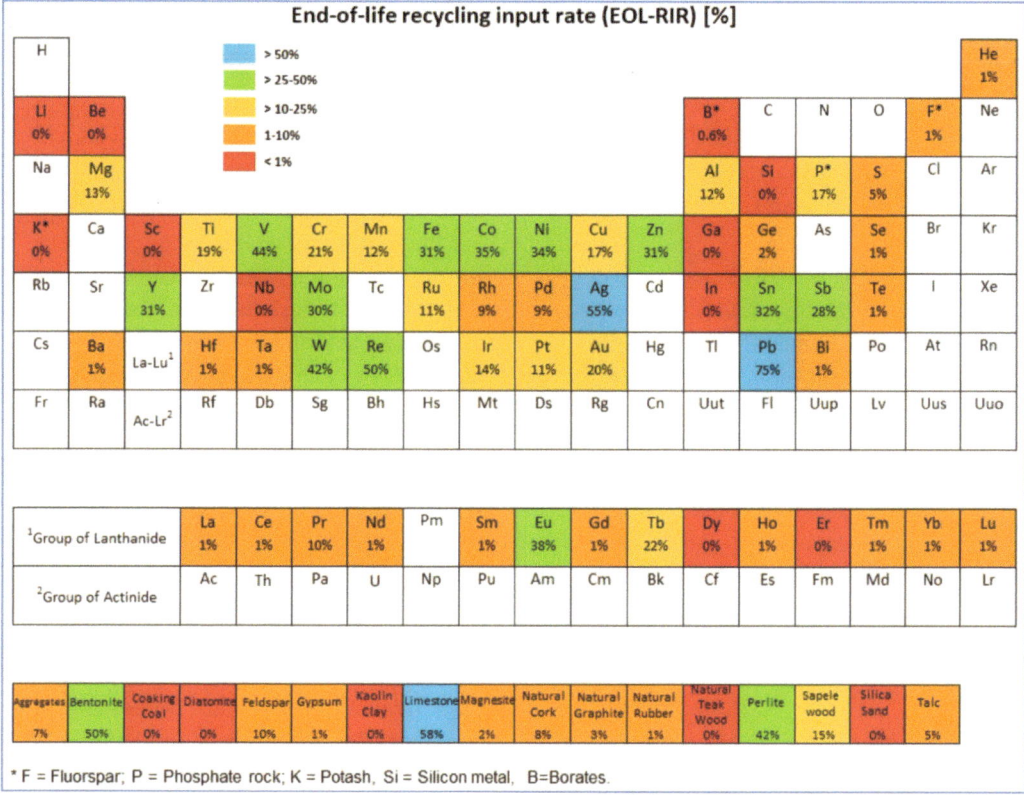

* F = Fluorspar; P = Phosphate rock; K = Potash, Si = Silicon metal, B=Borates.

Figure 11.3 End-of-life recycling input rates (EOL-RIR) for the EU-28 based on the MSA studies (when available) and used to draw the 2017 List of Critical Raw Materials for the EU

Source: Talens Peiro et al. (2018).

50% for commodity metals like iron (62%), aluminum (51%), and copper (61%) and precious metals like silver and gold. Even these numbers show large amounts of waste and buildup of stock, and this becomes worse when we look to rare-earth elements and other specialty metals like gallium and indium (e.g., used in solar cells and LCD screens), which have almost no recycling.

Low recycling rates are often due to limited collection of metals (or of products in which they are used). Low recycling rates are further due to technological limitations that make a number of materials unrecyclable

in particular applications. For example, flat panel displays (ranging from smartphones to TVs) have indium applied in a 100 nm-thick conductive indium oxide film, which can't be separated economically. Finally, low recycling rates can be attributed to design, which hampers the required separation of materials.

When a product is made from a single metal (e.g., aluminum cans, copper piping), recycling is straightforward because a large homogeneous batch of material can be collected. Often, as was evident from the electric kettle example, we use metals in specific alloys or in tiny amounts in close proximity to each other (e.g., in PCBs). This means that at the end of life, not just a single major metal needs to be recovered, but that elements need to be separated from each other. Here we encounter fundamental limitations, as not all metals can be simultaneously processed.

The most common way to recover metals is smelting at high temperature in a blast furnace ("pyrometallurgical" recovery). Depending on which metal is the majority (the "smelt stream"), some materials will easily be recovered, while others will dissolve and act (partly uncontrolled) as alloying elements, and others will be lost as oxides and end up as slag (waste material). Table 11.1 provides an overview of the compatibility of metals during this reprocessing, with columns for the three main smelting elements.

Table 11.1 shows that copper, aluminum, and iron should not be mixed, as they are not compatible with each other in recycling. This implies, for instance, that copper cannot be recovered from an iron smelt or aluminum smelt. So, fragments mixing both of these metals can never be fully recycled. However, copper is compatible with a number of other precious and specialty metals, implying that these metals *can* be recovered from a mixture with copper. Notably, this concerns metals that are frequently present in electronics, like tin, nickel, silver, gold, and palladium. Electronic printed circuit board assemblies are therefore preferably treated simultaneously with copper waste. Although metals like rare-earth elements and tantalum will be lost.

For example, consider a printed circuit board with electronic components that is connected to an aluminum heat spreader (this example

is discussed in more detail in Chapter 16, on the Design for Recycling). Fragments that contain parts of both heat spreader and electronics will lose the electronic materials when the fragment is separated into the aluminum smelt (Table 11.1's aluminum column shows their incompatibility); when the fragments end up in the copper smelt, the aluminum is lost (see Table 11.1's copper column). But in the latter case, Table 11.1 shows that other metals used in electronics, like gold, silver, tin, and lead, can be recovered.

Hydrometallurgical recovery can be an alternative to pyrometallurgy and can be more specific in the elements that are recovered. This is when a product or component is (partly) dissolved in a liquid, such as sulfuric acid, and subsequently the desired materials are precipitated in a number of chemical steps. This process is especially used for specific well-defined components, like batteries, as it puts higher demands on the input quality. It might also pose a solution for recovering indium from flat panel screens, dissolving it from the glass substrate, though this is still in development.

11.4 Designing with Metals

With the insights mentioned in this chapter, and using the Ecodesign Strategy Wheel, you can derive guidelines for designing with metals. Some examples are given here, but note that guidelines should be tuned to the product at hand and the context of use: choosing the right metal is not only part of the Ecodesign Strategy Wheel's "low-impact materials" group, but also directly relates to "clean manufacturing" "extended use," and "end of life." It can relate to others as well.

Table 11.1 Recycling compatibility and loss of metals. Green checkmarks mean the elements are generally recovered from a particular smelt, yellow dots mean they are likely to dissolve or form slag material and are not recovered, and red Xs mean the elements are generally lost.

Materials	Symbol	Recycling smelt stream			Examples of use
		Copper	Aluminum	Iron/Steel	
Ferrous metals					
Alloy steel	Fe	●	●	✓	Housing, mechanical parts, connections
Carbon steel	FeC_x	●	✗	✓	Structural components (housing, frame)
Stainless steel	$FeCr_xNi_y$	●	✗	✓	Food appliances, medical products
Non-ferrous metals					
Precious metals					
Copper	Cu	✓	●	✗	PCB conductor; wiring; thermal conductor; structural parts
Gold	Au	✓	✗	✗	Electronic contacts electroplating, bonding wires
Mercury	Hg	✓	●	✗	(restricted in EU) Fluorescent lamps, some cell batteries
Osmium	Os	✓	✗	✗	Ballpoint / fountain pen; electrical contacts; incandescent lamp
Palladium	Pd	✓	✗	✗	Fuel cell electrode
Platinum	Pt	✓	✗	✗	Electrode in high-end equipment(pacemaker), thermocouple
Silver	Ag	✓	✗	✗	Electronics (contacts, switches, antennas); printed electronics
Other metals					
Aluminum	Al	●	✓	●	Thermal conductor; structural components; wiring
Antimony	Sb	●	✗	✗	GaSb semiconductor: infrared LEDs & sensors; flame retardant
Bismuth	Bi	✓	✗	✗	Solder material; thermoelectric component
Cadmium	Cd	✓	✗	✗	(restricted in EU) Ni-Cd batteries; pigments (orange, yellow, red)
Chromium	Cr	✗	✗	●	Stainless steel alloy element
Cobalt	Co	✓	●	●	Electrode material in lithium-ion batteries
Lead	Pb	✓	✗	●	(restricted in EU) Solder; lead batteries
Lithium	Li	✗	✗	✗	Lithium ion batteries
Magnesium	Mg	●	●	●	Alloy element in aluminum; light-weight structural components
Nickel	Ni	✓	✗	●	Batteries, capacitors, switches, transformers; stainless steel alloy
Rare earth metals	RE	✗	●	●	High performance magnets
Silicon	Si	●	●	●	Semiconductor (ICs, PV cells); alloy element in aluminum
Tin	Sn	✓	●	✗	Solder; main alloy element in bronze
Titanium	Ti	✗	✗	●	Lightweight corrosion resistant structures; medical implants
Zinc	Zn	✓	✗	●	Main alloy element in brass

(Continued)

Table 11.1 (Continued)

| Materials | Symbol | Recycling smelt stream | | | Examples of use |
		Copper	Aluminum	Iron/Steel	
Other materials often used with metals and simultaneously recycled					
Polymers	$C_xH_yO_zN_p$	🟡	🟡	🟡	Housing, coatings, connected components
Rare earth oxides	REO	✖	✖	✖	Permanent magnets, battery alloys, fuel cell components
Indium oxide	In_2O_3	✔	✖	✖	Transparent conductor (e.g. in TBs, monitors, PV cells)
Tantalum oxide	Ta_2O_5	✖	✖	✖	High performance capacitors
Glass, silica	SiO_2	🟡	🟡	🟡	Window in products, filler in plastics
Aluminum oxide	Al_2O_3	🟡	🟡	✖	Electrical insulator, filler particle in plastics
Titanium oxide	TiO_2	✖	✖	✖	White, UV-reflecting pigment in plastics and coatings
Phosphorous	P	✖	✖	✖	Plasticizers, flame retardants in plastics
Bromine	Br	✔	✖	✖	(restricted in EU and USA) Flame retardant in plastics

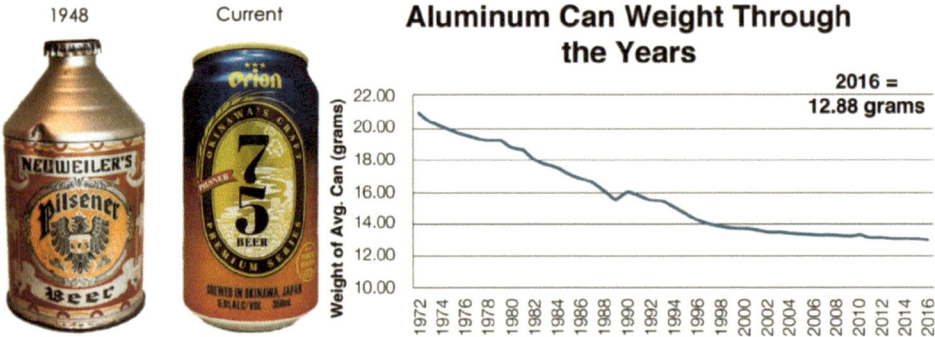

Figure 11.4 Mass of material used in aluminum cans has dropped 40% over time

Source: Wikimedia Commons and The Aluminum Association.

Here are rough guidelines for designing with metals:

- Minimize the amount of material used. For example, Figure 11.4 shows how aluminum drinking cans have cut material use almost in half through better engineering and design. This is often a win-win, as it also saves money.

- Use low-impact metals; e.g., steel and aluminum are usually preferred over metals like titanium. See Figure 11.5.

- Avoid replacing metallic parts by plastics coated with a metallic film, e.g., for reflection or electromagnetic shielding, as this considerably hampers recycling.

- Minimize the number of different metals, and/or make them easy to separate (see Chapter 16).

The chooser chart in Figure 11.5 is a quick guide to better or worse metals, balancing LCA impacts, potential health hazards, and social impacts of mining. It labels certain metals by popular properties.

CHOOSING SUSTAINABLE
METALS

— GOOD FOR —

Strength

Hardness

Low Density

Moisture
Resistance

Bearing
Surface

Bend & Cut

WHEN SANDING

Wear a respirator

Have good ventilation

WHEN WELDING

First clean off coatings

BETTER

Scrap Metal

Recycled Metal
(especially aluminum)

Mild Steel (Structural)

Mild Steel (Sheet)

Stainless Steel

Aluminum
Tin
Copper
Brass
Bronze

Tungsten
Titanium

Very High Environmental
Impacts or Health Risks:
Gold   Mercury
Silver   Cadmium
Platinum   Hexavalent
Lead   Chromium

WORSE

Figure 11.5 A chooser chart for low-impact metals, noting some functional properties

Source: based on Faludi (2015).

References

Faludi, J. (2015, August 14). Choosing greener metals. Instructables. Available at: https://www.instructables.com/Choosing-Greener-Metals/

Ferrá, F., Correia, A., Almeida, F. D., & Costa e Silva, E. (2024). Sustainable and optimized production in an aluminum extrusion process. Available at: www.preprints.org/manuscript/202401.1809

Talens Peiro, L., Nuss, P., Mathieux, F., & Blengini, G. (2018). Towards recycling indicators based on EU flows and raw materials system analysis data. Report #EUR29435 EN. Publications Office of the European Union.

How to Apply #11: Improve Metal Recyclability in Your Product
Time Estimate: 30 Minutes–2.5 Hours

Rethink the choice of metals in your product, to improve their likely recycling recovery and yield rates.

STEP 1: Identify Your Product's Metals and Recycling Streams
Time Estimate: 15–60 Minutes

Make a table of the metals in your product and categorize them by the main metal recycling streams of copper, aluminum, iron/steel, and lead.

What are your product's most important metals to recycle at end of life? Are they the ones with the most mass used in the product, or are some metals used less but still more important because they have much higher environmental impact, like gold? Optional: ideally use an LCA or other quantitative analysis you've done of your product.

STEP 2: Evaluate Material Recyclability and Potential Losses
Time Estimate: 10–20 Minutes

Use this chapter's Material Compatibility Table (Table 11.1) to identify any potential losses during the recycling process (red or yellow dots in Table 11.1) due to how metals are combined in the various parts of your product. Add a column in the table of your product's metals to track this with the same yellow or red dots. Especially focus on priority metals, but if you have time, consider all metals in your product.

STEP 3: Ideate More Recyclable Material Choices
Time Estimate: 10–30 Minutes

Are there metals in your design that will likely have recycling problems, given the other metals they're attached to in your product? If so, can you think of alternative material combinations that will provide higher recovery and yield rates in recycling? Brainstorm what materials you might substitute to ensure the high-priority metals in your product are recycled easily. (Ideally all metals in your product.) Note: your replacements don't have to be metals, they can be any materials with the right functionality.

Have at least 20 ideas or more. Be specific and concrete, it gets you more solutions from the same general ideas; have wild ideas, too, it's a brainstorm.

STEP 4: Ideate Non-Material Recyclability Improvements
Time Estimate: 10–30 Minutes

Can you think of ways to keep the same metals your product uses now, but attach or use them differently to provide higher recovery and yield rates in recycling?

Brainstorm new part attachments, subassembly organizations, user behavior nudges, business model variants, or anything other than material choice to ensure the high-priority metals in your product are recycled easily. (Ideally all metals in your product.) Note: Chapter 16 has many suggestions for attachment and separation.

Have at least 20 ideas or more. Be specific and concrete, but it's a brainstorm, so have wild crazy ideas, too.

STEP 5: Choose a Final Solution
Time Estimate: 5–30 Minutes

Choose one final solution, or combination of solutions, from your brainstorms. Use whatever decision criteria you like, but clearly and briefly describe it. Include an image of the new material or sketch of the new design solution.

Make a PDF showing:

* The table of metals in your product, including a column showing each one's recycling stream, which are priorities to recycle, and which have recycling problems.

* All your ideas (20+) on substituting materials.

* All your ideas (20+) on non-material solutions.

* Your winning idea to improve metal recycling recovery and yield, including an image or sketch of it (remember to attribute image sources!).

* A short 20–50-word description of the criteria or reasoning you used to choose the winning idea. Include cost, functionality, and environmental factors.

Checklist for Self-Assessment

To score your success on this exercise, see if you...

☐ *Made a table of the metals in your product.*

☐ *Identified priority metals to recycle.*

☐ *Listed the industrial recycling stream applicable to each metal.*

☐ *Identified potential recycling losses for each metal.*

☐ *Ideated 20+ alternative materials.*

☐ *Ideated 20+ non-material design solutions.*

☐ *Chose a winning idea to improve metal recycling, and explained it in 20–50 words, with an image.*

CHAPTER 12

Materials

Plastics

Ruud Balkenende,
Jeremy Faludi,
and Conny Bakker

Goals

- Apply sustainability considerations regarding use and recovery of plastic

- Identify end-of-life recovery pathways for plastics

- Understand how recycling affects the properties of plastics

DOI: 10.4324/9781003504672-14

Why It Matters

Plastics are ubiquitous, because their wide variety of properties and ease of shaping with mass manufacturing give them a huge number of applications. Unfortunately, their extensive use amplifies global warming and mining of fossil fuels, leads to health problems (mostly due to additives), and causes other environmental problems due to their resistance to degradation.

Summary

- Plastics are usually made from fossil fuels.

- Common polymers like PE, PP, PS, PA, PET, ABS, and PC are very stable. Their molecules don't decompose in the environment, but they may break down into "microplastics," which are increasingly concerning environmentally.

- Most plastics are recyclable, but the only plastics that are actually recycled often are PE, PET, and sometimes PP.

- Plastics are usually not pure polymers, but include a variety of additives, such as softeners, flame retardants, colorants, and/or fillers for mechanical modification. This complicates recycling and can add environmental impact.

- Mechanical recycling grinds up and melts plastics, which is less expensive and lower eco-impact but loses some mechanical performance.

- Chemical recycling breaks plastics down into monomers or further, then rebuilds the polymers to match primary mechanical properties, but is more expensive and has a higher eco-impact.

12.1 Polymers and Plastics in Products and Their Life Cycles

Polymers are used in many products, and often more than a single polymer is used. For example, the toothbrush in Figure 12.1: its handle combines relatively stiff and elastic plastics. The stiffer plastic is used to provide sufficient strength, the elastic plastic is used for the parts that are touched, for good grip and pleasant feel. Different colors are used for its aesthetic appearance. This part is made in a two-shot ("2k") injection molding process that offers reduced processing time and guarantees excellent adhesion between both plastics by molding them in the same mold. The bristles are made from nylon fibers which are strong and elastic, and are glued into the handle.

The toothbrushes in Figure 12.1 illustrate that plastics are versatile and relatively cheap, and therefore used in almost any type of product. Their properties are primarily related to the main constituent of a plastic: the polymer. A polymer is a large molecule composed of repeating units (monomers). The most common polymers used in products are commodity polymers like polyethylene (PE), polyethylene terephthalate (PET), polypropylene (PP), polystyrene (PS), and polyvinyl chloride (PVC), as shown in Figure 12.2. When higher standards are required for strength or other properties, engineering polymers like polycarbonate (PC), acrylonitrile butadiene styrene (ABS), polyamide (PA), and polyurethane (PU or PUR) are used. For extremely demanding situations, high-performance polymers like polyether ether ketone (PEEK) are used. Other specialty needs, like the rubberized

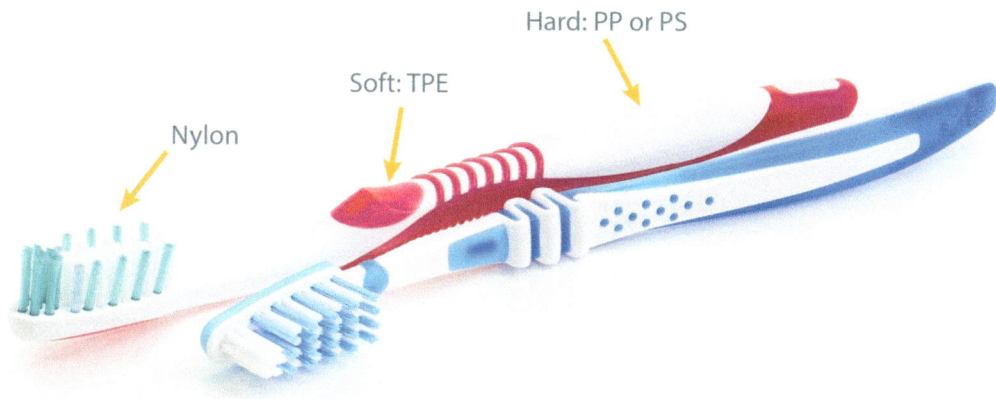

Figure 12.1 Three plastics in a toothbrush, used for their different properties

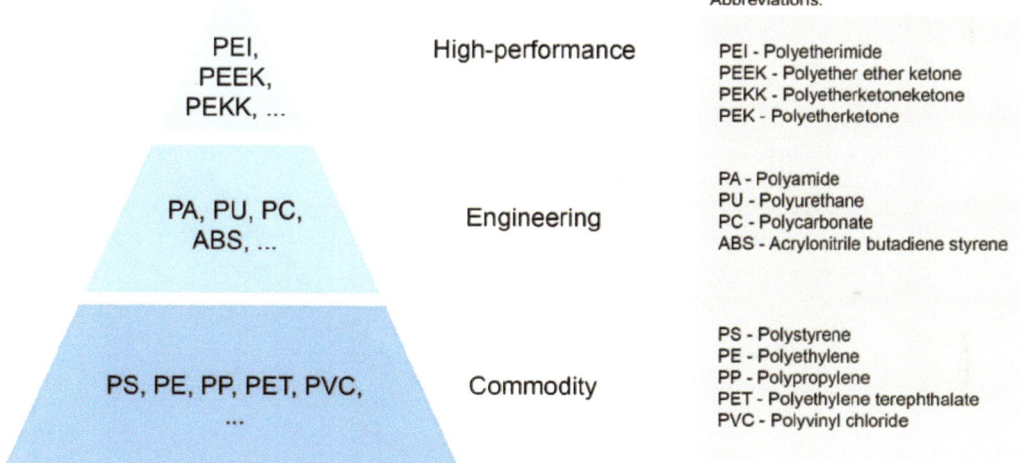

Abbreviations:

PEI - Polyetherimide
PEEK - Polyether ether ketone
PEKK - Polyetherketoneketone
PEK - Polyetherketone

PA - Polyamide
PU - Polyurethane
PC - Polycarbonate
ABS - Acrylonitrile butadiene styrene

PS - Polystyrene
PE - Polyethylene
PP - Polypropylene
PET - Polyethylene terephthalate
PVC - Polyvinyl chloride

Figure 12.2 Hierarchy of plastics based on their properties

grip in the toothbrushes, use thermoplastic elastomers (TPE) or other polymers. In most cases, the name of a plastic is directly derived from the polymer (although nylon, which is a PA, is an exception).

Plastics usually are not pure polymers, they often contain several additives. A randomly chosen plastic part may contain around 20 additives. These are chemicals blended into plastics to change their performance or appearance to better suit their intended applications. For example, plasticizers (to make parts more flexible), flame retardants, stabilizers (e.g., antioxidants and antimicrobial agents), colorants, and lubricants. About 5% of the average total mass of plastics put on the market consists of additives. In soft PVC, the amount of phthalate plasticizers added can be up to 50%. Many different chemicals are used as plastic additives. Their nature and

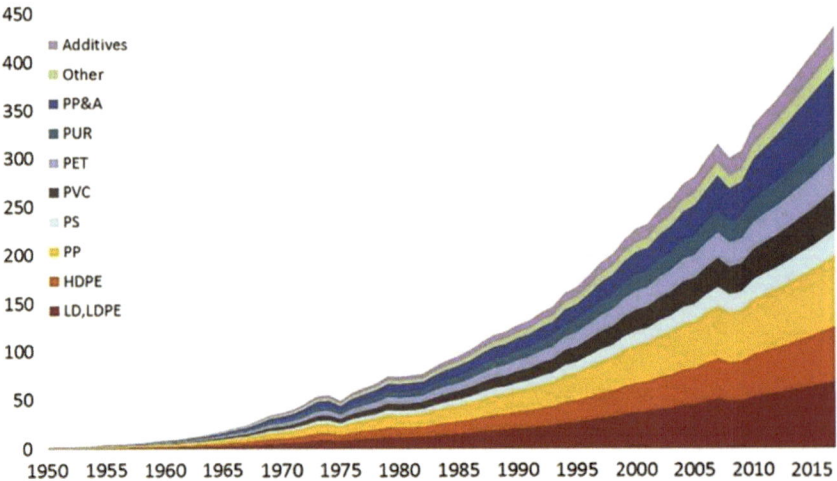

(a) Global annual primary plastics production (in Mt) by material type from 1950 to 2017.

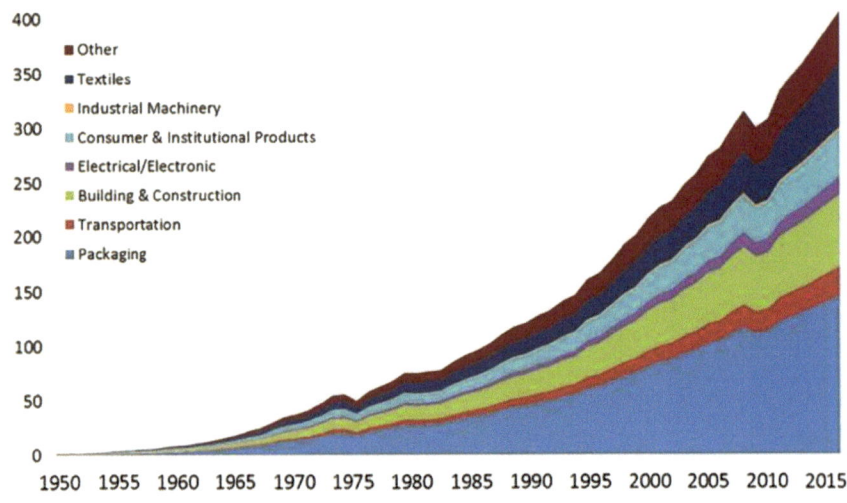

(b) Global plastics production: 1950 to 2015.

Figure 12.3 Increasing plastic use, and its industry applications, since 1950

Source: Geyer (2020).

concentration are usually only known to the plastic producer, and many of their health hazards have never been tested.

The development of synthetic polymers, and therefore also of plastics, is relatively recent—

it started early in the twentieth century, and mass production took off in the 1950s. Figure 12.3 shows the rapidly increasing annual production for major plastic types, with the applications in which they are used. With this huge and increasing volume of plastics,

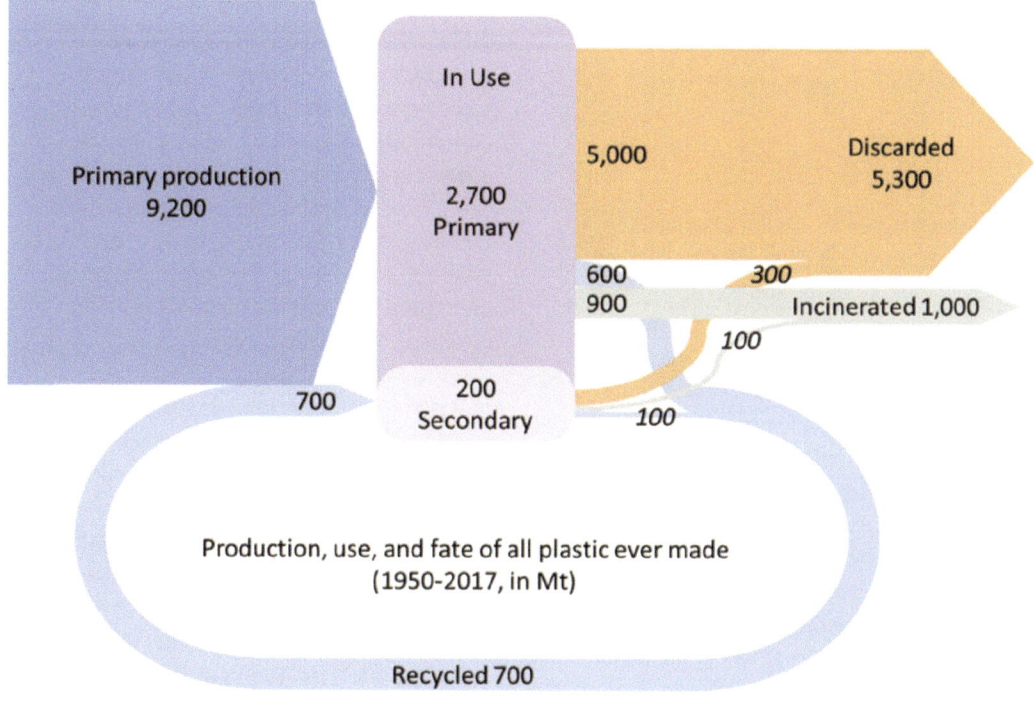

Figure 12.4 Cumulative production, use, and fate of all plastic ever made, from 1950 to 2017, in megatons (Mt)
Source: Geyer (2020).

the associated environmental problems are also now apparent on a global scale.

The lifetime of plastics in use varies from hours to weeks for plastic packaging, to years for plastics in durable products like small home appliances, to decades for plastics in buildings and infrastructure. The low cost and light weight of plastics have given rise to many single-use applications with long-lasting waste. You can avoid this by choosing different materials or by designing multiple-use products. This was strongly stimulated in the EU by the 2019 Single-use Plastics Directive, which bans a number of single-use plastic products (e.g., straws, cutlery, stirrers) when alternatives are easily available and affordable.

At their end of use, plastic products are largely discarded, implying incineration (43% in the EU, 50% in the Netherlands, 15% in the USA), landfill (25% in the EU, 0% in the Netherlands, 75% in the USA), or mismanaged in either the wrong waste stream or ending up in the environment. Only a limited fraction is actually recycled at end of life (33% in the EU, 50% in the Netherlands, 9% in the USA) (Law et al., 2020; PlasticsEurope, 2020). See Figure 12.4.

12.2 End of Life of Petrochemical Plastics

At the end of a product's life cycle, it's best to design for recovery operations that maintain product functionality (see Chapter 4). But here we will focus on the last-resort end-of-life options at the level of materials: recycling, incineration, landfill, and degradation in ambient conditions. See Figure 12.5.

12.2.1 Plastic Recycling

The major recycling route for plastics is mechanical recycling. This requires a good collection infrastructure and the ability to subsequently separate plastics. Currently, the only plastics commonly recycled globally are HDPE, LDPE, and PET. Other plastics are collected, but usually not separated adequately in high enough volumes to recycle. Details of the liberation and separation processes, usually shredding to obtain fragments and either infrared or density technology to separate these fragments, can be found in Chapter 15 on recycling processes. The toothbrush in Figure 12.1 illustrates some of the problems we encounter with much current product design. It uses three different plastics that are very hard to separate from each other, because they are co-molded and glued. Further, only commodity plastics are currently recycled at scale; even if separated, engineering plastics like TPE and nylon are often not recovered. Globally, about 15% of plastics are recycled, although this is likely an overestimation, as often people report the input to recycling facilities, not the resulting output.

In mechanical recycling, plastics are sorted based on their major polymer. After separation, the plastic is shredded further (producing the smaller particles needed for the subsequent processing steps), washed, ground into flakes, heated, and extruded into new pellets. The energy (and thus carbon footprint) associated with mechanical recycling are typically about 20% of the energy needed for primary plastic production.

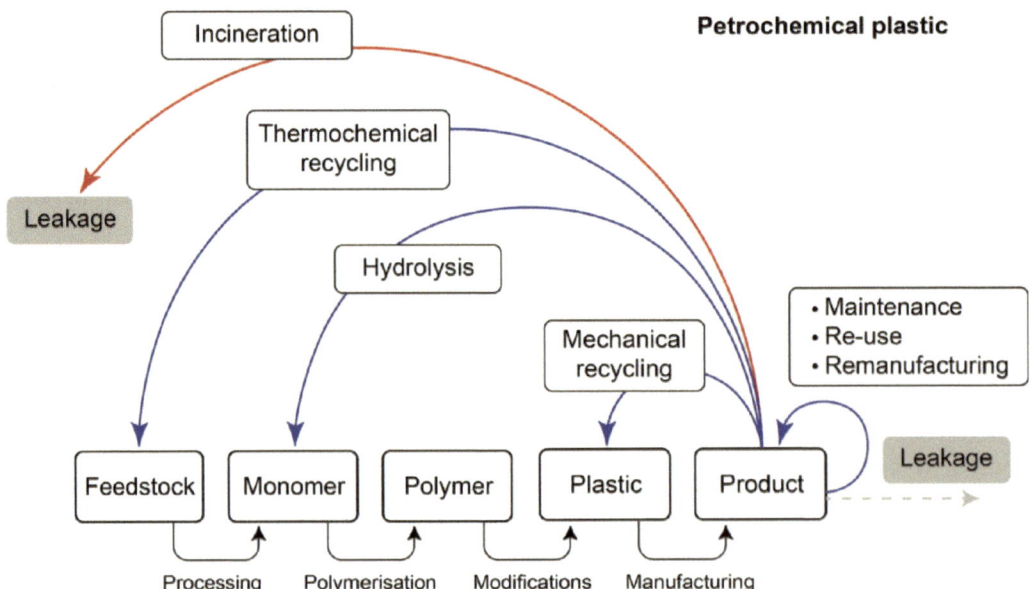

Figure 12.5 Recovery pathways for plastics, focusing on material recovery

Source: Adapted from Ritzen et al. (2023).

However, even with this sorting and cleaning, there are quality losses. Plastic grades vary in molecular weight (how long the polymer chain is) and additives to tune their properties. Many batches with different grades of a polymer are mixed during recycling, so the resulting recycled plastic will exhibit a larger distribution in molecular weight of the major polymer, and a larger variety of additives. A further complication is that separation processes are never perfect, implying that some contamination with other plastics is also likely. Finally, due to the shredding, washing, melting, etc., the molecular structure of the polymer can change through shortened polymer chains, oxidation, or a reaction with contaminants or additives present in the plastic. All these effects cause mechanically recycled plastic's properties to be less well defined than primary plastics, so you can only use it in applications with lower performance or quality standards. This is called "downcycling."

Chemical recycling comprises many methods, with widely varying results. They all have in common that they are still in an early stage in industry, not applied at large scale. One of the most investigated and applied routes is hydrolysis, in which the polymer is broken down into its monomers (the building blocks of the polymer) through a reaction with water. This is only applicable to "condensation polymers" like PET, PC, and PA; "addition polymers" like PE, PP, and PET are not suitable for this method. Hydrolysis is highly specific to a particular polymer and enables you to completely rebuild the polymer from the original building blocks, resulting in a polymer that cannot be distinguished from a primary polymer. It often requires just 50% of the energy of primary plastic production, but this value strongly depends on the nature of the plastic and the maturity of the processing, and might even be higher than for the primary plastic.

A different chemical route is thermochemical recycling, which is carried out at high temperatures (500–800°C) under reducing conditions to either form a fuel like oil or gas, or a mixture of small organic molecules that can be used as feedstock for subsequent chemical processes (which could include, but are not limited to, monomer manufacturing). The advantage of this method is that it can be applied to mixtures of (contaminated) plastics, the disadvantage is that a rather ill-defined feedstock results (quite similar to the feedstock resulting from cracking of crude oil).

12.2.2 Landfill and Incineration

Landfill or incineration of plastics are common waste management solutions when recycling is not done. Incineration implies the complete oxidation of a plastic to CO_2 and water. With plastics made from crude oil, this directly results in an increase of greenhouse gases, similar to burning the oil without becoming plastic first. If incineration is done with energy recovery, this might to some extent avoid emissions that would otherwise have been caused by burning additional fossil fuels, but this is a limited effect, as the caloric value of plastic waste is much lower than that of fossil fuels.

Landfilling might seem an opportunity to permanently store plastic waste and avoid the emissions associated with incineration. But in a landfill, plastics will slowly degrade (over the course of hundreds to thousands of years), still leading to CO_2 emissions; it also emits methane, which is a much stronger greenhouse gas. Also, additives and "microplastics" (microscopic particles of plastic) that escape can leach to soil and water, leading to negative environmental impacts.

12.2.3 Degradation in Ambient Conditions

A plastic wrapper carelessly tossed in the street will degrade under ambient conditions. This is a form of "leakage," implying that the material is lost for further use, pollutes the environment and might even degrade in a way that further increases its environmental impact. This is usually an uncontrolled process, in which a plastic slowly falls apart into smaller parts and microplastics due to weathering, sunlight exposure, temperature, humidity, and the composition of the plastic. Sometimes, plastics have additives that stimulate photodegradation (degradation under influence of light), hydro-degradation (under influence of water, comparable to the hydrolysis recycling route mentioned before but under ambient conditions and usually incomplete), or oxo-degradation (slow built-in chemical degradation). These terms might suggest complete degradation, but are misleading, as they likely just result in accelerated breakdown into small microparticles.

Biodegradation is sometimes an interesting alternative degradation pathway. This process might take place under ambient conditions in the environment, or under specifically created conditions like municipal compost facilities. Although some petrochemical plastics are biodegradable (PBS is an example), this property is more often observed with biobased plastics and will be discussed in Chapter 13 on bio-based plastics. However, here it is good to note that biodegradation under landfill conditions easily leads to the formation of methane, a strong greenhouse gas.

12.3 Impacts of Petrochemical Plastics

By far most plastics are of petrochemical origin, i.e., made from fossil fuels. Only a small fraction ($\sim$10%) is made from renewable biological resources (see Chapter 13). Plastics do result in some direct environmental benefits: As packaging material, their low weight results in reduced impacts due to transportation. Their preservation of food helps to reduce food waste (although a system based on more local supply and different behavior regarding storage would be better). Their resistance to degradation enables long-lived products. Alternative natural materials or biobased polymers would require increased land use, and might reduce biodiversity if they are not made from existing waste streams. Petrochemical plastics might slow this.

Overall, however, the abundant use of plastics is associated with large environmental impacts, most notably significant greenhouse gas emissions and pollution. Greenhouse gas emissions are due to the energy-intensive processing of the fossil fuels used as source polymers. This accounts at a global scale for about 1% of global warming gases. Further, about 4–8% of fossil fuels is used for plastic production.

The 30–50% of plastics that is incinerated at the end of life also contributes directly to global warming.

Plastic pollution has been found everywhere—in remote areas, at the poles, and in deep seas. This pollution is due to mismanaged plastic waste, either due to consumer behavior or to lack of a collection infrastructure. Also, some waste initially stored at landfills gets dispersed in the environment. Plastic pollution increases by about 10 million tons per year. These plastics accumulate in the environment, as most plastics take hundreds to thousands of years to degrade under a wide range of circumstances. This pollution directly affects wildlife through ingestion or entanglement. Slow degradation leads to the formation of microplastics, which have even been detected in human blood and brain tissue. Although the health risks are not yet well understood, this is considered highly worrisome.

The additives present in plastics pose an additional problem. Some additives leach out of the plastic and can be highly toxic to human health or to the environment. Such leaching might occur during the use phase (e.g., bisphenol-A leaching from polycarbonate baby bottles). But more common is leaching of additives into the environment at the end of life, where their eco-toxicity harms ecosystem health. Process chemicals are also a concern. For example, PVC needs chlorine during production, and might result in dioxin emission during incineration at end of life. Both substances are highly toxic but are not emitted to the environment during proper processing. The risks associated with production and end of life are not considered in the impact data of PVC, and it scores well in LCAs, but its use is nevertheless heavily disputed.

12.3.1 Using Recycled Plastics in Products

Plastics obtained through mechanical recycling have properties that are different from those of primary plastics and are usually considered to be less well-defined and of lower quality. This affects not only material properties (strength, impact resistance, chemical resistance, material shrinkage, leakage behavior), but also aesthetics (transparency, attainable colors, gloss levels) and molding flow properties. The latter may require adapting molds to avoid processing issues like blocked air venting, a blocked hot runner, or surface contamination due to outgassing.

Nevertheless, recycling plastics only makes sense if products are made with recycled plastics, and it is quite doable: recycled plastic is used in many high-quality products, from Patagonia jackets and Adidas shoes to Lander skateboards and Philips vacuum cleaners. Method brand cleaning products have used 100% recycled plastic bottles since 2008, and since 2012, their clear bottles are 100% recycled recovered coastal plastic by Plastic Bank's collection members (Figure 12.6). They collect and exchange plastic waste for money and social benefits, preventing the plastic from ending up in the ocean or a landfill and empowering local communities in places with poor waste management.

To design with recycled plastics, these strategies help:

- Use textured surfaces instead of uniform high gloss surfaces.

- Choose geometries for injection molded parts that allow easy flow paths and consider more venting parts.

- Reduce mechanical demands, e.g., thicker walls (it will likely still have lower eco-impact than thinner primary plastic parts).

- Use primary plastics only for very demanding parts.

The chooser chart in Figure 12.7 is a quick guide to better or worse plastics, balancing LCA impacts and health hazard assessments. It labels certain plastics by popular properties; all plastics in the chart can be injection molded.

Figure 12.6 A Plastic Bank member collecting plastics from a beach for recycling

Source: Plastic Bank.

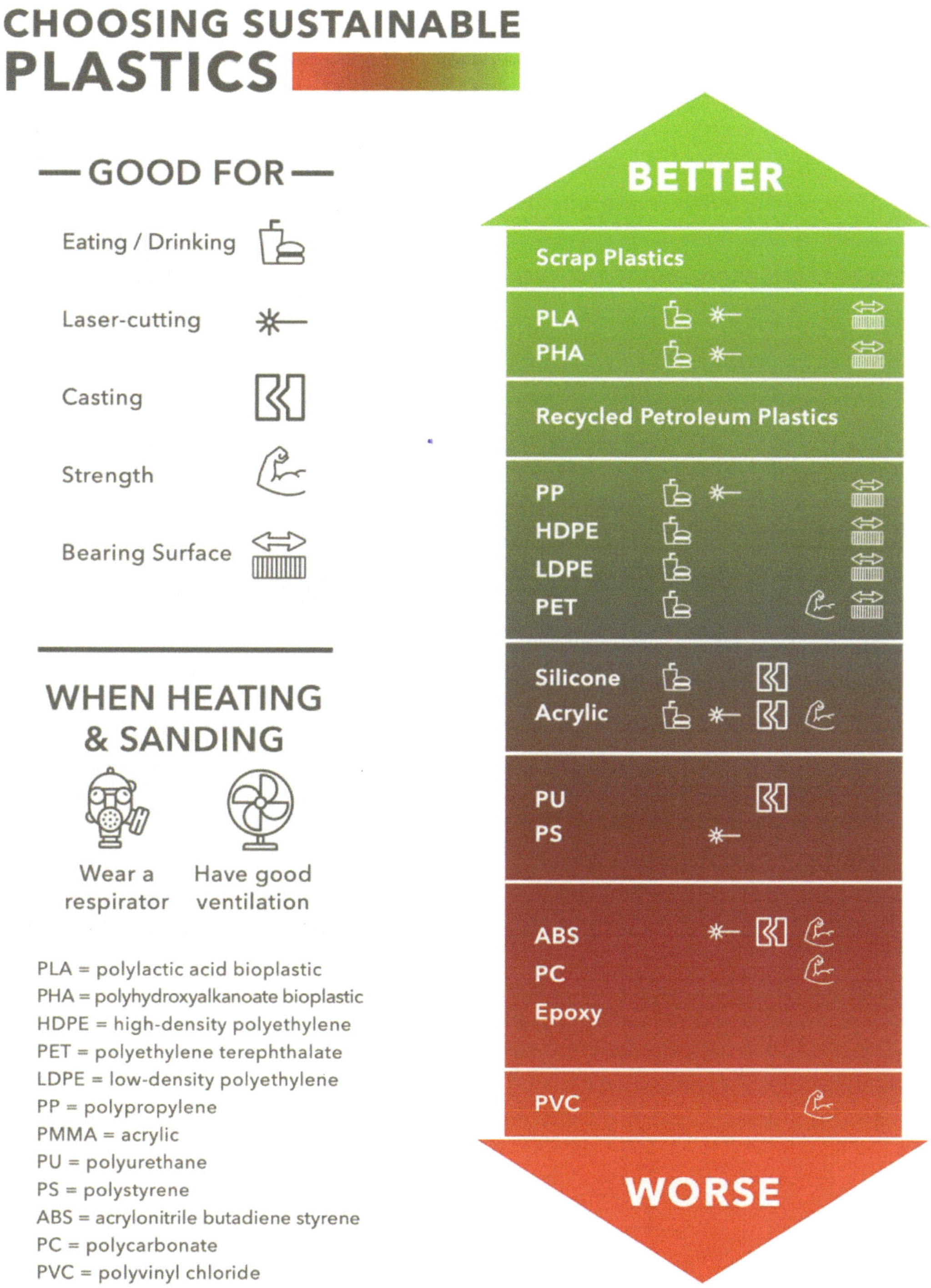

CHOOSING SUSTAINABLE
PLASTICS

— GOOD FOR —

Eating / Drinking

Laser-cutting

Casting

Strength

Bearing Surface

WHEN HEATING
& SANDING

Wear a   Have good
respirator   ventilation

PLA = polylactic acid bioplastic
PHA = polyhydroxyalkanoate bioplastic
HDPE = high-density polyethylene
PET = polyethylene terephthalate
LDPE = low-density polyethylene
PP = polypropylene
PMMA = acrylic
PU = polyurethane
PS = polystyrene
ABS = acrylonitrile butadiene styrene
PC = polycarbonate
PVC = polyvinyl chloride

BETTER

Scrap Plastics

PLA
PHA

Recycled Petroleum Plastics

PP
HDPE
LDPE
PET

Silicone
Acrylic

PU
PS

ABS
PC
Epoxy

PVC

WORSE

Figure 12.7 Quick reference chart for choosing greener plastics
Source: based on Faludi (2015).

Resources and References

Resources for Further Study

- European Commission. (2018). A European strategy for plastics in a circular economy. 52018DC0028. European Commission. https://eur-lex.europa.eu/legal-content/EN/TXT/?uri=COM%3A2018%3A28%3AFIN.

- Rossi, M., & Lent, T. (2006). Creating safe and healthy spaces: Selecting materials that support healing. *Designing the 21st Century Hospital*, 55.

- Henry, B., Laitala, K., & Grimstad Klepp, I. (2019). Microfibres from apparel and home textiles: Prospects for including microplastics in environmental sustainability assessment. *Science of the Total Environment* 652: 483–494.

References

Faludi, J. (2015, August 14). Choosing greener plastics. Instructables. Available at: https://www.instructables.com/Choosing-Greener-Plastics/

Geyer, R. (2020). Production, use, and fate of synthetic polymers. In *Plastic Waste and Recycling* (pp. 13–32). Elsevier.

Law, K. L., Starr, N., Siegler, T. R., Jambeck, J. R., Mallos, N. J., & Leonard, G. H. (2020, October 30). The United States' contribution of plastic waste to land and ocean. *Science Advances*, 6(44).

PlasticsEurope. (2020). Plastics – the facts 2020. Available at: https://plasticseurope.org/nl/knowledge-hub/plastics-the-facts-2020-2/

Ritzen, L., Sprecher, B., Bakker, C., & Balkenende, R. (2023). Bio-based plastics in a circular economy: A review of recovery pathways and implications for product design. *Resources, Conservation and Recycling*, 199, 107268.

How to Apply #12: Source Recycled Plastics for Your Product
Time Estimate: 1–5 Hours

Find sources to buy recycled plastics that can replace primary plastics in your product. Evaluate the feasibility of switching over from a design and cost standpoint.

STEP 1: Identify and List the Plastics in Your Product
Time Estimate: 5–30 Minutes

Make a table of each plastic part, the type of plastic, and the mass (can be rough estimates). Identify which part(s) you'd like to investigate recycled alternatives for. Optional: ideally use LCAs, including both materials and manufacturing methods, to choose the plastic parts with highest environmental impact, or aim to switch all your plastic use to recycled sourcing.

STEP 2: Research Suppliers of Recycled Plastics
Time Estimate: 30 Minutes–2 Hours

Hunt for one or two companies from whom you might buy recycled stock of the plastic(s) you chose in Step 1. You can find companies in online catalogs and databases from the Resources for Further Study in Chapter 10, or search engines, blogs, physical material libraries, wherever you want to look. If you can't find companies that sell recycled versions of your plastic, find the closest recycled plastic you can.

List the company's relevant product offering(s), prices, and the company's website or other contact information. Also note any information on mechanical or aesthetic performance that's important to your application. If the details you need aren't on their website, call them. For prices, get a quote for a realistic production-level amount of material (e.g., for a popular consumer product selling 100,000 units/year with 0.1 kg of the material per product, you might ask for a price on 10,000 kg; for a low-to-medium sales volume product, you might ask for 1/10th or 1/100th as much).

STEP 3: Evaluate Feasibility and Redesign as Needed
Time Estimate: 15–90 Minutes

Make short lists of pros and cons (maybe 5–20 words each) of using these recycled plastics. Is it feasible? Will it impact design, cost, or operations, versus primary materials? Are there design changes that would make it more feasible? (For instance, if the recycled plastic is 20% lower strength than the same primary plastic, you might increase wall thicknesses by 10%, or might perform the calculations required for the exact geometry changes needed.)

STEP 4 (Optional): Quantify Impacts
Time Estimate: 15–60 Minutes

If you started with an LCA of your product and its plastics, run a quick screening LCA to see how much this design option would lower impacts compared to the original design. Show the graph comparison, and briefly write (5–50 words) whether replacing the primary plastics with recycled plastics is good to pursue, given the design or cost changes you estimated, or whether other design interventions would give better results for the whole system. (For example, replacing metals or reducing product energy use.)

Checklist for Self-Assessment

To score your success on this exercise, see if you…

☐   *Made a table of plastic parts.*

☐   *Chose the best part(s) for recycled plastics, or chose all of them.*

☐   *Listed one or two suppliers of recycled plastics.*

☐   *Listed data on available recycled plastics for your chosen part(s).*

☐   *Made a list of pros and cons of using the recycled material, including required design changes.*

☐   *(Optional) quantified the impact of the design change using a quick screening LCA.*

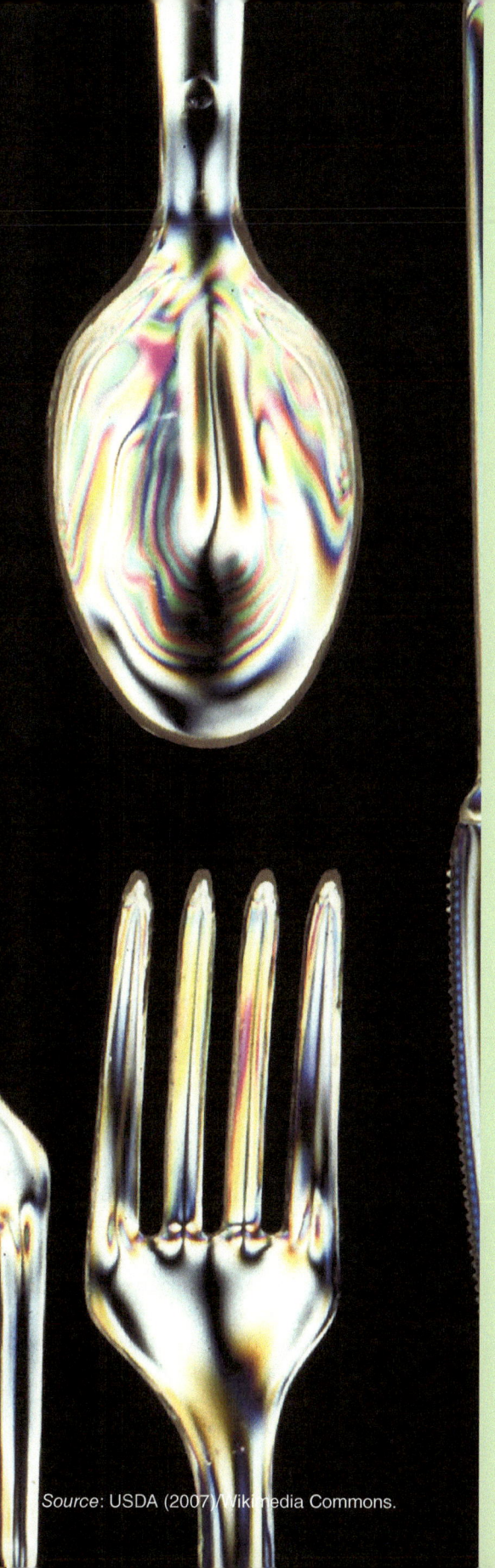

CHAPTER 13
Materials

Bio-Based Plastics

Ruud Balkenende,
Jeremy Faludi,
and Conny Bakker

Goals

- Distinguish between different types of bio-based plastics, their sourcing, and end-of-life recovery pathways

- Explain the difference between bio-based plastics and biodegradable plastics

DOI: 10.4324/9781003504672-15

Why It Matters

If we want to reduce our use of fossil fuels, the use of petrochemical plastics should also be reduced. Bio-based plastics form an alternative to petrochemical plastics. But they have their own challenges when it comes to sourcing of the materials, they might have different properties necessitating redesign of products, and they might need to be recovered differently at the end of life.

Summary

- Bio-based plastics are sourced from plants or micro-organisms. This should not be confused with biodegradability, which refers to potential end-of-life treatments involving degradation under the influence of micro-organisms.

- Not all bio-based plastics are biodegradable, and some petrochemical plastics are biodegradable.

- Bio-based plastics generally cause less greenhouse gas emissions than petrochemical plastics. However, their use of land, fertilizers, pesticides, and sometimes competition with food production, introduces other environmental and social impacts.

- "Drop-in bio-based plastics" are identical to conventional petrochemical plastics from a design perspective. They can be used interchangeably.

- "Dedicated bio-based plastics" use bio-based monomers to form plastics with distinct new properties.

- Recovery pathways for bio-based plastics are largely similar to those for conventional plastics.

- Composting and incineration of bio-based plastics are carbon-neutral, unlike for fossil-based plastics.

13.1 Bio-based Polymers

Bio-based plastics are made from polymers produced from renewable resources such as biomass. They are carbon-neutral, implying that the atmospheric carbon dioxide concentration does not increase even after their incineration. This is because the carbon they release into the air is carbon they recently sequestered out of the air by growing, as opposed to fossil fuels, which grew by pulling carbon out of the air millions of years ago. Bio-based plastics complement natural and renewable materials such as wood, cork, cotton, and wool, offering a wider and better-defined range of properties than these natural materials.

Bio-based polymers are still mostly used in low-end applications like packaging, disposable catering supplies, and agricultural films, with PLA (polylactic acid) as a well-known example. An example showing the combination of a bio-based plastic and a natural material is the shoe shown in Figure 13.1. The sole is made from bio-TPU (thermoplastic polyurethane) derived from corn, and the top is made from cotton. A shoe sole puts relatively high demands on wear resistance and flexibility of the material, illustrating that bio-based plastics are also suitable for durable engineering applications.

The use of a renewable feedstock is necessary to move away from fossil fuels. However, this alone does not necessarily make bio-based materials sustainable. The production of bio-based plastics has been criticized for competing with resources for food production. As a result, the "first-generation" feedstocks that consisted of edible seeds or corn are increasingly being replaced by second, third, and even fourth generation production methods.

Figure 13.1 Reebok cotton + corn sneaker
Source: Windle (2021).

These use non-edible, fast-growing crops or agricultural by-products (second generation), algae biomass (third generation) or, as a new development on the horizon, "fourth generation" genetically modified algae and other microbes as feedstock.

The processing also adds to the impact. In synthesizing raw bio-materials into chemicals for polymer production, a wide range of industrial processing techniques are involved. And just like petrochemical polymers, bio-based polymers will need additives such as antioxidants, plasticizers, colorants, and fillers, to adapt their physical properties and improve their suitability for a particular use. This might complicate end-of-life recovery.

13.1.1 Dedicated Bio-based Polymers

Dedicated bio-based polymers like PLA do not have an identical fossil-based counterpart. They are made via a dedicated pathway in which polymers are built from monomers directly based on organic building blocks like sugars and organic acids. These bio-based polymers do not resemble existing petrochemical polymers in structure, and exhibit different properties. Figure 13.2 gives an impression of the range of monomers that are available and the versatility of the polymers that can be obtained.

The best-known example is polylactic acid (PLA). PLA is a polyester obtained from the polymerization of lactic acid, obtained from fermentation of sugar or starch. It is a rigid, transparent, glossy, industrially biodegradable polymer. Initially, this polymer was mainly used for bio-medical applications. With dropping prices, it is now used at large scale as packaging material and for other disposables. Another example is polyethylene furanoate (PEF), a polyester made completely from bio-based building blocks. Although still in a pilot stage, it is comparable to polyethylene terephthalate (PET) in application range and manufacturability. Its gas barrier and thermal properties are superior to PET, which makes it attractive for applications in the packaging industry for beverages.

13.1.2 Drop-In Bio-based Polymers

A drop-in bio-based plastic is chemically identical to the petrochemical alternative but made from biomass instead of fossil oil or gas. They, therefore, are a direct substitute for conventional polymers, and can be used instead of their petrochemical equivalents without technology changes or equipment investments. Drop-ins are usually made via bio-ethanol, which is converted to ethylene; the starting point for many conventional

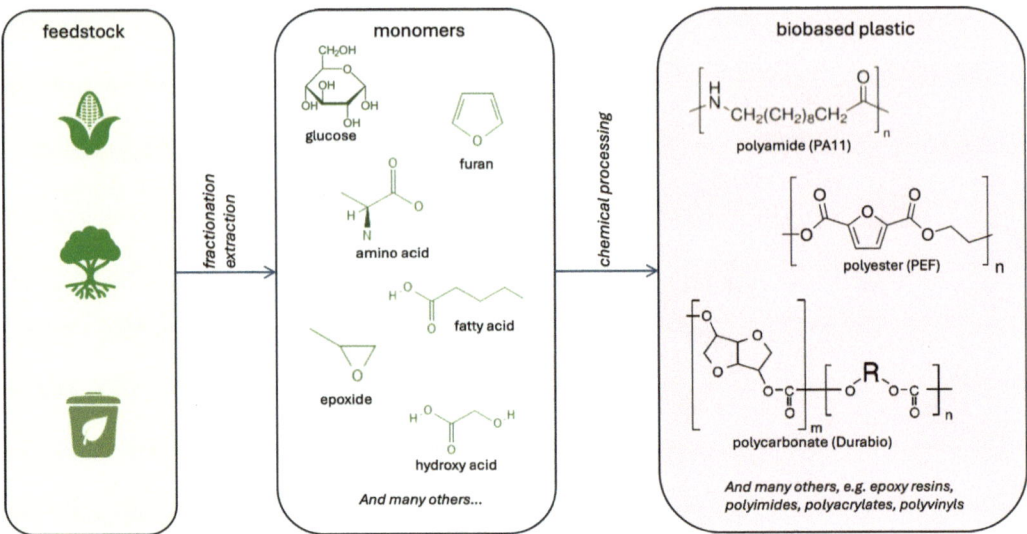

Figure 13.2 The range of polymers that can be obtained from biomass feedstocks

Source: Cywar et al. (2022).

plastics. Examples include bio-polyethylene (bio-PE), bio-propylene (bio-PP), and bio-polyethylene terephthalate (bio-PET). As drop-ins do not add new properties, they must compete on price with conventional polymers, which makes them less attractive from an economic point of view.

Even partially bio-based polymers exist. Plastics are also called bio-based if part of their monomers originate from renewable sources. Bio-PET, for example, has bio-ethylene-glycol as one of its monomers, while the other monomer is terephthalic acid, which is often made from fossil sources, implying that bio-PET consists of plant-based material for 30% by weight, and only 20% by counting carbon atoms. However, this field is developing rapidly, and recently also bio-terephthalic acid has become available.

13.1.3 Polymers and Their Applications

The polymer pyramid shown in Figure 13.3 indicates which plastics are suitable when a certain performance is required. Figure 13.3 shows that at the commodity level, there are already many bio-based alternatives for commonly used polymers.

At the engineering grade, there are also some replacements, but at the high-performance level there are currently few practical replacements for petrochemical polymers. Figure 13.4 shows global production of bio-based plastics for various applications.

Bio-based polymers are mainly used in packaging (starch blends, PLA, bio-PET, bio-PE) and textiles (bio-PTT). With the

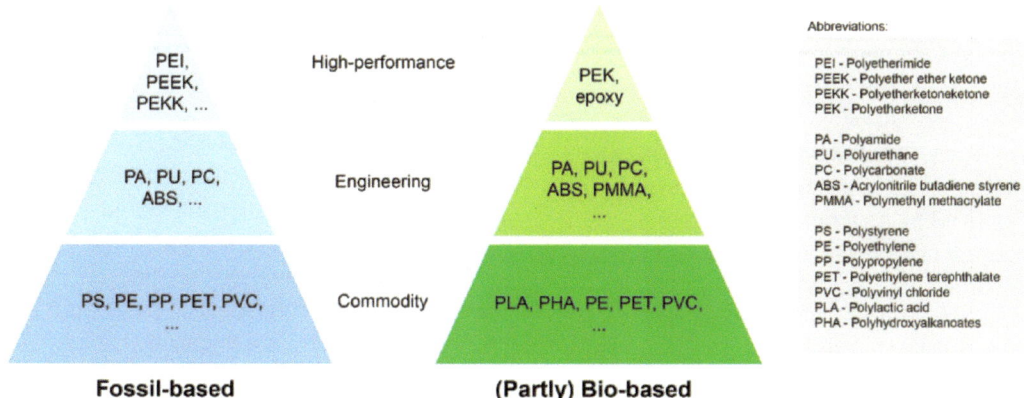

Figure 13.3 Polymer pyramid for fossil-fuel based polymers and bio-based polymers
Source: European Bioplastics, nova-institute (2023).

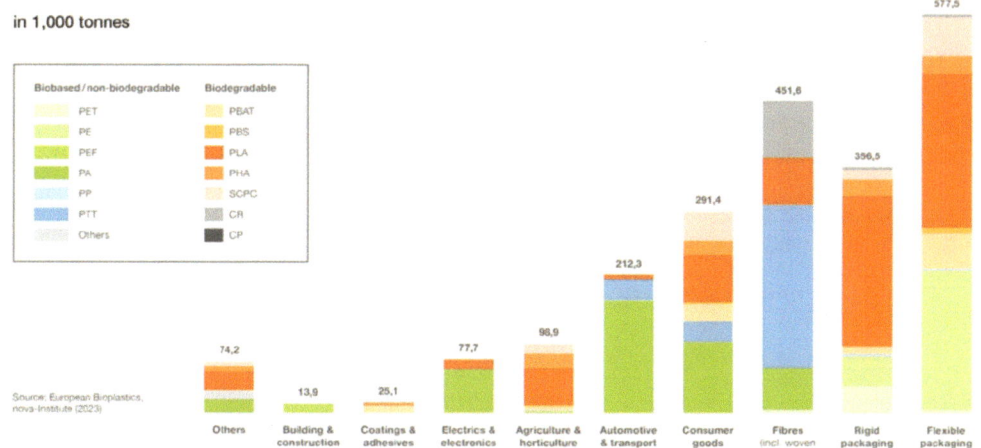

Figure 13.4 Global production capacities of bioplastics in 2022, forecast to 2028

exception of PLA and starch-based blends, dedicated bio-based polymers are still emerging materials from a material science perspective, and also to designers, who need to learn about their properties in relation to applications. Bio-based plastics are being explored in engineering applications, but currently drop-in bio-based polymers are most common.

13.2 Recovery Pathways for Bio-based Plastics

The recovery pathways for petrochemical plastics also apply to bio-based plastics (Figure 13.5). As for plastics in general, recovery of products and components is preferred over recovery pathways that only retrieve the material. However, in contrast to petrochemical plastics, for bio-based plastics, complete decomposition through incineration or composting is acceptable, as the CO_2 that is emitted is part of a short carbon cycle, it doesn't come from ancient geological sources.

13.2.1 Recycling of Bio-based Plastics

Recycling of bio-based plastics is similar to petrochemical plastics, and can be carried out through mechanical or chemical recycling, shown in Figure 13.5. Mechanical recycling consists of well-known physical steps, such as the shredding, washing, separating, drying, regranulation, and compounding of thermoplastics. The properties of these recycled plastics are usually downgraded when compared to primary plastics. Dedicated bio-based plastics can be chemically recycled, as they are usually "condensation polymers." Chemical recycling is a depolymerization process (hydrolysis) that converts plastic back into monomers, making it possible to remake the pure polymer. More information about these processes can be found in Chapter 15.

Although mechanical and chemical recycling are possible, they are only done occasionally. This is related to the low volume of bio-based plastics currently used, and difficulties in separating these plastics from the bulk of petrochemical plastics. The exception here is the drop-in bio-based plastics, which

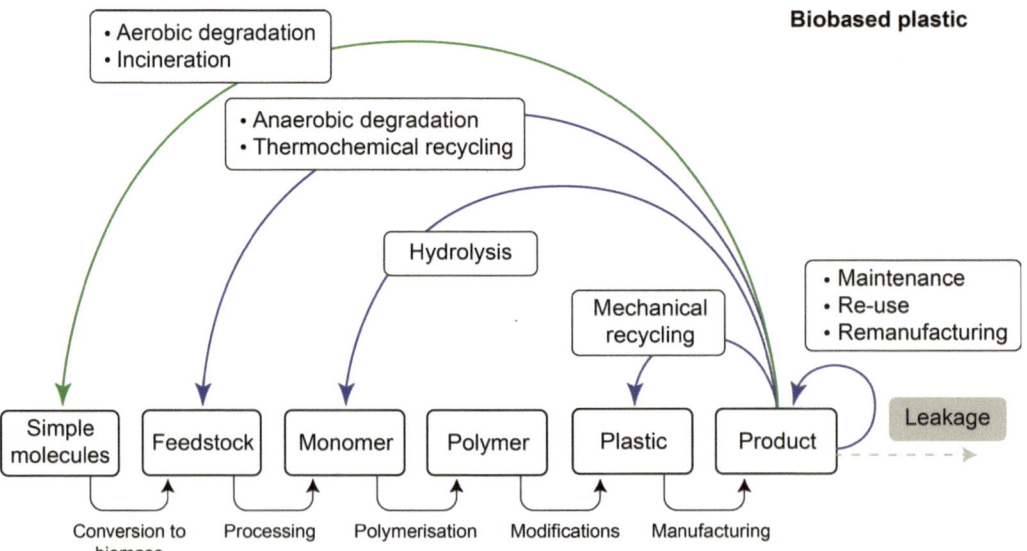

Figure 13.5 Recovery pathways for bioplastics, focusing on material recovery

Source: Adapted from Ritzen et al. (2023).

are chemically indistinguishable from their petrochemical counterparts, and will be recycled with those plastics. So, for example, bio-PE will become mixed with PE during recycling.

13.2.2 Biodegradation

The term "biodegradability" in common speech is used for degradation mechanisms that are actually very different, shown in Figure 13.5, and not all of them are environmentally benign. First, let's clearly distinguish "biodegradation" from "degradation under environmental conditions." Degradation under environmental conditions is a process by which a material disintegrates. It can, for instance, take place because of natural daylight ("photodegradation") or because special additives were included to help the plastic break down into small fragments over time ("oxo-degradable" plastics). These processes result in micro-plastics.

Biodegradation, in contrast, results from the action of naturally occurring micro-organisms such as bacteria, fungi, and algae. Biodegradation breaks the polymer chains, lowering the molecular mass of the polymer that constitutes the plastic. In the presence of sufficient oxygen, micro-organisms cause "aerobic" degradation, converting biodegradable plastic completely into water and CO_2. Without oxygen, "anaerobic" degradation gives rise to production of biogas, mainly methane, CO and CO_2. Under controlled industrial conditions, biogas can be used as fuel. Under uncontrolled conditions, this process is harmful, as methane is a potent greenhouse gas (about 28 times as effective as CO_2), so this degradation is *not* carbon-neutral.

Composting is a form of aerobic biodegradation that implies "complete oxidation," meaning it results in nothing but carbon dioxide, water, and compost (biomass) that contains valuable nutrients and is used as a soil improver. Note that biodegradable polymers, unlike natural organic waste, don't contain fibers and nutrients and will not leave compost. Composting is often misunderstood as a process that happens "naturally" when a biodegradable polymer ends up somewhere in the open environment. However, few biodegradable polymers break down successfully at ambient temperatures in a reasonable timeframe. Most can only be composted under carefully controlled conditions (e.g., a temperature of 58°C), in industrial composting plants, meaning they will not be compostable for most users or when left in the environment. The EU and US standards for determining compostability (EN 13432/ASTM D6400) require at least 90% biodegradation in six months. Other standards have specified the requirements for home composting as at least 90% decomposition within 12 months at ambient temperature (NF T 51–800). Also, for home composting, a relatively high temperature of at least 32°C is still required for optimal digestion by micro-organisms. When specifying biodegradable plastics, be sure to require one of these certifications; ideally home-compostable certification.

Even in the case of home-compostable plastics, the risk of biodegradation under anaerobic conditions or at too low temperature cannot be completely excluded. Unmanaged biodegradation, caused by the uncontrolled release of biodegradable plastics in the environment, should therefore be avoided, as it can lead to slow and incomplete biodegradation, and methane emissions.

13.2.3 Incineration

With bio-based plastics, incineration is carbon-neutral, whereas for petrochemical plastics it leads to an increased carbon footprint. Incineration, especially if combined with energy recovery, is, therefore, a feasible end-of-life pathway for bio-based plastics.

13.2.4 Bio-based Versus Biodegradable

It is important to distinguish bio-based clearly from biodegradable. A common misunderstanding is that plastics produced from renewable feedstock are biodegradable, but this is not necessarily the case. The term "bio-based" indicates the renewable origin of the carbon source used for manufacturing the plastic. In contrast, "biodegradable" points to a potential degradation mechanism at end of life, which is unrelated to the carbon source. Biodegradable plastics can be made from fossil fuels as well as bio-based sources (see Figure 13.6 for examples). Biodegradation thus doesn't relate to the production phase of a material, but to its disposal phase.

13.3 Impact of Bio-based Plastics

The renewable nature of the sources for bio-based plastics is an advantage when it comes to global warming and resource depletion. Also, recovery options with bio-based materials are clearly mapped and offer the advantage of incineration being climate-neutral. But the environmental impact of their production, use and recovery still comes with considerable uncertainty. A life cycle assessment (LCA) of bio-based

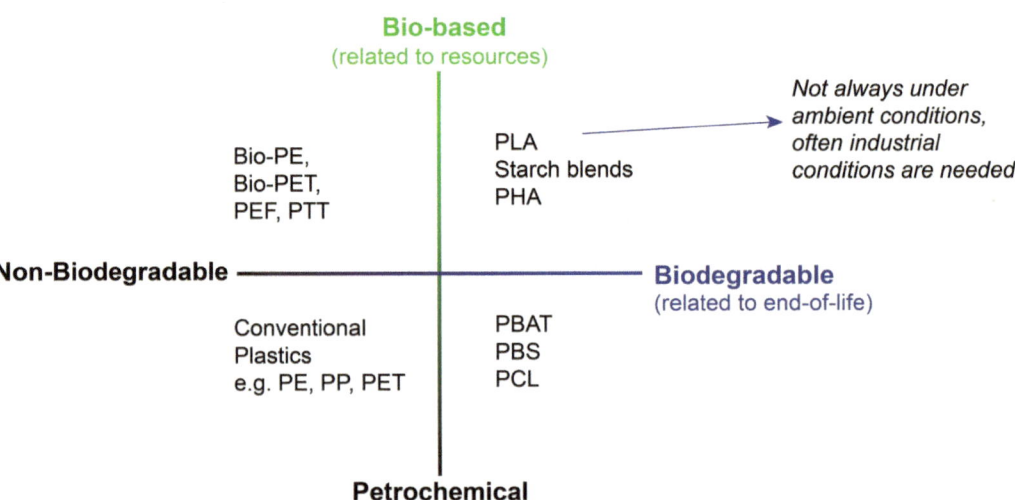

Figure 13.6 Plastic source and biodegradability categories and their examples

plastics has proven to be complicated, because most of them are relatively novel, so there is little data available for these materials. An additional complication is the variety of sources and large local differences in cultivation impact.

Carbon is not the only consideration for bioplastics. Growing crops for bio-based plastics leads to large increases in land and water use. Production of bioplastics often causes more eutrophication and acidification than petrochemical plastics, and also marine eco-toxity is usually higher, though the carbon benefits are often worth the tradeoff. Social impacts are less clear and vary more by production location than plastic type, implying that the impact of bio-based plastics is strongly dependent on the nature of the crop source and where it's grown (Ritzen et al., 2024).

Resources and References

Resources for Further Study

- The Biopolymer Database by Altair Engineering GmbH: https://www.materialdatacenter.com/bo/main (commercial biopolymers listing engineering data, applications, and companies to order from)

- Materiom database of bio-based materials: https://materiom.org (do-it-yourself recipes to make materials, with example products).

References

Cywar, R. M., Rorrer, N. A., Hoyt, C. B., Beckham, G. T., & Chen, E. Y. X. (2022). Bio-based polymers with performance-advantaged properties. *Nature Reviews Materials*, 7(2), 83–103.

European Bioplastics, nova-institute. (2023). Market. European bioplastics. Available at: https://www.european-bioplastics.org/market/

Ritzen, L., Sprecher, B., Bakker, C., & Balkenende, R. (2023). Bio-based plastics in a circular economy: A review of recovery pathways and implications for product design. *Resources, Conservation and Recycling*, 199, 107268.

Ritzen, L., Sprecher, B., Bakker, C., & Balkenende, R. (2024). Sustainability of bio-based polyethylene: The influence of biomass sourcing and end-of-life. *Journal of Industrial Ecology*, 28(6), 1684–1698.

Windle, J. (2021, April 23). Reebok cotton + corn sustainable sneakers. Jeans a Teacup. Available at: https://www.jeansandateacup.com/reebok-cotton-corn-sustainable-sneakers/

How to Apply #13: Evaluate Replacing a Plastic Part with Bio-based and Biodegradable Materials
Time Estimate: 1–3.5 Hours

Investigate replacing a plastic part in your product with bioplastic, a composite, or other biomaterial. Most bioplastics are new for durable applications; they aren't always a greener option, but in the right circumstances they can be a good choice. However, plastics have only been used for about 80 years, so traditional biomaterials like wood or molded paper pulp may also be viable. Here, you'll generate some ideas and evaluate their performance for your product.

STEP 1: Identify Plastic Components to Be Replaced
Time Estimate: 15–30 Minutes

Make a list of the plastic parts in your product, including types of plastic and rough masses (you may already have this table from the exercise in Chapter 12). Identify for which part(s) you'd like to investigate bio-based and biodegradable ("B&B") alternatives. Consider:

- Are certain parts consumable/disposable? Biodegradability can be a benefit here.

- Are some parts more obviously replaceable with B&B plastics? (For instance, less stringent requirements for strength, heat resistance, or other functionality.)

- Do some parts especially communicate with customers, helping create an emotional connection if they know it's B&B plastic?

- Optional: ideally use LCA including both materials and manufacturing methods to identify the plastic parts with highest environmental impact.

Your final decision of which plastic part(s) to replace with B&B plastics may balance all of these.

STEP 2: Find Suppliers of Biomaterials
Time Estimate: 30 Minutes–2 Hours

Hunt for biomaterials you can buy for your part—ideally both bio-based and biodegradable, but perhaps just one or the other, depending on your part's performance needs. They may be bioplastics, or composites, or other biomaterials. You can find companies in online catalogs and databases mentioned in the "Resources for Further Study" list in this chapter or Chapter 10, or search engines, blogs, physical material libraries, wherever you want to look. Find at least five materials from two companies.

List the materials you found, each with a URL or reference for where you found them.

Narrow down the materials to one or two top candidates, by investigating each material to find whatever mechanical or aesthetic performance information you need (e.g., elasticity, transparency, waterproofness, etc.). If they don't have enough details online, call them. Consider:

- What functional properties do these plastics have versus what your parts need?
- Are the polymer feedstocks renewable and responsible?
- Are the polymers compostable or biodegradable at end of life? Under what conditions? Will your product be disposed of in those conditions?

These considerations may make you decide bioplastics aren't the best choice for your product. Whether they are or aren't clearly a good choice, continue through the rest of the exercise to see how you might turn them into a good choice.

Once you've narrowed down to one or two best options, get a price quote. Get the quote for a realistic production-level amount of material (e.g., for a popular consumer product selling 100,000 units/year with 0.1 kg of the material per product, you might ask for a price on 10,000 kg; for a low-to-medium sales volume product, you might ask for 1/10th or 1/100th as much).

Write down your final material choice(s) with brief notes on why (5–20 words). Include the price, notes on functionality or aesthetics, and the company website or other contact information.

STEP 3: Design Considerations and Use-Phase Design Changes
Time Estimate: 15–30 Minutes

To use the new material you found for this part, should you change the design? Consider:

- Does the new material have unique properties that could improve your product's functionality or desirability?
- Are there design or geometry changes necessitated by different physical or aesthetic properties? (Such as thicker walls to account for lower strength.)
- Should the product's user recognize the part as biomaterial? If so, would you make any aesthetic changes to tell the story?

Sketch or briefly write how you would change the design or not, and why.

STEP 4: End-of-Life Considerations
Time Estimate: 15 Minutes–1 Hour

If you were to use a bioplastic or other biomaterial for these components, what design changes or business/operations changes (such as taking products back at end of life) would help ensure the product is disposed of properly at the end of life? Consider:

- What's the expected end of life for your product and its materials? How will the new materials affect that? Will they contaminate standard plastic recycling streams, or be easily separated and composted, or have their own recycling infrastructure, or other?

- Is there a way to make it easy to separate components (see Chapter 16)?

- Is there a way to communicate your intended end-of-life recovery pathway to users?

Sketch or briefly write your intended end-of-life pathway, with brief notes on why (5–20 words). If these considerations make you decide biomaterials aren't the best choice for your product, say so and say why.

Checklist for Self-Assessment

To score your success on this exercise, see if you...

- ☐ *Listed plastic parts in your product, and chose which part(s) to investigate replacing with biomaterials.*

- ☐ *Listed 5+ bioplastics or other biomaterials from 2+ companies, citing sources.*

- ☐ *Chose a best material and said why, listing its price and relevant functional or aesthetic properties. (Or listed why no biomaterial is a wise choice in this case.)*

- ☐ *Sketched or briefly described design changes to accommodate the new material in product use phase.*

- ☐ *Sketched or briefly described design changes to accommodate the new material at end of life.*

CHAPTER 14

Product Lifetime Extension

*Ruud Balkenende,
Jeremy Faludi,
and Conny Bakker*

Goals

- Explain the rationale for product lifetime extension
- Apply principles of design for durability
- Apply principles of design for repair, upgrade, refurbishment, or remanufacturing

DOI: 10.4324/9781003504672-16

Why It Matters

Keeping products functional for a prolonged time is often the most effective way to reduce environmental impacts associated with mining, manufacturing, and end of life, as no new products need to be made. Maintenance, reuse, and repair prolong the use phase of a product, while refurbishment, remanufacturing, and parts harvesting enable a new life cycle for products or components. To enable these recovery pathways, the product should be built to last. Diagnosis of the status of its components, disassembly, replacement of components, and reassembly should be enabled to facilitate repair and remanufacturing. Legislation increasingly recognizes the importance of lifetime prolongation.

Summary

- Product lifetime extension includes (in order of preference): durability (including maintenance), reuse, repair or upgrade, refurbishment, and remanufacturing.
- Design for repair, upgrade, and remanufacturing includes how products are diagnosed, disassembled, components are repaired or replaced, and products are reassembled and tested.
- Design for repair, upgrade, and remanufacturing shares many product architecture strategies.
- Repair needs business models providing affordable and available spare parts, while remanufacturing needs reverse logistics and a long-term strategic approach to designing product generations/platforms.
- Design for repair/remanufacturing requires:
  - **easy diagnosis**: clearly trace lost/ decreased functionality to particular (sets of) components.
  - **easy disassembly**: products must be disassembled quickly and simply, and "priority parts" are accessible.
  - **easy reassembly**: connections can be reused and can be established manually.

14.1 Product Lifetime Extension in a Circular Economy

The circular economy aims to avoid waste by closing the loops of product and material flows. For products, this can be done at different levels. Often, closing loops is associated with recycling. However, if you only recover and reprocess materials and manufacture them into new products, you lose all the energy, effort, and financial value put into the product during the manufacturing stage. From both an economic and environmental perspective, it's better for you to recover whole products and components, to maintain functionality and value. Therefore, in addition to designing for recycling (materials recovery), products should also be designed for longer use, reuse, repair, and remanufacturing (all implying product-level recovery) as well as parts harvesting (component-level recovery). In fact, these are higher priorities. The Value Hill visualizes these priorities of retaining value in Figure 4.2 in Chapter 4.

Designing products that last, and that can be repaired, refurbished, and remanufactured easily and economically, has received increasing attention in the past few years. There is a growing societal interest, especially in repair, and most notably self-repair, stirred by consumers and grassroots associations which aim to repair their products (see iFixit's repair manifesto in Figure 14.1). Also at a political level, awareness has been growing. Improving the reparability of consumer products is one of the measures in the European Commission's Ecodesign for Sustainable Products Regulation (European Commission, 2024) to arrive at more environmentally sustainable

Figure 14.1 iFixit's repair manifesto

and circular products. Right-to-repair laws are also sweeping many US states.

Repairing, refurbishing, and remanufacturing products require your product architecture to facilitate these activities, as well as associated service and business models (Bocken et al., 2016). This will ensure that users as well as businesses adopt new approaches for dealing with products at the end of their functional lives. To design for these factors, see Chapters 20 and 21. This chapter focuses on product architecture strategies.

14.2 Design for Durability

Durability refers to long physical reliability of a product. Examples are the development of products that can take a lot of wear and tear without breaking down. Reliability of components, ease of maintenance, and material choices are important features for products that are intended to be durable. Examples can be found in many products for professional use, ranging from washing machines to furnaces and power tools. Some consumer products have similar reliable design aspects. For instance, washing machines produced by Miele are built to have an average lifetime of 18.5 years compared to an industry average of 11.7 years (Bakker et al., 2014).

Designing for durability implies preventing the occurrence of common causes for failure. We can distinguish mechanical, thermal, electrical, chemical, and radiation-induced failures. Failures usually occur if a product suffers continuous high exposure or large variations in exposure to a specific cause or a combination of causes. Table 14.1 lists potential failure

causes and provides examples of the damage that might be caused.

To prevent failure from happening, you can apply a variety of design strategies, either at the product level or the component level: prevent, block, distribute, dissipate, and endure. Here are summaries of each:

- **Prevent** means not exposing vulnerable components to failure causes. For example, make indicators for proper use and maintenance, separate sensitive components from components that produce high exposure (e.g., decouple electronics from vibrating components through dampers), and design ergonomically (e.g., to avoid dropping).

- **Block** means shielding vulnerable components from failure causes. For example, add surface coating to resist UV exposure, thermal insulation to protect temperature sensitive materials, watertight seals to avoid water ingress, and provide breakers (e.g., overheat shutdown, electronic fuse).

- **Distribute** means distributing stress from a failure cause to avoid peak loads. For example, with mechanical loads, follow lines of stress; with thermal loads, add mass as a heat sink or add thermally conductive material as a heat spreader.

- **Dissipate** means moving failure causes to the surrounding environment. For example, dissipate thermal loads through passive (large surface area) or active (water flow) cooling. A damper might absorb vibration and shocks.

- **Endure** means designing the product to withstand the failure causes. For example, select robust materials and components, or over-specify dimensions and properties.

Table 14.1 Overview of failure causes and examples of associated failure modes

Failure cause		Possible failure modes
Mechanical	Bending	• Cracking (rapid, fatigue or creep)
	Vibration	• Plastic deformation
	Impact	• Disconnection of components
	Wear	• Degradation of material (microplowing, microcutting, microcracking) • Degradation of nearby electronics due to overheating • Product appearance deterioration
Thermal	Overheating	• Deformation and other degradation of plastics • Degradation of temperature sensitive components • Thermal runaway leading to circuit-board damage
	Heat shock	• Local deformation due to thermal expansion leading to component failure • Material damage • Rapid degradation or total failure of temperature sensitive components • Thermal runaway leading to circuit-board damage • Dislodgement of soldered components
	Cold shock	• Joint failure due to thermal mismatch • Cracking through mechanical stress • Detachment of adhesives due to embrittlement • Local deformation and failure due to bending in component interconnections • Brittle fracture
Electrical	Electrostatic discharge	• Malfunction of voltage-sensitive components (e.g., transistors)
	Electrical overstress	• Complete destruction of the sensitive components • Fire hazard
Chemical	Water (POWER ON)	• Short circuiting • Malfunctioning due to increased leakage current
	Water (POWER OFF)	• Permanent damage to electronics due to galvanic corrosion of metals • Short circuit pathways due to galvanic corrosion leading to failures when powered • Absorption by polymers of electronics resulting to changing electrical properties • Swelling and degradation of materials impacting mechanical robustness • Corrosion resulting in degradation of appearance
	Acids, bases	• Acceleration of water related failures
	Lubrication, cleaning agents	• Absorption by PCB laminate and polymers leading to weakening and cracks • Damage of internal structure of capacitors leading to failure • Degradation of product appearance
	Humidity	• Corrosion weakening the overall structural integrity and performance • Corrosion leading to leakage current, contact resistance and short circuit • Corrosion leading to degradation of appearance
Radiation	Ultraviolet (UV)	• Degradation of mechanical properties of polymers • Degradation of product appearance

Designing for robustness/durability can use multiple strategies. For instance, a failure caused by overheating could be prevented by using components that are less sensitive to heat, using components that produce less heat, using features that spread heat, and/or insulating sensitive components. In practice, a balanced approach between different design strategies is likely. Table 14.2 gives an overview of design principles you can apply. Note that these principles might simultaneously fit several of the above-mentioned strategies.

However, longevity depends not only on the engineering aspects of a product. These basically define the technical lifetime of the product. But often the economic lifetime of a product is considerably shorter (Bakker et al., 2014). This is often because new products have more functionality, or the user's needs change, rendering old products obsolete; see Figure 21.5 in Chapter 21 to see why people

Table 14.2 Design principles for physical product durability

Design principle	Explanatory examples
Design simplicity	Using basic operating principles that reduce the number of (moving) parts required
Material & component selection	Matching type and grade of components and materials to their functional requirements and use environment
Over-specification/over-dimensioning	Ensuring the load on the part will not exceed the load that the part can handle. Dimensioning interfaces of moving parts to provide smooth running.
Enclosing	Protecting sensitive electronics from water by using a watertight cover; airtight cooling system in refrigerator
Breaker	Interrupting operation to protect against failure, e.g., electric overcurrent by providing a fuse, computer shutting down before overheating
Surface treatment	Select surface treatments to prevent degradation from weathering or wear, or to age gracefully
Redundancy	Duplication of critical components
Expendable parts/materials	Inexpensive parts designed to wear out during use, protecting more expensive parts, e.g., brake pads or lubricant
Decoupling	Separating components from each other's influence, e.g., positioning capacitors away from hot transistors, or damping vibration through a shock absorber
Controlled tolerances	Choosing tolerances of parts or between parts to reduce internal variations in load, thus reducing fatigue from stress cycles
Proper use/maintenance encouragement	Avoid excess stresses and wear through user care and handling, e.g., indicator for oil replenishment in a car
Condition maintenance	Features to stay within safe operating conditions, e.g., cooling of computer chip, load balancing in a washing machine, throttle valve against overpressure
Ergonomic design	Features to avoid dropping and otherwise assure proper handling, e.g., hand grip for transporting a vacuum cleaner

throw away various products. For changing functionality or user needs, you can design for upgradability, discussed later.

Another reason for throwing away products is changing fashion. For this, you can create an emotional attachment to a product, as described in Chapter 20. Attachment is determined by multiple aspects, like memories, enjoyment, usability, appearance, and reliability (Page, 2014). Self-expression and group affiliation have been identified as other important factors (Mugge, 2017). With materials that age gracefully, like a well-worn leather jacket, the user might experience a shared history with the product (Mugge et al., 2008; Rognoli & Karana, 2013). This storytelling can build personal connection. However, attachment is generally difficult to control, and these type of design interventions are likely to be limited to certain product categories (e.g., watches) and to people with certain personalities or interests (Mugge et al., 2005).

Designing for durability is not the best strategy for all products—for example, food packaging that is not easily collected for reuse at end of life might be better designed for recycling or composting. Other products might be better designed for repair or other life extension. But durability is a very valuable strategy for most products.

14.3 Design for Lifetime Extension

A product's life can also be extended through maintenance, repair, technical upgrading, refurbishment, or remanufacturing. Definitions and examples are provided in Table 14.3.

14.3.1 Maintenance

Maintenance, repair, and upgrading have many aspects in common, with maintenance aiming to keep a product in working condition, repair (sometimes called "corrective maintenance") to restore the product to working condition after a failure, and upgrading to improve performance. Design for maintenance means including features to physically ease maintenance activities (e.g., a car with convenient screw-cap access to replace motor oil). It may also mean including features to psychologically encourage maintenance (e.g., a car dashboard light reminding users to change the oil).

14.3.2 Repair

Repairability of many consumer products has considerably decreased over the past decade due to factors like miniaturization of electronics and increased complexity of products, but also due to design interventions that, consciously or unconsciously, hurt the user's ability to diagnose faults, and limit the accessibility of parts. An example of the latter are deeply hidden screws and glued parts. Design for repair guidelines usually focus on aspects like component standardization, modularity, and the ease of disassembly.

Easy repair doesn't need easy access to all components, it should focus on "priority parts": the parts essential for the product's performance and most likely to break down or deteriorate. (When also designing for remanufacturing or recycling, the definition of priority parts expands to also include parts with high economic value to recover, and hazardous parts like batteries or chemicals that shouldn't go into shredders.) A

Table 14.3 Definitions of reuse, repair, etc.

Recovery operation	Definition	Examples
Reuse	Using the whole product again.	• Second-hand clothing
Maintenance	Preventive action to retain reliable functionality.	• Replacing motor oil in a car • "Descaling" of mineral deposits from a coffee machine before it malfunctions
Repair	Corrective action to restore functional performance of a part (Linton & Jayaraman, 2005)	• Replacing a broken part • Descaling a coffee machine after it malfunctions
Upgrade	Extending use by improving the product's quality, value, effectiveness, or performance	• Replacing computer RAM that still functions with bigger RAM to expand functionality.
Refurbish	Restoring a product to an acceptable level of functionality (often by third parties) (Hatcher et al., 2014)	• Smartphone that is inspected and cleaned, replacing parts that are broken or severely underperforming
Remanufacture	Restoring a product to a condition the same as or better than a new product (British Standards Institute)	• Smartphone that is built anew from a combination of used, repaired, and new parts, with at least the same specification as the original parts
Parts harvesting	Retrieving components, modules, or parts from products, to use them as spare parts for servicing, maintenance, and repair. These parts might be refurbished or remanufactured.	• Common in in the automotive industry, for ICT equipment, and professional machinery to reuse still functioning parts

"disassembly map" (Figure 14.2) can be used to show the number of steps and the difficulty of reaching various components. And often neglected, but equally important, the ease of reassembly should also be considered. Especially preventing damage to connection points, which hampers reinstalling parts.

Smartphones are a typical example of products that are very hard to repair (at least by non-professionals), due to glue, soldered components, and the high degree of integration (e.g., combined screen and motherboard). However, Fairphone, with its modular build and easy accessibility of the components, shows that you can design such products to let users easily replace a broken part (Figure 14.3).

Diagnosing faults is also key to repair. Fault diagnosis is the process of identifying and characterizing a fault when a failure occurs. It is, therefore, an essential step to take before product repair. However, most products are not designed for an easy diagnosis process, as feedback to the user is often absent or hard to understand, and access to components is difficult (both visually and manually). This makes repair an obscure process, especially for ordinary users, and may limit their willingness to repair, as the required effort and cost are not clear. Providing error codes or other clear signals may greatly help diagnosis, but may complicate the build of the product. As successful diagnosis often requires access to internal components, ease of disassembly and reassembly is also helpful here.

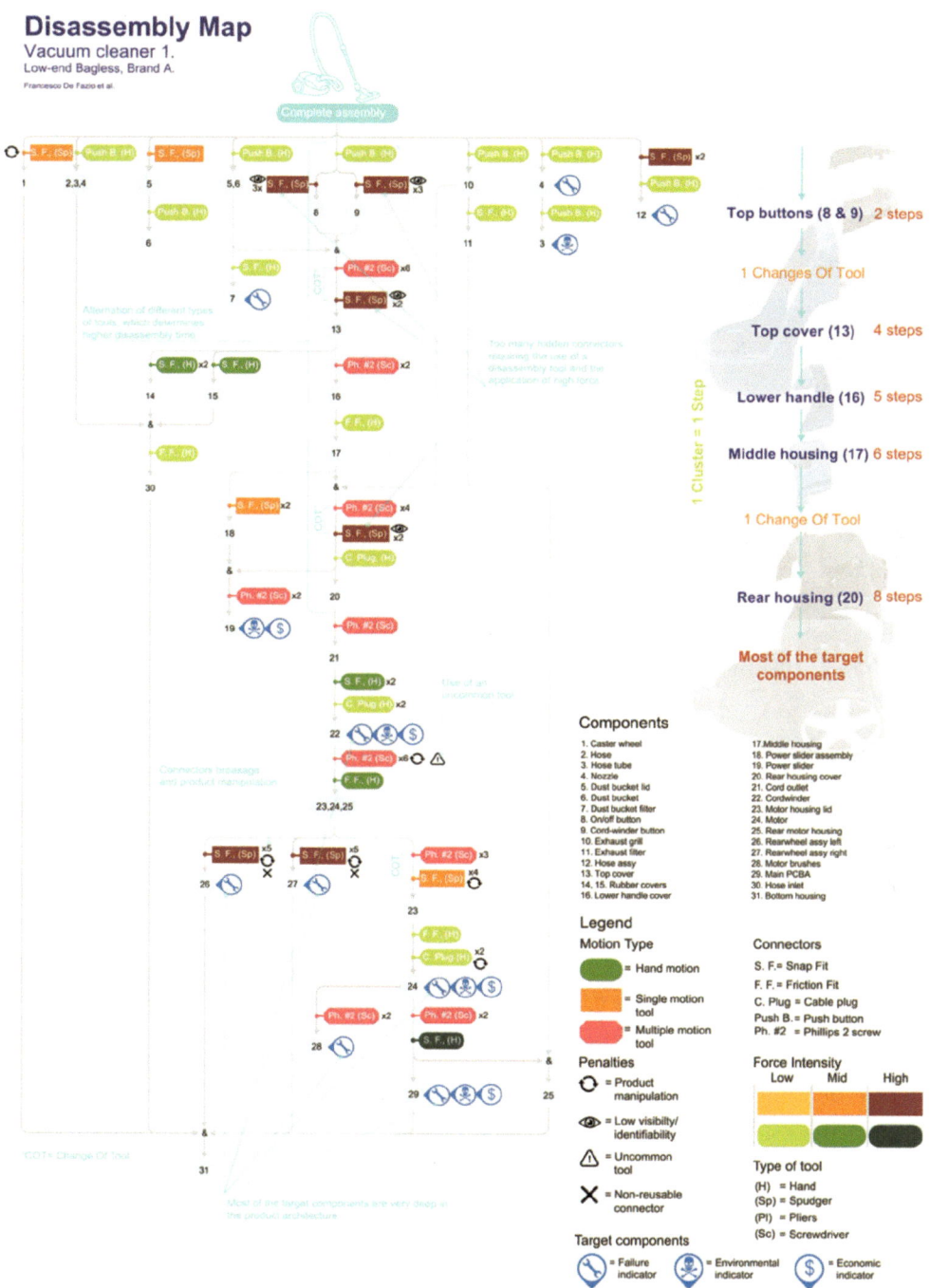

Figure 14.2 Disassembly map for a vacuum cleaner. Vertical depth shows the number of disassembly steps to reach components (numbered circles). Colored boxes show tools and actions required. Blue icons show priority parts ("P"), hazards (skull and crossbones), and financial value ("$$").

Source: De Fazio et al. (2021).

Figure 14.3 A Fairphone 2 is easy to disassemble for repair and upgrade

Source: iFixit.com

14.3.3 Upgrade

Upgradability is the ability of a product to continue being useful under changing conditions by improving the quality, value, and effectiveness or performance (Linton & Jayaraman, 2005). Technological developments leading to improved product features are often the reason for buying a new product. This has been evident in information and communication technology developments over the past decade, where new generation of smartphones and tablets followed each other rapidly. This can, to some extent, be prevented by making products upgradable.

Designing for upgrade requires product architecture similar to design for repair: easy disassembly and modularity. However, it also requires thinking ahead in time:

you need to consider technological roadmaps and expected changes in user needs. Technologically, the definitions of your product's functional modules and standardized interfaces for them are critical. Products are not designed individually, but based on a "platform" forming the core of multiple generations of a product, with the possibility to exchange modules that are expected to improve most over time. For example, desktop computers that allow for upgrades in chips and memory, and allow for expanded functionality like multiple monitors. The platform is the computer chassis and motherboard, which can last for several generations of memory, CPU, and peripherals. This especially asks for a careful consideration of modularity, accessibility, and interfacing (Kasarda et al., 2007; Rashid et al., 2013).

14.3.4 Refurbish/Remanufacture

Refurbishment and remanufacturing, although in daily practice often used interchangeably, result in different product qualifications (see definitions above). Regardless, designing for them both is almost the same. For both operations, a product is thoroughly inspected, cleaned, worn parts are replaced, and the reassembled product is tested extensively. Usually, these products come with warranties. The main difference is the higher quality assurance of remanufactured products. Parts harvesting is closely connected to refurbishment and remanufacturing, again with shared design principles, but with the focus on parts instead of the whole product.

To design for refurbishment or remanufacturing, use strategies listed above for component durability, disassembly and reassembly, and accessibility; also design for cleaning, reverse logistics, and marketing (Shu & Flowers, 1999; Nasr &Thurston, 2006). Useful design concepts are modularization and platform design (Hatcher et al., 2014). All these concepts are familiar in traditional product design, but get a different focus in remanufacturing. For example, modularization and platform design are usually intended to improve manufacturing efficiency and cost; while in remanufacturing, the focus is on efficient disassembly and process organization. Design for remanufacturing involves decisions like standardization of parts, and selection of materials and fasteners.

The role of the user is especially important in the case of refurbishment, where user acceptance of refurbished products is of critical importance. Reducing the perceived risk of low quality is important. This can be achieved through informing users and by increasing the benefits offered by refurbished products. Suggested approaches are to enable improvement of product aesthetics by easily resurfacing vulnerable parts (Mugge, 2017), or to give new users information on the product's previous usage history (van Weelden et al., 2016).

14.3.5 Combining Strategies

Ideally, you would design your product for longer life in all the ways listed above, but time and effort and money are limited. Like designing for durability, you need to use your own judgment about whether to prioritize design for repair, upgrade, remanufacturing, or other strategies. As always, it's best to use an evidence-based approach, looking at data on how and when your products fail, how often those failures are repaired versus ending the product's life, and how well your products are recovered at end of life, etc. As mentioned above, product architecture is not enough, you also need a business model and support systems (e.g., collection infrastructure or a repair network). However, to solve the product architecture part of lifetime extension, Figure 14.4 sums up these design strategies in a checklist.

Design for repair & reincarnation checklist

Access a product's components... 〉 **to keep the product alive longer...** 〉 **and give its parts multiple lives**

Design for Disassembly
Ensure products are easy to take apart quickly.

Parts
- ☐ Minimize the number of parts.
- ☐ Simplify structure and form.
- ☐ Use ferromagnetic materials to enable sorting and disassembly.

Tools and Fasteners
- ☐ Require only a few standard tools.
- ☐ Avoid requiring tools for the most common actions.
- ☐ Minimize the number and variety of fasteners.
- ☐ Use intuitive snap-fits, clips, or sliding connections.
- ☐ Design connections that are visually and physically accessible.
- ☐ Access fasteners from the same axis.
- ☐ Hold multiple parts with one fastener.
- ☐ Use coarse threaded screws for speed; use nuts and bolts for strength.
- ☐ Use human-scale fasteners.
- ☐ Use hand-strength press-fits instead of tight press-fits.
- ☐ Avoid glues, and use only glues that are easily soluble or heat reversible.
- ☐ Ensure fasteners are adequate for structural integrity.
- ☐ Use fasteners that will hold up over repeated use.

Documents
- ☐ Embed clear, graphical disassembly instructions into the product.
- ☐ Document materials and methods for deconstruction for the user.

Design for Repair
Easily fix failures through disassembly and:

Product Architecture
- ☐ Use modular assemblies that enable the replacement of discrete components.
- ☐ Ensure easy access to parts likely to need maintenance.
- ☐ Use self-locating parts.
- ☐ Use robust connectors.
- ☐ Label and color-code parts to enable troubleshooting.
- ☐ Standardize parts between product lines and across generations.

Documents
- ☐ Make technical documentation freely available or open-sourced.
- ☐ Include parts list and part numbers.
- ☐ Create user interfaces and troubleshooting tools to diagnose problems.

Business
- ☐ Make repair and service options clear to customers.
- ☐ Consider repair-friendly warranty terms.
- ☐ Make replacement parts available and affordable.

Design for Upgrade
Keep products relevant and useful longer.

Product Arch.
- ☐ Use standard-size modular parts to enable interchangeability and customization.
- ☐ Design easy access to parts likely to become obsolete.
- ☐ Use standard, cross-platform connections (for example, USB).

Tools
- ☐ Build diagnostic tools to help users understand the components that are limiting performance.

Design for Remanufacturing
Enable reuse of old components in new products.

Business
- ☐ Design collection infrastructure with smooth touchpoints between the company and users.
- ☐ Design a quality-control system for testing returned components.
- ☐ Use a take-back program, product-service-system, or other enabling business model.

Figure 14.4 Quick reference guide to design for repair, upgrade, and remanufacturing

Resources and References

Resources for Further Study

* Videos on design for durability, disassembly, repair, and upgrade in the VentureWell Tools for Design and Sustainability website: https://venturewell.org/tools_for_design/design-lifetime-sharing/

References

Bakker, C., Wang, F., Huisman, J., & den Hollander, M. (2014). Products that go round: Exploring product life extension through design. *Journal of Cleaner Production*, 69, 10–16.

Bocken, N. M. P., de Pauw, I., Bakker, C., & van der Grinten, B. (2016). Product design and business model strategies for a circular economy. *Journal of Industrial and Production Engineering*, 33(5), 308–320.

De Fazio, F., Bakker, C., Flipsen, B., & Balkenende, R. (2021). The Disassembly Map: A new method to enhance design for product repairability. *Journal of Cleaner Production*, 320, 128552.

European Commission. (2022). Sustainable Product Policy Initiative. Available at: en.euractiv.eu/wp-content/uploads/sites/2/...

European Commission. (2024). *Ecodesign for Sustainable Products Regulation, 2024/1781*. Available at http://data.europa.eu/eli/reg/2024/1781/oj/eng.

Hatcher, G. D., Ijomah, W. L., & Windmill, J. F. C. (2014). A network model to assist design for remanufacture integration into the design process. *Journal of Cleaner Production*, 64, 244–253.

Kasarda, M. E., Terpenny, J. P., Inman, D., Precoda, K. R., Jelesko, J., Sahin, A., & Park, J. (2007). Design for adaptability (DFAD): A new concept for achieving sustainable design. *Robotics and Computer-Integrated Manufacturing*, 23(6), 727–734.

Linton, J. D., & Jayaraman, V. (2005). A framework for identifying differences and similarities in the managerial competencies associated with different modes of product life extension. *International Journal of Production Research*, 43(9), 1807–1829.

Mugge, R. (2017). A consumer's perspective on the circular economy. In J. Chapman (Ed.), *Routledge handbook of sustainable product design*. Routledge.

Mugge, R., Schifferstein, H. N. J., & Schoormans, J. P. L. (2005). Product attachment and product lifetime: The role of personality congruity and fashion. In K. M. Ekstrom & H. Brembeck (Eds.), *E - European advances in consumer research* (vol. 7, pp. 460–467). Association for Consumer Research.

Mugge, R., Schoormans, J. P., & Schifferstein, H. N. (2008). Product attachment: Design strategies to stimulate the emotional bonding to products. In H. N Schifferstein, & P. Hekkers (Eds.), *Product experience* (pp. 425–440). Elsevier.

Nasr, N. & Thurston, M. (2006). Remanufacturing: A key enabler to sustainable product systems. In *Proceedings of LCE2006*, pp. 15–18.

Page, T. (2014). Product attachment and replacement: Implications for sustainable design. *International Journal of Sustainable Design*, 2(3), 265.

Rashid, A., Asif, F. M., Krajnik, P., & Nicolescu, C. M. (2013). Resource Conservative Manufacturing: An essential change in business and technology paradigm for sustainable manufacturing. *Journal of Cleaner Production*, 57, 166–177.

Rognoli, V., & Karana, E. (2013). Toward a new materials aesthetic based on imperfection and graceful aging. In E. Karana, O. Pedgley, & V. Rognoli (Eds.), *Materials experience* (pp. 145–154). Butterworth-Heinemann.

Shu, L. H., & Flowers, W. C. (1999). Application of a design-for-remanufacture framework to the selection of product life-cycle fastening and joining methods. *Robotics and Computer-Integrated Manufacturing*, 15(3), 179–190.

van Weelden, E., Mugge, R., & Bakker, C. (2016). Paving the way towards circular consumption: Exploring consumer acceptance of refurbished mobile phones in the Dutch market. *Journal of Cleaner Production*, 113, 743–754.

How to Apply #14: Design for Durability
Time Estimate: 30 Minutes–4 Hours

Generate design alternatives to extend your product's useful life. This is a deeper dive into design strategies you may have considered in Chapter 4. You may want to revisit your work from that exercise.

STEP 1: Generate Design Ideas to Extend the Use Phase
Time Estimate: 15–60 Minutes

Review Table 14.2, "Design principles for physical product durability." In addition to these, also consider the two cultural durability points mentioned in the text:

• timeless aesthetics, not driven by fashion

• flexible use for changing user needs and scenarios

Brainstorm design alternatives by generating at least 5+ ideas for each topic, 50+ ideas total. If you're really certain a principle doesn't apply, you can skip it, but stretch yourself to try to think of applications. Especially push for specific and concrete ideas, but it's a brainstorm, all wild crazy ideas are okay. Take note of which category of design strategies feels most important for your product, and why.

STEP 2: Choose, Develop, and Explain Your Top Idea
Time Estimate: 15–60 Minutes

Choose your favorite design idea from your brainstorm, and develop it more. Sketch any changes in form. If the product looks the same, with just material substitution, write down the new material and what it replaces.

Write briefly (5–20 words) why you chose this design direction, including what usage changes you expect (if any), and what environmental impact improvements you expect.

STEP 3: Plan to Avoid Unintended Consequences and Encourage Intended Consequences
Time Estimate: 10–20 Minutes

Sometimes a design change to make something more durable can end up leading to more environmental impact, if the product is still thrown away early in a disposable culture. Can you think of any risks or unintended consequences from your design change? Is there anything you can do to mitigate these risks to make sure your design change has the desired effects? Consider:

- Are you asking users to change behaviors?

- Are you relying on other systems or services during the use phase or end of life?

- How can you positively drive users to reuse or share? (This can include ideas from Chapters 20 and 21.)

BONUS STEP: (Optional): Generate Design Ideas for Post-Use (Repair, Upgrade, and Remanufacturing)
Time Estimate: 30 Minutes–1.5 Hours

Repeat the steps above, but brainstorming on the Design for Product Lifetime quick reference guide (Figure 14.4), specifically the lists of design for repair, upgrade, or remanufacturing principles. This generates and develops design ideas for the post-use phase of your product.

Checklist for Self-Assessment

To score your success on this exercise, see if you...

- ☐ *Generated 5+ ideas for each strategy in the design for physical product durability table, plus the two cultural durability principles; 50+ ideas total.*

- ☐ *Chose a top idea and briefly described why, with a sketch or brief description of it.*

- ☐ *Listed possible unintended consequences from your top idea's design changes, and strategies to help ensure your design has the intended effects.*

- ☐ *(Bonus, optional): Followed a similar process to develop design ideas for the post-use phase of your product's life (repair, upgrade, or remanufacture).*

CHAPTER 15

Material Recovery

Recycling Processes

*Ruud Balkenende,
Jeremy Faludi,
and Conny Bakker*

Goals

- Describe the recycling process of products/materials

- Explain how the recycling of products is affected by design

- Recognize metrics for recycling effectiveness, such as recovery yield and grade

DOI: 10.4324/9781003504672-17

Why It Matters

Recycling of materials is necessary when the functionality of products can no longer be maintained or restored. Through recycling, we reduce environmental impacts by avoiding the damage and energy consumption of mining of new materials. Also, emissions of hazardous materials at the end-of-life stage are prevented. From an economic perspective, it prevents losing scarce or critical materials. Design for recycling is needed to enable high recovery yields of materials, which is a necessity to reach increasingly demanding recycling targets.

Summary

- Recycling processes liberate materials from products, usually in a destructive way.

- Materials must be separated into homogeneous material fractions to obtain high recovery yields and avoid downcycling.

- Copper (Cu), iron (Fe), and aluminum (Al) cannot be recycled simultaneously. When one is recovered from a mixture, the others are lost. Several other metals can be recovered when mixed with Cu, Fe, or Al.

- Electronic circuit boards usually consist of materials with high impact and high economic value, like copper, gold (Au), silver (Ag), and palladium (Pd). Fortunately, these materials can be processed simultaneously, in the Cu pyrometallurgical recovery process.

- Recycling is sometimes referred to as urban mining, so it uses mining terms like "recovery yield" (quantity of obtained material) and "grade" (purity of obtained material).

Recycling aims to recover materials. How this is done depends on the nature and the source of the material. Figure 15.1 depicts the recycling process as part of the product life cycle. From a product design perspective, we can distinguish between post-industrial recycled material, which originates from manufacturing waste, and post-consumer recycled material, which is obtained from used products. The purity of post-industrial recycled material is usually like that of primary material. Post-consumer waste may vary widely in quality, depending on its level of contamination, the quality of recycling processes and the final reprocessing step. From the perspective of sustainable design, we are mostly interested in the recyclability of materials from used products and the use of post-consumer recycled materials in new products.

Different materials need their own processes to be recycled: glass and paper need to be treated separately, plastics should not be mixed, and metals like aluminum, iron, and copper cannot be recovered simultaneously. This means that the materials that together form the product need to be separated from each other at the end-of-life stage. Note also that recycling cannot be carried out on a small scale, as efficiency of scale requires that many different products are treated simultaneously in a bulk process. For the separation of materials from products, manual labor is considered too expensive, and usually brute force is used. Products are shredded into small fragments, and those fragments are subsequently subjected to detection and separation processes that distribute the materials in different "fractions" (streams of similar material). These material fractions are then further processed to

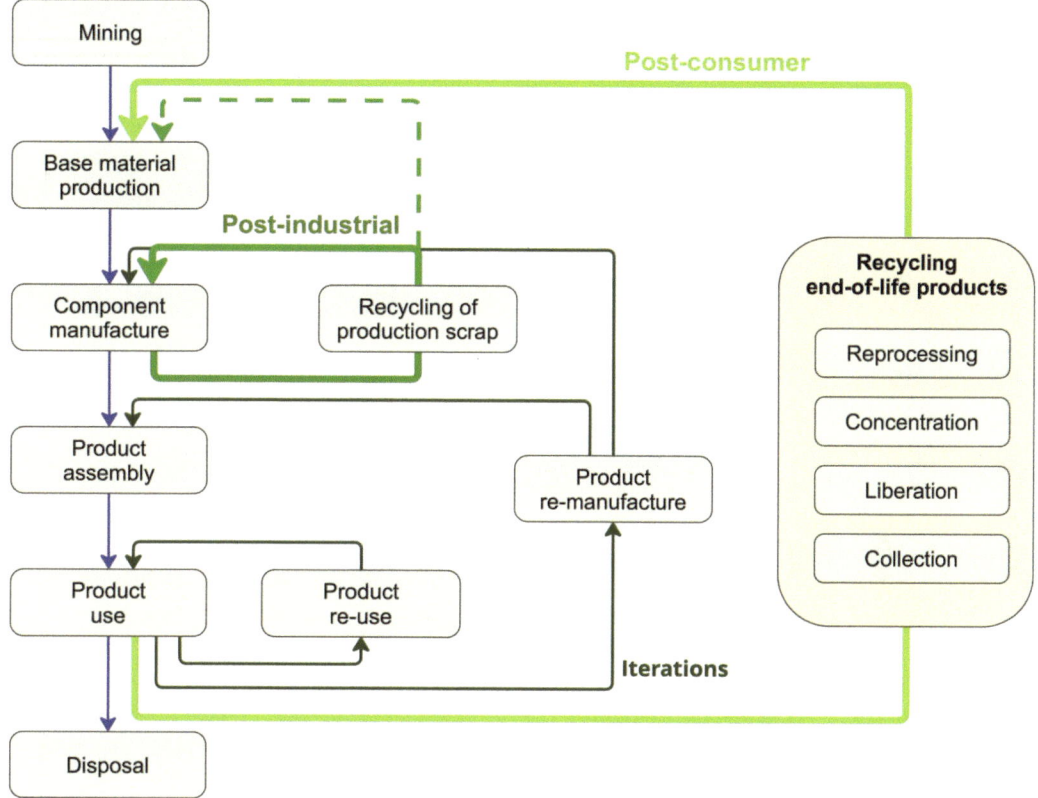

Figure 15.1 The recycling process as part of the product life cycle

Source: Based on Tempelman et al. (2014).

obtain materials that can be used again in manufacturing.

Here we will take recycling of electronic products as an example, as these highlight some specific issues, while the process is representative for many different product categories. Let's focus on the steps in the recycling process to explore how the design of a product influences the results of these steps. We use LED lamps as an example.

Step 1: Collection

For many products, dedicated collection systems exist. In Europe, paper, glass, and batteries have already been separately collected for decades. More recently, dedicated systems have also been set up for organic waste, plastic packaging (sometimes combined with Tetrapak drink packaging and cans), textiles and shoes, and electronics. This is mostly done though curbside collection, roadside collection points, in stores, or at municipal waste collection points (Figure 15.2). Sometimes a financial deposit system is used to stimulate high return rates.

Electronic waste should be collected as a separate waste stream. In 2019, the EU raised its recycling target from 45% by weight to 65% (this percentage is based on

Figure 15.2 Different collection systems

the products sold in the previous 3 years). In the Netherlands, various channels for disposing of obsolete electronic products are available; the most common are returning to retail stores or municipal collection points. The product's design and architecture as such have no direct effect on the collection rate of electronics, but aspects like clear communication, preferably directly on the product or packaging, and communication about the collection infrastructure, are essential and can be considered part of the service associated with a product.

Step 2: Liberation

Liberation is a process of freeing connected materials from each other. This is usually a destructive mechanical bulk process of shredding or crushing. However, manual liberation of some materials might be necessary to keep hazardous substances out of mechanical equipment (e.g., shredding a battery could burn down a recycling plant). This means that batteries, refrigerator coolant, and frying pan oil, for example, are first removed manually in a "depollution" step. During this initial step, easy-to-grab high value materials like power cables might also be manually removed.

Shredding the product is an irreversible destructive step in which the products are cut or broken into small fragments. Figure 15.3 shows examples of typical equipment used for this operation. By combining the shredder with a sieve, more uniformly

Dual shaft crusher   **Hammer mill**   **Auger Crusher**

Figure 15.3 Typical equipment used for liberation

shaped fragments are obtained, which helps subsequent separation.

The results of shredding of LED lamps are shown in Figure 15.4. Some fragments are, as intended, homogeneous in composition. However, many fragments still consist of multiple materials, which cannot be recycled simultaneously. Analysis of such mixed fragments shows that the incomplete liberation of the materials is due to the connections that join different parts.

Different types of connections have different separability. For example, friction fits and snap fits are liberated easily by both shredding and manual disassembly.

By contrast, glued or welded parts are inseparable either by shredding or manual dismantling. See Table 16.2 in Chapter 16 for which connection types to prefer and avoid.

Step 3: Concentration/Separation

The homogeneous and mixed fragments that result after recycling are separated to obtain fractions that are concentrated in the desired material. Separation is done based on physical properties of the materials (Figure 15.5).

In a typical separation sequence for electronic products, first, magnetic materials

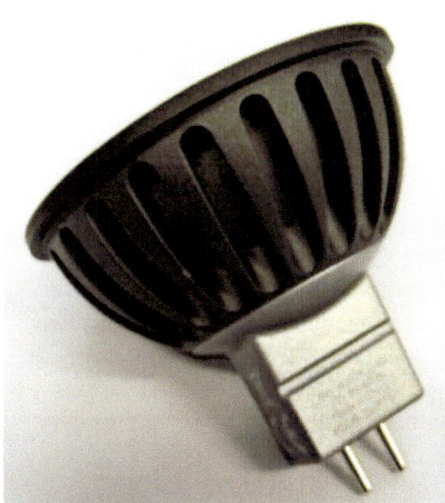

Figure 15.4 Standard spotlight and the resulting fragments after shredding
Source: Aerts et al. (2014).

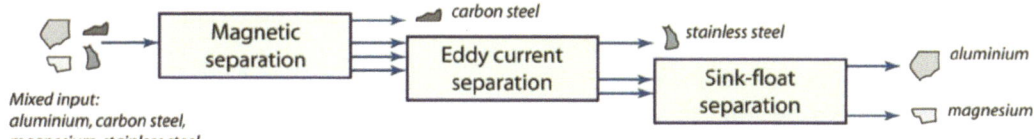

Figure 15.5 Metals separation using a sequence of separation machines
Source: Tempelman et al. (2014).

are separated using an electromagnetic rotating drum. Typically, steel is separated in this step. But note that also permanent magnets (e.g., originating from transformers or speakers) and mixed fragments that contain steel are likely to end up in the magnetic fraction. This will contaminate the steel fraction with other materials and reduce the recovery of those other materials.

To separate non-magnetic conductive materials from non-conductive materials, eddy current separation is a common next step. A rapidly switching magnetic field induces so-called "eddy currents" in the conductive fragments, and this current in turn induces a magnetic moment in the fragment, which causes repulsion of the conductive fragment. The higher the conductivity, the greater the magnetic repulsion. In this way, non-magnetic conductive materials like copper and aluminum are separated. If particle sizes are sufficiently uniform, aluminum and copper can also be separated from each other.

Remaining materials can be separated based on density. This can be done using "float tables," in which water with varying salt concentrations is used. Wind sifting (blowing air at materials) is another method for this type of separation. Major polymers like PE, PP, PA, ABS, PC, PET can be separated in this way.

If mainly polymers are present, for example, in the case of recycling of plastic packaging waste, infrared detection is a more accurate way to separate plastics. With this technique, a "molecular fingerprint" of the plastic is obtained. Plastics are identified and are then separated into a specific fraction by targeted air puffs.

Step 4: Reprocessing

The material fractions that are obtained after separation are subsequently reprocessed to obtain materials that can be used again in products. More information on reprocessing of metals can be found in Chapters 11 and 12.

15.2 Recycling as Urban Mining

Like mining, recycling aims to extract valuable materials. The end-of-life products that are recycled can be considered to be ore that needs to be processed to extract the desired material fractions. "Tailings" means waste material resulting from this process. Due to the similarity with mining, the recycling of materials from waste is sometimes referred to as "urban mining."

As with mining of ores, the "recovery yield" and "grade" of the resulting material are of interest, as shown in Figure 15.6. The recovery yield measures how much of the desired material from the input is extracted into the output. The grade measures the purity of the output fraction and indicates the level of contamination with undesired materials.

$$\text{Recovery yield (material X)} = \frac{\text{mass of material X in fraction X}}{\text{total mass of material X in input}}$$

$$\text{Grade (fraction X)} = \frac{\text{mass of material X in fraction X}}{\text{total mass of fraction X}}$$

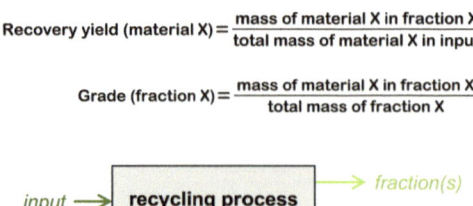

Figure 15.6 Schematic of the recycling process, with equations for yield and grade

A higher recovery yield and grade usually imply that more energy is needed for the liberation and separation processes, which also makes the process more expensive. But here, designers to a large extent, can determine the ease of recycling of the materials used in a product through material choice and product architecture, see Chapter 16.

Recycling yield ideally would represent the yield of the entire recycling process—the fraction of material that is recycled from obsolete products. But different organizations measure different things. Be sure to check what is measured when quoting figures. Especially note that, in some cases, the recycling rate includes incineration with energy recovery (also euphemistically referred to as "thermal recycling"). However, incineration is not at all related to material recovery, and you should avoid the term thermal recycling.

Recycling rates of paper and cardboard are fairly high worldwide, with a 60% global average, a minimum of 37% in Africa, and a maximum of over 73% in Europe (EPRC, 2022). Glass is recycled at a global average of roughly 20–30%, but 75% in Europe. In 2019, plastic packaging materials were recycled less than 20% globally, but 41% in Europe. In 2020, the EU's actual electronics recycling rate was 43%, far below the target of 65% (Forti et al., 2020). Note that these data are by weight, not differentiating between materials based on their environmental impact or economic value. Finally, as these numbers are inputs *into* recycling facilities, they do not account for the considerable losses during the subsequent liberation, separation, and reprocessing. Actual recovery yields for many metals are less than 10% (see Figure 11.3 in

Chapter 11). The limited degree of recycling is also clear from the low level at which recycled materials are used in new products.

15.3 Environmental Impact of Recycling

The environmental impacts of waste are obvious. Waste in landfills or the environment is not inert. Biodegradable materials decomposing anaerobically lead to methane gas that has a high global warming potential. Plastics degrade and form micro-plastics that have adverse effects on food chains globally. Toxic materials, like heavy metals and polyaromatic compounds, leach from products and are emitted into groundwater and air. Leakage from a cell phone battery can contaminate 600 m^3 of groundwater. Incineration causes air pollution, CO_2, and sulfur dioxide emissions, leading to global warming and acid rain, and might produce toxic substances like dioxins. Recycling eliminates or lowers these impacts, which are sometimes not taken into account in cradle-to-grave LCAs.

Recycling also lowers impacts and emissions by reducing the need for the production of primary materials, saving energy and water, while avoiding the burden on natural resources. Although recycling processes of course also need energy and lead to emissions, this is usually significantly less than needed to produce primary materials. Recycling a glass bottle means a 20% reduction in air pollution and a 50% reduction in water pollution compared to making a bottle from raw materials. Recycling 1 ton of copper means that typically mining of 150–200 tons of copper ore and 4 tons of

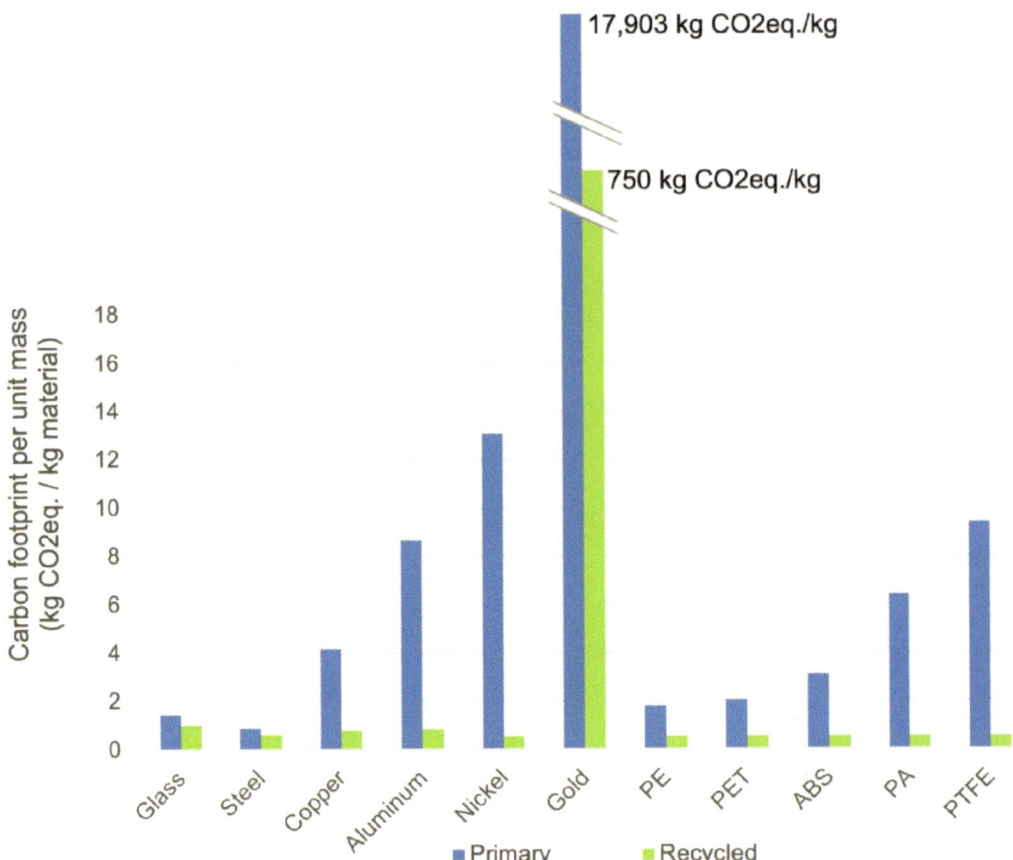

Figure 15.7 Carbon footprints of some materials, primary versus recycled.

Source: Data from Idemat (Vogtlander et al., 2024).

coal are avoided. Life cycle assessment inventory data distinguish between primary and recycled materials, which provides a direct indication of the avoided impacts when using recycled materials. Figure 15.7 shows an LCA comparing production of 1 kg of several different materials, both primary and secondary.

References

Aerts, M., Felix, J., Huisman, J., & Balkenende, R. (2014). Lamp redesign: shredding before selling. Going Green – CARE Innovation 2014 conference, November, Vienna, Austria.

EPRC. (2022). Monitoring Report 2021: European Declaration on Paper Recycling 2021–2030. European Paper Recycling Council (EPRC).

Forti, V., Baldé, P.C., Kuehr, R., & Bel, G. (2020). The Global E-Waste Monitor 2020: Quantities, flows and the circular economy potential. United Nations University & International Solid Waste Association.

Tempelman, E., Shercliff, H., & van Eyben, B. N. (2014). *Manufacturing and design: Understanding the principles of how things are made.* Elsevier.

Vogtlander, J. G., et al. (2024). Idemat LCA database. Available at: EcoCostValue.com.

How to Apply #15: Investigate Local Recycling
Time Estimate: 1–6 Hours

While many products are distributed globally, the first step to understanding global waste treatment is to experience local facilities. Learn about recycling facilities in your area, and visit one in person, if possible. Talk with the people there about how you might improve recovery of your product.

STEP 1: Map Waste Processing Facilities in Your Area
Time Estimate: 15–60 Minutes

Find where waste goes in your local region, including recycling of metals, plastics, and e-waste, as well as compost, incineration, and/or landfill. "Local" can mean your city, state, or other region; whatever radius is required to find at least one processing facility for each of these waste streams. You can find facilities through municipal government websites, other online searches, or whatever means you need. You don't need to find facilities smelting metals or melting plastics to make granulate; try facilities that sort, liberate, and separate are enough. Informal facilities (such as neighborhood gardens taking compost) are okay, but official facilities are better.

Mark each of these facilities on a map.

STEP 2: Visit a Recycler in Your Area
Time Estimate: 2–4 Hours

Schedule a tour of your local recycling facility. (Any facility, whether plastic or metal or e-waste, will do; you could even visit multiple facilities.) If they don't normally host tours, see if you can make it worth their while, perhaps by gathering a group of people; or try different facilities. If you can't get a visit, see if you can get a phone conversation.

On the tour, take notes to list which materials are recycled and not recycled. Include notes on different kinds of the same polymers as well (e.g., some facilities separate white PET, clear PET, and others). Notice how waste is sorted and how different materials are liberated and separated.

Ask questions relevant to how your product could be recycled there, or how your material selection could improve recyclability. Consider:

- What materials do you hope will be recycled?
- What is the percent yield of separating that material in this facility?

- What is the grade of that material output from this facility?

- Which waste streams did these materials go through to reach a facility like this one?

- Where do the materials go after this facility, on their way to become new products?

- Any other questions you can think of to make your product easier to recycle.

Write down your list of what's recycled and summarize the answers to your questions.

STEP 3: Assess Priorities to Improve Your Product's Recyclability
Time Estimate: 15–60 Minutes

List the likely biggest difficulties for recycling your product, based on your discussion with the recyclers. You don't need to generate solutions yet—you'll do that in Chapter 16. Note whether the main problems are related to material choices, part connections (if your product has detailed enough design to discuss connections), manual disassembly, economics, or other factors. Or if your current design is already perfect for recycling, say why.

Checklist for Self-Assessment

To score your success on this exercise, see if you…

- ☐ *Mapped waste treatment facilities in your area, including recycling, compost, incineration, and/or landfill facilities.*

- ☐ *Visited a recycler to either have a tour or talk about your product design.*

- ☐ *Listed what materials they recycle and the answers to your questions.*

- ☐ *Listed the likely biggest difficulties for recycling your product, based on your visit notes.*

CHAPTER 16

Material Recovery

Design for Recycling

*Ruud Balkenende
and Jeremy Faludi*

Goals

- Explain the rationale for Design for Recycling guidelines

- Improve a product's recyclability by applying Design for Recycling guidelines

DOI: 10.4324/9781003504672-18

Why It Matters

Recycling is a prerequisite when the functionality of products can no longer be maintained or restored. Through recycling, we reduce the environmental impact by avoiding the damage and energy consumption due to mining new materials, and emissions of hazardous materials at the end-of-life stage. It also prevents losing scarce or critical materials. Design of products affects the result of the recycling process and determines to a large extent which materials can be separated from each other, which is essential for high recycling yields.

Summary

- Design for Recycling requires knowledge of recycling processes that are applicable for a particular product.

- Choose materials that are recyclable and are actually recycled from the waste stream in which the products end up. Ideally, also use 100% recycled material.

- Make hazardous components (e.g., batteries) and materials (e.g., oil) easy to remove without contaminating other parts.

- Choose connection types that easily break down during liberation, to enable separation of the materials when shredded or manually disassembled. For example, avoid glues and coatings, prefer snap-fit or lock connections.

Application of these design guidelines offers a variety of solutions that can significantly improve the recyclability and environmental performance of products.

Recycling becomes easier when the materials that constitute the product can be turned into homogeneous fractions, from which new materials can be reprocessed. This implies that during the (destructive) liberation phase, the product should ideally fall apart in fragments consisting of only a single material. A further prerequisite is of course that these materials are recyclable. Also fragments consisting of mixed materials can be recycled as long as the materials are compatible in the recycling process.

Design for recycling guidelines are summarized in Table 16.1. Recycling usually focuses on materials, but collection of products, breaking down of connections between materials, and dealing with specific components (often containing hazardous materials) are just as important.

Collection requires an infrastructure for taking back products and getting them to collection points. For example, curbside collection, dedicated collection points in stores, municipal collection points, and deposit systems. These might focus on very specific products and materials, as in the case of dedicated collection of glass, paper, and textile, or bottle deposit systems. However, collection can also cover broad ranges of products, as in the Dutch plastic-metal-drink-packaging (PMD) system. Some product classes have multiple dedicated pathways. Electronics, for instance, can be disposed of at many electronics stores and municipal collection points. Collection thus also is a first phase in separation of products and materials. For that purpose, it is important that it is easy for consumers

Table 16.1 Design for recycling guidelines

Design for recycling guidelines	
Collection	• Enable easy material identification (e.g., recycling codes) • Provide suitable infrastructure (e.g., collection points, take-back programs)
Materials	• Use commonly collected and recycled materials • Ideally specify 100% recycled materials when sourcing • Minimize the number of different materials • Avoid difficult-to-separate combinations of materials • Avoid paints/coatings • Avoid hazardous substances (e.g., toxic, flammable)

Connections	• Avoid fixed connections
	• If hazardous materials are used, make easily and quickly removable
	• Break down (by shredding) to:
	  • pieces with uniform composition
	  • pieces of relatively large size (> l cm)

Special (electronics, batteries, hazards)	• Enable easy and fast detection of materials • Remove part in one piece • Separate for dedicated recycling

to recognize in which category a product should be disposed, while also during automated separation in later stages ease of identification is important. As an example, the different standardized shape of many batteries allows for optical recognition of battery types.

For materials, using recyclable materials is obvious; using recycled materials as much as possible is even better, because it not only verifies that the material is indeed recyclable, but also ensures that the design is suited for the potentially different specifications of recycled material. Further, it helps create a market for the use of recycled materials. The main materials used in products are metals and plastics. Metals are in general relatively easy to recycle, but some are recycled much more than others (e.g., steel and aluminum, see Figure 11.3 in Chapter 11). Common plastics such as PE, PP, PS, and ABS are relatively easy to recycle in principle, unless they have a high content of additives or fillers. However, globally, the only plastics commonly recycled are PE and PET. Many other plastics can be recycled, but currently occur in too small volumes in the waste stream to make recycling economically viable.

Minimizing the number of different materials minimizes losses due to separation and liberation of different materials; so does avoiding combinations of materials that are difficult to separate. For example, it's easy to separate mild steel from aluminum or plastics, because steel is magnetic but the others aren't. Gold and silver can be recycled without mechanical separation from copper, but not steel. For a table of metal recycling compatibilities, see Table 11.1 in Chapter 11.

Coatings should be avoided, if possible, as in most cases they pollute the material fractions. Coatings that involve painting or metal deposition on plastics should especially be avoided. Such coatings not only contaminate the plastic material fraction but can also be detrimental to the separation process that is often based on density. If the weight increases by more than 1%, it can hurt separation. Oxidation layers on metals (e.g., anodized aluminum, blackened steel) sometimes avoid these problems, but not always.

Specific attention is needed for hazardous substances. These should preferably not be used at all, but if they are present, easy access and removal of hazardous components should be enabled. Examples of such components are gas discharge lamps containing mercury (also sometimes present as backlight in displays); batteries containing lithium, cadmium, mercury, or lead; printed circuit boards containing lead (in solder) and often many other problematic materials. Further, hazardous materials might be used as additives to plastics to improve flame retardance (brominated compounds) or to tune elasticity (phthalates). The latter cases are often harder to deal with as these additives are present in a diluted form in materials that are sometimes abundantly used, and often without the designer being aware of the nature of the additives used. In general, hazardous materials should be easy to remove quickly in a manual step. If this is not possible, avoid their presence in the product.

Connections should break down during liberation; otherwise, fragments consisting of multiple materials will be formed, which will inevitably lead to contamination of the material fractions. Examples of the main types of connection (form closure, force closure, form and force closure, and material enclosure) are given in Figure 16.1. The problems with mixed metal fragments have already been discussed in Table 11.1 in Chapter 11. Steel, stainless steel, aluminum, copper, and many other metals used in electronics should not remain connected, otherwise losses are inevitable. Also, different plastics should be separated from each other and from metallic parts. Breaking down the connections between different materials is thus essential.

Table 16.2 shows an overview of connections and their ease of disconnection in manual dismantling and mechanical disintegration (shredding). As recycling predominantly takes place through shredding, connections should be designed to break down during mechanical disintegration. The exception to designing for shredding are components that need specific attention, for instance, because of hazardous substances (e.g., batteries). In those cases, rapid manual dismantling of those components is mandatory. Manual dismantling is also needed for product repair or remanufacturing. To design for manual disassembly, see Chapter 14.

Some connections are easily liberated (e.g., press fits or turn locks); you should prefer these in your designs. Some connections are almost impossible to break down (e.g., glue connections); you should avoid these. This includes single parts made of multiple materials in a way that prohibits separation, like overmolding or 2K molding (different plastics injected into the same mold). Some connections depend on the design (e.g., screw-together parts) will frequently not fall apart.

16.2 Design for Recycling Case Study

The design guidelines listed in Table 16.1 focus on the materials used, but especially

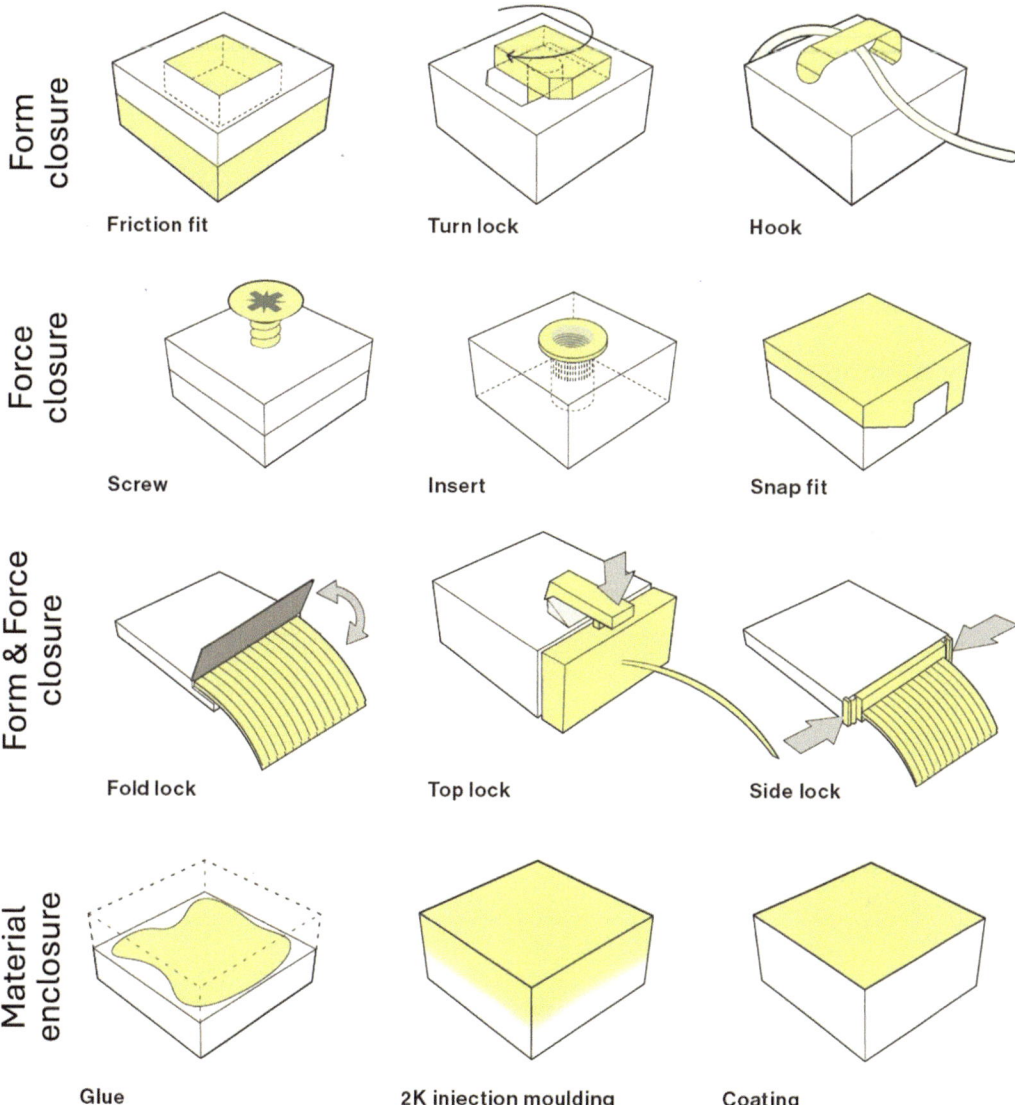

Form closure

Friction fit Turn lock Hook

Force closure

Screw Insert Snap fit

Form & Force closure

Fold lock Top lock Side lock

Material enclosure

Glue 2K injection moulding Coating

Figure 16.1 Examples of different types of connections between components
Source: Versloot (2024).

on the way in which materials and parts are connected, with as a key aim the ability to enable separation of the electronics at the end of life. Redesign of a LED spotlight will be used to show the effect of different design for recycling approaches on the recyclability of the lamps.

16.2.1 Recyclability of the Original Product

The standard spotlight is depicted in Figure 16.2(a). The lamp contained a relatively large aluminum heat spreader that was connected to parts made of engineering

Table 16.2 Compatibility of part joining methods with material liberation methods

Connection		Disconnection	
		✓ good ● design dependent ✗ bad	
Class	Specific connection	Manual dismantling	Mechanical disintegration
Form closure	Friction fit	✓	✓
	Clamping hook	✓	✓
	Turn lock	✓	✓
	Clicking	✓	✓
Form & Force closure	Fold lock	✓	●
	Side lock	✓	●
	Top lock	✓	●
Force closure	Snap fit	✓	✓
	Magnetic connection	✓	✓
	Screw	✓	●
	Nut and bolt	✓	●
	Rivet	✗	●
	Folding	✗	●
	Crimping	✗	●
	Clinching	✗	●
	Heat staking	✗	●
Material enclosure	Pressure sensitive adhesives	✓	●
	Hot-melt adhesives	●	✗
	Irreversible glue	✗	✗
	Cements	✗	✗
	Soldering	●	✗
	Welding (including brazing)	✗	✗
	2K molding	✗	✗
	Coating	✗	✗

Note: ✓ good; ● design-dependent; ✗ bad.

polymers. These polymers had high percentages of glass fiber fill and contained additives like flame retardants. The printed circuit board assemblies (PCBAs), one with the LEDs and the other with the driver electronics, were connected to the heat spreader by screws. When shredding this lamp, the polymer parts that were fairly brittle broke down and could be separated. However, due to the presence of fillers and additives, it was not possible to recycle them and they would be incinerated or landfilled. The heat spreader broke into smaller fragments but remained connected to the PCBAs, as the screw connections did not break down. This implies that either the electronics would end up in the aluminum fraction and be lost, or that the aluminum itself would end up in the copper fraction, where it would be lost, and also hurt the recovery of materials from the electronics.

Figure 16.2 A "standard" LED spotlight (MR16): (a) as a CAD model; (b) the manufactured lamp; and (c) fragments resulting from the liberation process, showing many mixed material fragments

Source: Aerts et al. (2014).

16.2.2 Redesign 1: Applying Fracture Lines

The fragmentation of the heat spreader in Figure 16.2(b) shows the brittleness of this die-cast aluminum part. This lamp redesign example uses this fracturing behavior to improve liberation during shredding. The way this part fractured could be controlled to a large extent by applying fracture lines in the heat spreader, as shown in Figure 16.3(a). The fracture lines were defined so that they also passed through or just along the screw holes, so the heat spreader's fragmentation would release the screw connections, implying that the PCBAs were no longer connected to the heat spreader fragments. Figure 16.3(b) shows the result of a shredding test on these lamps, yielding pure fractions of aluminum heat spreader fragments and PCBAs. It demonstrates that understanding how design affects part fragmentation can greatly improve actual recyclability without adding cost or significantly changing the production process.

Figure 16.3 Fragments resulting from the liberation process for a lamp with the fracture lines in redesign #1. (a) heat spreader fragments, showing a quarter fragment that was clearly fractured along the laser-inscribed fracture lines (note also the screw hole); (b) fragments resulting from other parts than the heat spreader, nicely disconnected from all other parts

Source: Aerts et al. (2014).

16.2.3 Redesign 2: Material Selection and Stacked Design

The above approach optimized the original design. When a complete redesign is possible, new design aspects can be introduced. The second redesign of the spotlight eliminated non-recyclable materials

Figure 16.4 LED lamps with radically improved separation from redesign #2. (a) exploded view of the lamp; (b) the lamp as manufactured; (c) the result of the liberation process without potting material; (d) the result of the liberation process for lamps filled with potting material.

Source: Aerts et al. (2014).

as much as possible, while connections were limited to a single directly accessible one. This lamp was mainly made of stacked deep-drawn aluminum parts. The lamp and an exploded view are shown in Figure 16.4(a) and Figure 16.4(b). In this lamp, the deep-drawn aluminum parts replaced parts made of engineering plastics and the die-cast aluminum heat spreader. Furthermore, all parts were simply stacked and pressed together only by one ring, directly accessible at the top of the lamp. This lamp could therefore easily be disassembled manually.

However, LED lamps are relatively small electronic products that are too inexpensive for manual dismantling to be economical. They most likely end in a shredding process as described above. This should therefore also be taken into account in the lamp design. A small-scale shredding test showed that the lamp folded due to the high compliance of the thin aluminum parts (Figure 16.4(c)). By introducing "potting" (filling the hollow interior of the lamp with sand or silicone, both easy to remove from the electronics and other parts), this was prevented, resulting in excellent separation of the electronics when shredding the lamps (Figure 16.4(d)).

16.2.4 Redesign 3: Lower Impact with Glass Housing

When focusing primarily on environmental impact instead of recyclability, you can make different design choices. Glass has a lower environmental impact to produce than metals and engineering polymers. The third redesign used glass as the housing material, but otherwise mostly followed the stacked approach described in the previous redesign. To account for sufficient heat dissipation, a small aluminum insert was still needed. An exploded view is shown in Figure 16.5(a).

A recycling test of this lamp showed that the glass shattered, while all other parts remained almost intact and could be easily separated (see Figure 16.5(c)). In practice, glass is usually not recovered from an electronics waste stream, thus lowering the recycled weight. Further, the economic residual materials value of this lamp was reduced, as it no longer contained so much aluminum to recover, making it less attractive to recyclers. This redesign thus decreased the recyclability on a weight basis compared to the previous designs. However, the environmental footprint of manufacturing this lamp improved significantly by replacing

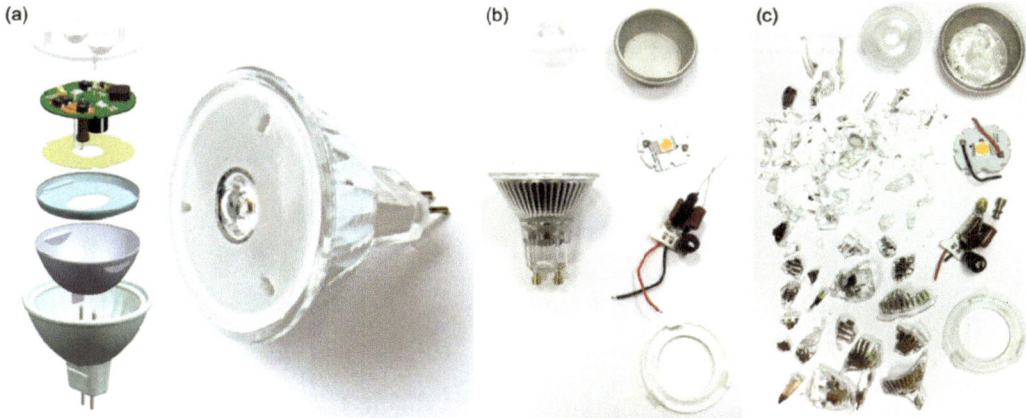

Figure 16.5 The LED lamp redesigned with glass housing to prioritize environmental impact as well as recycling: (a) exploded view of the lamp; (b) the assembled lamp; (c) the parts of the lamp before assembly; (d) fragments resulting after the liberation process

Source: Aerts et al. (2014).

aluminum with glass (Aerts et al., 2014). This shows that a lower environmental impact does not always need a higher recycling rate, although recycling all materials would be preferred, of course.

16.3 Comparing the Redesigns

The net environmental burden for recycling the redesigned lamps was calculated using life cycle assessment. These calculations considered the fragmentation of the product and the implication of mixed fragments on the possible recovery yield of the materials present. They also considered the energy input in collection, separation and recovery (Huisman, 2003).

The results are shown in Figure 16.6. From an environmental perspective, the potential loss of especially copper and gold present in the electronics is significant, whereas

these materials would be hardly visible in a weight-based approach. Comparing the "standard" lamp with the lamp with fracture lines shows how greatly fragmentation can reduce the occurrence of mixed fragments, resulting in better separation and recovery. This is especially the case for retrieving the PCBAs as a separate fraction, as by proper liberation the recovery of the most valuable elements is maximized. This stresses that recovery and environmental impact are strongly affected if parts cannot be properly separated, and therefore end in a waste stream from which their constituting materials cannot be recovered.

Interestingly, the lamp with the stacked design scored slightly worse than the standard lamp, although designed to improve recyclability. This was caused by the folding that occurred in the liberation step, which encapsulated the PCBAs and strongly reduced their separation, as is evident from the large copper and gold contribution to the impact (indicating that these materials

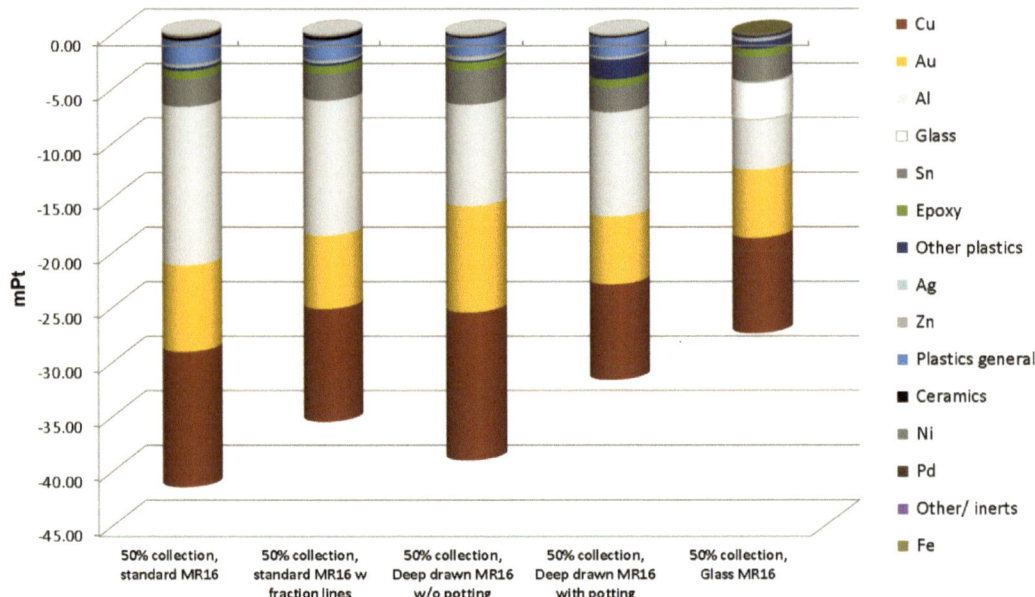

Figure 16.6 LCA results showing environmental impacts of the different design options over their entire lifecycle of a lamp (MR16) from manufacturing to recycling. Note that these values originate from LCA software that puts negative values on higher impacts.

Source: Aerts et al. (2014).

were lost). When potting was used to avoid the folding, recovery was much better, so the impacts became much smaller.

Figure 16.6 further shows that the glass-based LED lamp has the lowest environmental burden, although the difference with the stacked lamp with potting is small. As explained above, this is not due

to the higher recycling rate (on a weight basis), but to the good separation of the electronics after liberation in combination with the use of low impact materials. Finally, the environmental impact calculations show the importance of measuring impacts of different designs to verify their effectiveness, not assuming they perform better because of the design intentions.

References

Aerts, M., Felix, J., Huisman, J., & Balkenende, R. (2014). Lamp redesign: shredding before selling. CARE Innovation 2014, November, Vienna, Austria.

Huisman, J. (2003). *The QWERTY/EE concept: Quantifying recyclability and eco-efficiency for end-of-life treatment of consumer electronic products* [PhD dissertation]. Delft University of Technology.

Versloot, D. (2024). *Design for recycling of electronic products: Study on smart TVs* [Master's thesis]. Delft University of Technology.

How to Apply #16: Redesign for Recycling
Time Estimate: 1–6 Hours

How can you redesign your product's part connections and materials to enable easier liberation and separation of materials, either by users or recyclers? This requires you to already have a fairly detailed design.

STEP 1: Identify Your Product's Recycling Problems
Time Estimate: 15–60 Minutes

Make a table of all the parts in your product, with columns listing each one's material, the other parts it is attached to, and the method of attachment, like Table 16.3.

Notes to help fill out your table:
- You don't need to list each connection for both parts (e.g., say the outer housing is press-fit to the legs and say the legs are press-fit to the housing). As long as each connection is listed once, that's enough.

- Printed circuit boards and the components mounted to them can be considered single parts, as they'll be recycled together.

- For any parts made of plastic, metal, glass, wood, or others, list each material component as its own part.

- For complex products, you don't need to list every single part, but list the top ten you expect to be the highest priority, considering both the environmental value and economic value of recovering them, as well as hazardous materials like flammable batteries.

- Optional: ideally base priorities on real LCA, cost, and hazard data for your product.

Table 16.3 Example part connections

Part	Material	Connection method	Mounted to
Outer housing	Polypropylene	(None)	(None)
Legs	Aluminum	Press fit	Outer housing
Electronics assembly	PCB with mounted ICs	Screws	Legs
		Solder	Screen
Screen	LCD screen	Snap-fit	Outer housing
(etc.)	(etc.)	(etc.)	(etc.)

STEP 2: Evaluate Part Recyclability and Potential Losses
Time Estimate: 10–20 Minutes

Use Table 16.1, "Compatibility of part joining methods with material liberation methods," to identify likely losses during the recycling process (red or yellow dots) due to how parts are joined to other parts of different materials. Consider whether each part connection in your table will be disassembled manually by a person (such as batteries), or disintegrated by shredding the product (almost everything).

For metal parts that don't easily separate from each other, also use the Material Compatibility Table (Table 11.1 in Chapter 11). If two metal parts aren't separated but are compatible in the same recycling stream, they still score green rather than yellow or red.

Add two columns to your table from Step 1: one column for "manual vs. shredded", and one column for "likely losses" (green/yellow/red from Table 11.1 in Chapter 11).

STEP 3: Ideate More Recyclable Connections or Material Choices
Time Estimate: 10–30 Minutes

Given your results in Step 2, what parts in your product will likely have recycling problems? Of these, choose one or two priority parts to improve their recyclability. For these priority part(s), brainstorm to improve their recycling liberation and separation, either by changing connection methods or by material choice.

Have at least 30 ideas or more, with at least 5 ideas on connection methods and at least 5 ideas on material choice. Be specific and concrete, it gets you more solutions from the same general ideas; also have wild ideas, it's a brainstorm. If the solutions for your priority parts are simple and obvious, with no downsides, brainstorm on improving the lower priority parts in your table, to reach the 30+ ideas.

Optional: If you have time, brainstorm to improve the recyclability of all problematic parts.

STEP 4: Choose a Final Solution
Time Estimate: 5–30 Minutes

Choose one winning solution for each part, from your brainstorms. Use whatever decision criteria you like, but clearly and briefly (20–50 words) describe your reasoning; it's okay to combine solutions that synergize. Develop the winning idea(s) further as needed, with a sketch of the new joining geometry or list of material replacements or both.

Checklist for Self-Assessment

To score your success on this exercise, see if you…

☐   *Made a table of parts in your product, with columns for material, connection method, part it is mounted to, manual disassembly or shredding, and likely loss in recycling.*

☐   *Identified priority parts to improve their recycling.*

☐   *Listed 30+ new ideas, including 5+ on material choice and 5+ on connections.*

☐   *Sketched or briefly described the winning idea(s) you'll move forward with.*

☐   *Explained the reasons for choosing the winning idea(s), in 20–50 words.*

CHAPTER 17
Energy Literacy

*Jeremy Faludi,
Ruud Balkenende,
and Conny Bakker*

Goals

- Explain how energy use and its environmental impacts are calculated

- Calculate a product's energy use for different power modes

- Calculate system energy efficiency

- Calculate energy payback time for simple renewable energy generation

- Identify opportunities for energy efficiency, effectiveness, and sustainable energy production

DOI: 10.4324/9781003504672-19

Why It Matters

Energy is indispensable for life and modern living, but burning fossil fuels causes over 73% of the world's greenhouse gas emissions (Ritchie et al., 2020), as well as most particulate pollution, acidification, and resource depletion. Using energy more effectively in the systems you design can greatly reduce these impacts, and perhaps even drive positive impacts. To do this, you must understand the basic concepts of energy, power, efficiency, and effectiveness.

Summary

- Energy is the ability to do work; power is the rate of energy use.

- No conversion is 100% efficient, and when energy is converted many times, the inefficiencies multiply. Thus, your product's energy (mechanical motion, cooling, light, or electronics) may require five or ten times the amount of fossil fuel energy to be mined and burned.

- Energy used to refine and manufacture materials is called "embodied energy."

- You can calculate energy "payback" times of renewable energy systems by finding when the energy required to produce and operate them is surpassed by their avoidance of burning fossil fuel energy.

- Improving energy efficiency is nice, but improving environmental impact is what matters. This energy "effectiveness" can be reached by improving efficiency, cleaner energy sources, and/or critically rethinking where and why energy is needed.

- The whole world must transition to renewable clean energy. You can design products and systems to drive that transition.

17.1 Energy and Power Basics

Energy is "the ability to do work," whether that work is moving an apple (mechanical energy), cooling the apple to a desired temperature (thermal energy), or digesting the apple (chemical energy), or other activities. Energy is measured in joules (J). To get an intuition of a Joule, lift an apple from the floor to a desk: lifting a mass of 0.1kg a height of 1m against gravity's acceleration of 9.8m/s^2 requires about 1 Joule (J) of energy. Lifting a 100kg person one meter requires about a kilojoule (kJ) of mechanical energy. An AA battery holds about 10 kJ of chemical energy. It takes about 10 kJ to cool an apple from room temperature to refrigerator temperature; the refrigerator uses electrical energy to pump heat energy out of its insulated box and into the room (Figure 17.1).

Cooling an apple once does not keep it cool forever. Heat leaks through the fridge's insulation and out the door when it's opened. To stay cool, the fridge keeps using energy

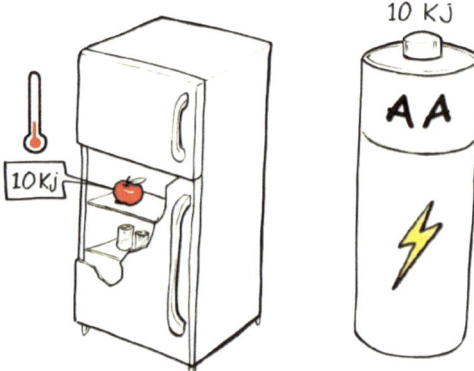

Figure 17.1 It takes 10 kJ to cool an apple to refrigerator temperature—the same as an AA battery's capacity

Source: based on Faludi et al. (2011).

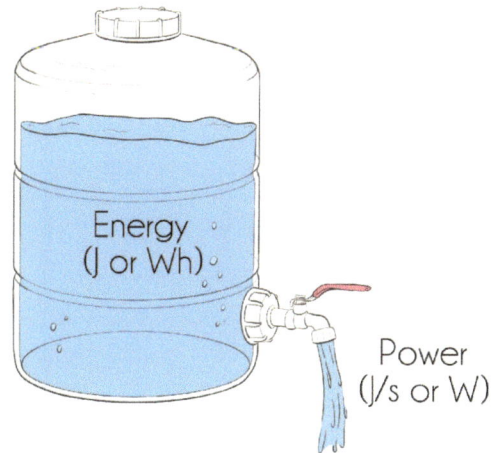

Figure 17.2 Energy is capacity, power is rate of energy flow

over time. This is power, measured in watts (W), defined as one joule per second. If your refrigerator demands 100 watts for 100 seconds, it uses 10kJ of energy; it uses the same amount of energy if it demands 500 watts for 20 seconds. Thus, saving power may not save energy—it just loses energy more slowly. Given enough time, a dripping faucet can empty a tank of water just as completely as a fast-pouring faucet, as in Figure 17.2. Note that a watt-hour (Wh) is not a unit of power—it is power multiplied by time, making it another unit of energy (3600 Joules, because there are 3600 seconds in an hour).

Table 17.1 lists some common equations for energy and power in different forms. You don't need to know these equations per se, but it's good to have a general instinct for what matters a lot or a little for different forms of energy. For example, kinetic energy depends on velocity squared, so ten times the speed means 100 times the energy. Electrical energy doesn't depend on current, but electrical power does. For thermal energy and power, it's not just the mass and

temperature difference that matter, but also the heat capacity of the material, which can vary greatly between materials.

There's also a distinction between "sensible" heat (changing the temperature) and "latent" heat (changing phase, i.e. freezing/thawing or boiling/condensing). It takes 4.2 J of energy to raise the temperature of 1 g of water by 1°C, but it takes 2,260 J of energy to turn 1 g water at 100°C to steam at 100°C. That's over four times the amount of energy it takes to heat water from 0° to 100°!

17.2 Efficiency and Losses

A refrigerator cools food with electricity by using a motor to pump coolant through coils inside and outside the refrigerator body, which pumps heat from inside to outside. It's the same as an air conditioner or other heat pump. The electricity is converted to mechanical motion to fluid flow to a temperature change. In turn, the electricity running the fridge might have come from a solar panel on the roof, or from burning coal that was mined thousands of kilometers away. Some energy is "lost" in each of these conversions. Technically, by the first law of thermodynamics, energy can never be created or destroyed, but when energy is converted into things that we can't use, like noise, vibration, and low-grade heat, we call it lost and our system's energy efficiency decreases. For example, heating water to boil into steam to spin a turbine is inefficient because boiling water converts a lot of energy to latent heat in the steam, which doesn't turn into mechanical motion of steam pushing the turbine. One way of increasing system efficiency is recovering lost energy, as in "cogeneration" plants, which output

Table 17.1 Energy and power equations

Type	Energy "E" (Joules) and Power "P" (Watts)
Mechanical	$E_{potential} = mgh$ $E_{kinetic} = \frac{1}{2} mv^2$ $P_{kinetic} = Fv$ Where: m = mass (kilogram) h = height (meter) g = gravitational acceleration (meter / second^2) v = velocity (meter / second) F = force (kilogram • meter / second^2)
Electrical	$E = CV$ $P = IV$ Where: C = charge (Coulomb) I = current (Amp) = Coulomb / second V = voltage (Volt) = (kilogram • meter^2) / (Amp • second^3)
Thermal	$E = mc_p\Delta T$ (sensible) $E = mL$ (latent) $P = \dot{m}c_p\Delta T$ (sensible) $P = \dot{m}L$ (latent) Where: m = mass (kilogram), c_p = specific heat capacity (Joule / kilogram °Celsius) L = specific latent heat (Joule / kilogram) ΔT = temperature change (°Celsius) $\dot{m}$ = mass flow rate (kilogram / second)

both electricity from the spinning turbine and also the leftover heat in the steam, to be used as heat. Table 17.2 lists various energy efficiencies of conversions between different kinds of energy.

17.3 System Efficiency

If your power comes from coal, the coal is first mined and transported to a power plant. This costs little energy relative to the energy in the coal, so it's about 97% efficient. The power plant burns the coal to boil water into steam, which turns a turbine with an electric alternator, generating electricity. This converts the coal's chemical energy into heat energy, then kinetic energy, then electrical energy. Their combined efficiency is about 35%, so 2/3 of the coal's energy is lost as leaked heat and particles of incompletely burned soot in the air. The electricity then travels to your home over high-voltage power

Table 17.2 Various conversion types and their efficiencies

Conversion process	Conversion type	Energy efficiency
Electricity generation		
Gas turbine	Chemical to thermal to electrical	30–43%
Gas turbine plus steam turbine (combined cycle)	Chemical to thermal to electrical	up to 60%
Gas turbine cogeneration	Chemical to thermal to electrical and thermal	up to 64%
Water turbine	Gravity potential to electrical	up to 90%
Wind turbine	Kinetic to electrical	up to 59%
Solar PV cell	Radiative to electrical	6–40% (technology-dependent, 15–20% most often)
Fuel cell	Chemical to electrical	40–60%
Electricity storage		
Lithium-ion battery	Chemical to electrical/reversible	80–90%
Nickel-metal hydride battery	Chemical to electrical/reversible	66%
Lead-acid battery	Chemical to electrical/reversible	50–95%
Engine/motor		
Electric motor	Electrical to kinetic	80–97% (> 200 W) 50–90% (10–200 W) 30–60% (< 10 W)
Combustion engine (car)	Chemical to kinetic	21–36%
Turbofan	Chemical to kinetic	20–40%
Appliance		
Household refrigerator	Electrical to thermal	low-end systems ~20%% high-end systems 40–50%
Incandescent light bulb	Electrical to radiative	0.7–5.1%
Light-emitting diode (LED)	Electrical to radiative	4.2–53%
Electric heater	Electrical to thermal	~100%
Others		
Electrolysis of water	Electrical to chemical	50–70%

Source: Paoli & Cullen (2020); Wikipedia (2022).

lines; this is 93% efficient. Finally, perhaps 50% of your fridge's electricity use actually cools the food inside. The other half is lost converting from electricity to mechanical to heat energy movement, leaking heat through the fridge's insulation, and leaking cold air plus letting in warm air when the door is opened. (Small amounts are also used by fridge electronics and lighting.) These steps are shown in Figure 17.3.

Efficiencies always multiply, so the whole system's efficiency is simply all the individual efficiencies multiplied together:

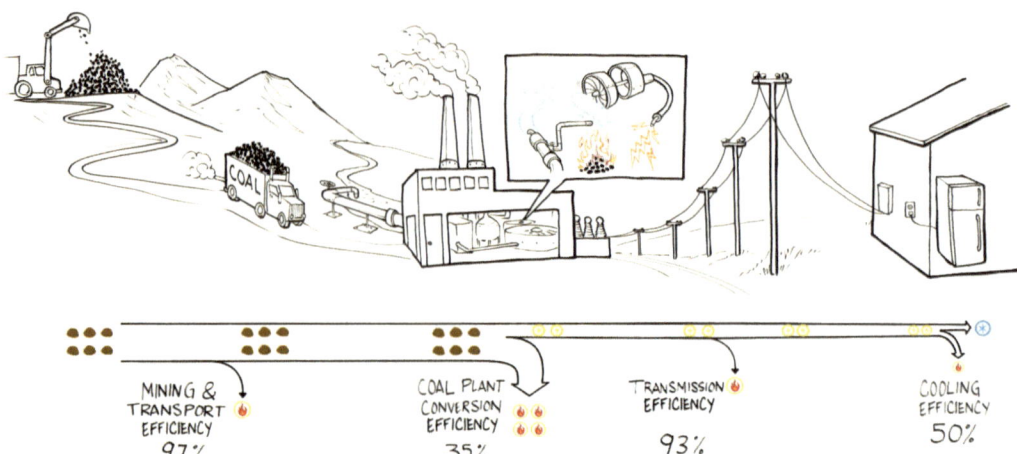

Figure 17.3 The many energy conversions between mining coal and the refrigerator cause 5/6 of the energy to be lost

Source: based on Faludi et al. (2011).

(97%)(35%)•(93%)•(50%) = 16% system efficiency

Because the whole system is only 16% efficient, the 10 kJ used to cool one apple becomes 60 kJ of coal mining. But if you redesign the fridge with better insulation, or ways to open it without losing cool air, then every unit of cooling energy you save the end user saves six units of energy upstream!

17.4 Environmental Impacts and the Energy Transition

Energy makes the world go round—it is vital to human activity, industry, and the economy. Cheap ubiquitous energy has lifted people out of poverty and driven the high standards of living we have today. However, most energy today is obtained from fossil fuels, which cause huge health and environmental problems—three-quarters of global greenhouse gas emissions come from burning fossil fuels for energy (see Figure 17.4). Fossil fuels cause similar percentages of global particulate emissions, smog, and terrestrial acidification. In addition to the environmental problems of fossil fuels, the limited supplies of oil and gas drive wealth and power inequalities, including wars (Kruse, 2013; Ross, 2012).

To bring modern society into environmental balance with the planetary boundaries and into social balance with the UN Sustainable Development Goals, we must transition to renewable and clean energy worldwide. It's expensive to transition, and renewable energy generation still causes some impacts (largely mining materials, including rare minerals for magnets in wind turbines, solar panels, batteries, and the wires connecting them all). Therefore, it's important to save energy in the products and services we design, but the top goal is not energy

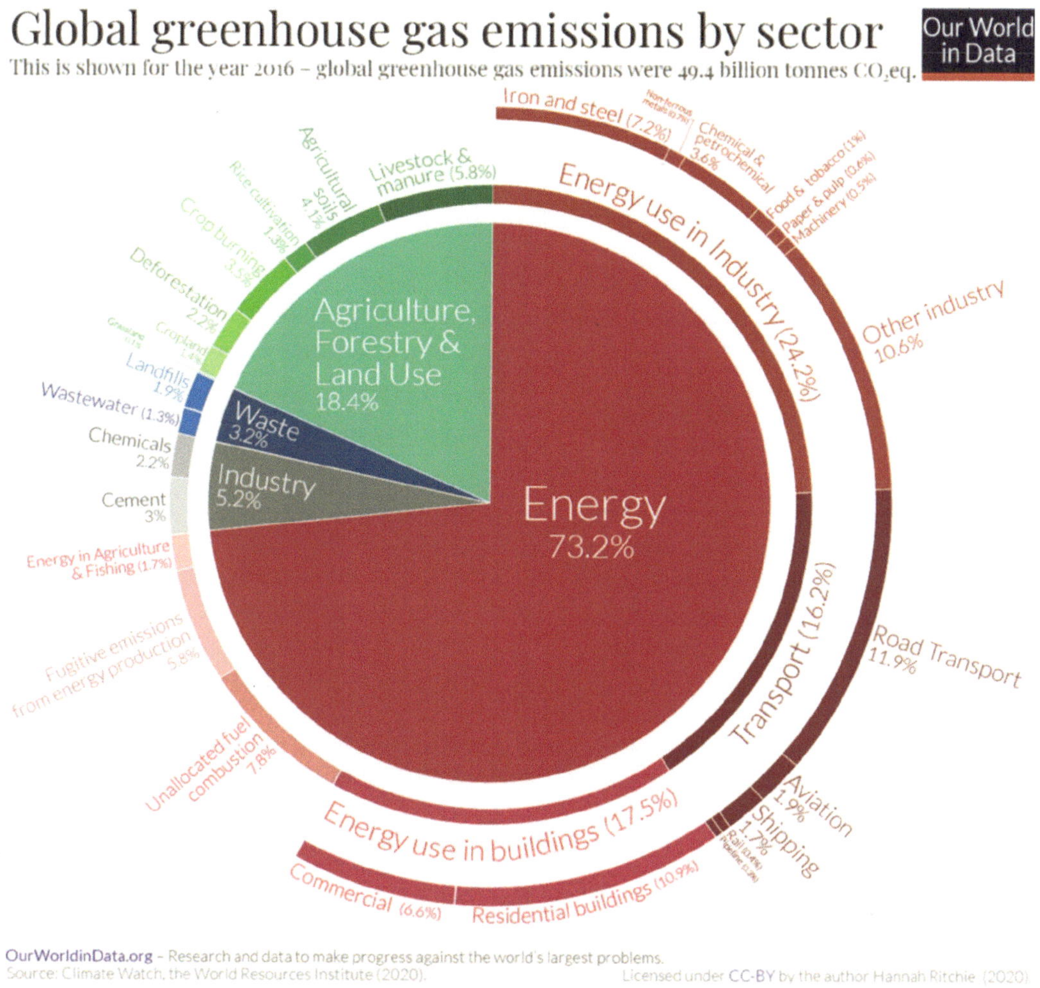

Global greenhouse gas emissions by sector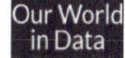

This is shown for the year 2016 – global greenhouse gas emissions were 49.4 billion tonnes CO_2eq.

Figure 17.4 Global greenhouse emissions by sector, 2016

Source: Ritchie et al. (2020).

efficiency, it's improving the environmental and social impacts of energy.

For example, when warming your home, even a 100% efficient conversion of chemical energy to heat by burning propane still causes more pollution, climate change, and resource extraction than a solar photovoltaic (PV) panel's 20% efficient conversion of sunlight to electricity and heating through resistive wires. According to the Idemat 2021 LCA database (Vogtlander et al., 2021), the solar PV electric heating cuts greenhouse emissions by 70% and cuts comprehensive ReCiPe point impacts by 64% compared to burning gas. The ReCiPe savings are not as good as greenhouse savings because of the material mining required for the equipment, but is still a huge saving.

17.5 Energy Effectiveness

Efficiency is doing things right; effectiveness is doing the right things. That means energy effectiveness might include energy efficiency, sourcing clean energy, using the product less, or reframing the problem to meet user needs in different ways that use no energy at all. The latter is sometimes called "modal shift" (Akenji et al., 2019).

For example, in car design, efficiency might mean re-engineering it for lighter weight and less drag, using less energy for the same function. Clean sourcing might mean replacing the gas-powered engine with an electric motor, fed by solar power from the grid. Critical re-thinking might mean considering that people drive rather than walking or biking because they live far from where they work or play, so instead of redesigning the vehicle, redesigning neighborhoods so people can live nearby and walk or bike. (Shifting modes to eliminate energy demand.)

"Tunneling through the cost barrier" is a concept by Amory Lovins at Rocky Mountain Institute (Hawken et al., 1999) to escape the cost problems of traditional efficiency. The "cost barrier" is the diminishing returns of buying more efficient components—often costs go up exponentially versus efficiency gains, and at some point the gains in efficiency are no longer worth the extra cost. However, sometimes extreme measures (and higher costs) in one component can eliminate the need for other components, thus "tunneling" through the wall of diminishing returns to save cost at the whole system level (Figure 17.5).

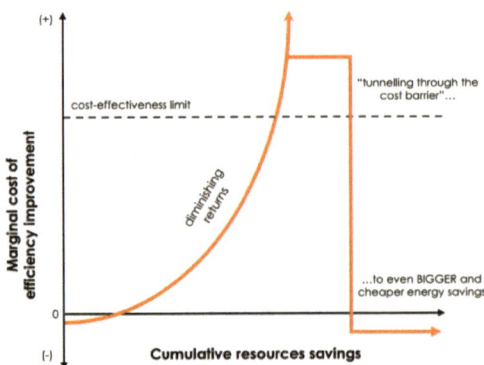

Figure 17.5 Tunneling through the cost barrier
Source: Based on Hawken et al. (1999).

For example, the original Rocky Mountain Institute headquarters building used several times more insulation and thermal mass than usual in its walls, roof, and floor. Normally, adding more insulation adds cost with diminishing returns on investment, but this building's extreme over-insulation made it so energy efficient that it didn't need a boiler or heating ducts, heating itself only with sunlight through windows and the body heat of people in the building on most days. This tunneled through the cost barrier by simultaneously saving much more energy and decreasing construction cost (Hawken et al., 1999). Such buildings are called "passive houses." This can work in products and manufacturing lines also. Servers in data centers cause enormous amounts of waste heat, which then requires enormous amounts of energy to cool. Roughly half of data center energy use is actual computing, half is cooling. Improvements in server efficiency therefore get doubled, because every watt of computing energy reduction is also one watt of cooling energy reduction (Shehabi et al., 2018).

17.6 Embodied Energy and Payback Time

Renewable energy always sounds good, but sometimes can actually be worse than existing fossil fuel grid energy. It takes energy to manufacture products, including solar panels and batteries. The energy put into mining, refining, manufacturing, and transporting a physical object is called its "embodied energy." You can see if it makes environmental sense to power a product with its own solar panel and battery rather than a fossil-powered electrical grid by comparing the energy embodied in the extra solar hardware versus the cumulative energy demand of the product and the whole upstream electricity system shown in Figure 17.3.

The results of this calculation depend on time—the product's solar power system causes a big impact initially, but then generates power "for free" during use, while the grid-powered device continues to cause environmental impacts with every hour of operation. A system reaches "energy payback time" or "energy return on investment" (EROI) when its embodied energy is lower than the cumulative energy demand of grid power. For large-scale systems like rooftop photovoltaics (PV) to power a home, payback time is within 1.5–3 years, and then environmental impact savings accrue for the rest of the PV system's life—usually 30+ years (Figure 17.6).

While many people report energy payback times, as mentioned above, the important thing is environmental impacts. Therefore it's more relevant to calculate carbon payback times, ReCiPe payback times, or other actual

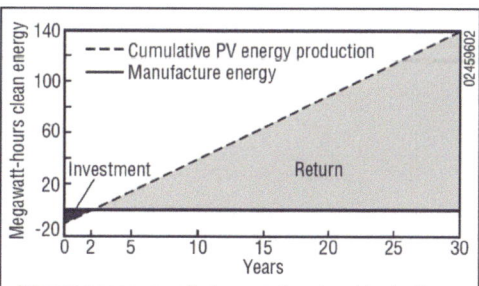

PV systems can repay their energy investment in about 2 years. During its 28 remaining years of assumed operation, a PV system that meets half of an average household's electrical use would eliminate half a ton of sulfur dioxide and one-third of a ton of nitrogen-oxides pollution. The carbon-dioxide emissions avoided would offset the operation of two cars for those 28 years.

Figure 17.6 Typical embodied energy payback time for rooftop solar power
Source: NREL (2004).

environmental impact metrics. This is an LCA that considers time.

For example, let's calculate carbon payback of redesigning a computer keyboard from being powered by the grid through a cable, versus solar-powered with a battery, as in Figure 17.7. For simplicity, assume the only hardware difference between the solar keyboard and a normal one is the addition of a PV cell 56 cm^2 in size, and a lithium ion battery weighing 5g. (Ignore wires and electronics.) On average, the wired keyboard uses 0.7 W of power for 8 hrs/day every day of the year, using average Dutch domestic electricity, every day. Table 17.3 shows the impact comparison.

In this scenario, Table 17.3 shows that the system pays back the embodied impacts of production in slightly over a year. A keyboard is likely to last much longer than that, so it's beneficial to switch from grid to on-board solar power. However, if the keyboard were

Figure 17.7 A solar-powered keyboard—when will its avoided carbon emissions pay back the impacts of its PV panels?

Table 17.3 Comparing impacts of solar to normal keyboard, with data from Idemat 2021 LCA database (Vogtlander et al., 2021)

Wired keyboard	Grid impact intensity ("Electricity low voltage, domestic use general"): 0.12 kg CO_2/MJ Electricity use: (0.7 W) • (8 hr/day) • (365 days/yr) • (0.0036 MJ/ Wh) = 7.36 MJ/yr → impact = (0.12 kg CO_2/MJ) • (7.36 MJ/yr) = **0.87 kg CO_2eq./yr**
Solar keyboard	PV panel impact intensity: 130.2 kg CO_2eq./m² PV panel size: 56 cm² = 0.0056 m² → impact = (130.1 kg CO_2eq./m²) • (0.0056 m²) = 0.73 kg CO_2eq. Battery impact intensity: 46.07 kg CO_2eq./kg Battery size: 5 g = 0.005 kg → impact = (46.07 kg CO_2eq./kg) • (0.005 kg) = 0.23 kg CO_2eq. → Total impact = 0.73 + 0.23 = **0.96 kg CO_2eq.**
Payback time	(PV system production impact) / (grid electricity impact / year) = (0.96 kg CO_2eq.) / (0.87 kg CO2eq./yr) = **1.1 year**

Table 17.4 Rough values for estimating energy or carbon payback times

Material or electricity source	Carbon intensity
Solar PV panel, single crystal (Generates: 200 W/m² outdoor full sun, 20 W/m² outdoor cloudy skies, 1 W/m² from indoor electric lights)	130.1 kg CO_2 eq./m²
Printed circuit board (including ICs)	260.4 kg CO_2 eq./kg
Lithium-ion $LiCoO_2$ laptop battery (180 Wh/kg)	46.1 kg CO_2 eq./kg
Grid electricity, Western Europe average	0.13 kg CO_2 eq./MJ
Grid electricity, North America average	0.15 kg CO_2 eq./MJ
Grid electricity, China average	0.17 kg CO_2 eq./MJ
Grid electricity, India average	0.20 kg CO_2 eq./MJ
Grid electricity, South America and Africa average	0.15 kg CO_2 eq./MJ

Source: Verwaal (2019); Teehan & Kandlikar (2012).

only used for one hour per day, the payback time would be 8.8 years, so it likely would not be beneficial.

You can calculate any scenario like this. If the renewable energy option also causes some impacts over time (perhaps from replacing batteries, or requiring maintenance), you must include those. If you graphed this like Figure 17.6, it would simply mean the renewable option is no longer a flat line, but has a slope that intersects the grid power line later. If your product use isn't constant over time but varies greatly, you might choose a different functional unit that's relevant. For example, you might measure washing machines by kg CO_2 eq. per wash, or measure cars by kg CO_2 eq. per km driven. If your product operates in a different region, use the carbon intensity of grid electricity for that region (Table 17.4).

Resources and References

Resources for Further Study

- MacKay, D. J. C. (2009). Sustainable energy – Without the hot air. Chapters 1, 2. Available at: https://www.withouthotair.com

- Hawkens, P., Lovins, A., & Lovins, L. H. (1999). *Natural capitalism: Creating the next industrial revolution*. Little, Brown and Company.

References

Akenji, L., Lettenmeier, M., Koide, R., Toivio, V., & Amellina, A. (2019). 1.5-degree lifestyles: Targets and options for reducing lifestyle carbon footprints. Institute for Global Environmental Strategies, Aalto University.

Faludi, J., Mentor, A., Sachs, J., & Danby, D. (2011). Autodesk Sustainability Workshop. Available at: https://venturewell.org/tools_for_design/energy-effectiveness/

Hawken, P., Lovins, A., & Lovins, L. H. (1999). *Natural capitalism: Creating the next industrial revolution*. Little, Brown and Company.

Kruse, F. (2013). *Oil politics: The West and its desire for energy security since 1950*. Anchor Academic Publishing.

NREL (National Renewable Energy Laboratory). (2004). PV FAQs: What is the energy payback for PV? (DOE/GO-102004-1847). Available at: https://www.nrel.gov/docs/fy04osti/35489.pdf

Paoli, L., & Cullen, J. (2020). Technical limits for energy conversion efficiency. *Energy*, 192, 1–26.

Ritchie, H., Roser, M., & Rosado, P. (2020). Emissions by sector. OurWorldInData.org. Available at: https:// ourworldindata.org/emissions-by-sector (accessed September 21, 2022).

Ross, M. (2012). *The oil curse: How petroleum wealth shapes the development of nations*. Princeton University Press.

Shehabi, A., Smith, S. J., Masanet, E., & Koomey, J. (2018). Data center growth in the United States: Decoupling the demand for services from electricity use. *Environmental Research Letters*, 13(12), 124030.

Teehan, P., & Kandlikar, M. (2012). Sources of variation in life cycle assessments of desktop computers. *Journal of Industrial Ecology*, 16, S182–S194.

Verwaal, M. (2019). The Small PV Systems Design Guide: A short introduction in the design of small solar powered products. TU Delft course curriculum.

Vogtlander, J. G., et al. (2021). Idemat LCA database. Available at: EcoCostValue.com.

Wikipedia (2022). Energy conversion efficiency. Wikimedia Foundation. Available at: https:// en.wikipedia. org/wiki/Energy_conversion_efficiency

How to Apply #17: Assess Priorities for Energy Improvements
Time Estimate: 1–3 Hours

If your product uses energy during its life, when does it use energy, and what specific components use most of the energy? Or if your product's energy impacts are indirect (like clothes being washed and dried), what part of the system uses the most energy?

Embodied Energy
0.01 Wh/min of service

Idle Energy
0.08 Wh/min of service

Cooking Energy
0.07 Wh/min of service

STEP 1: Estimate Your Product System's Biggest Uses of Energy
Time Estimate: 0.5–1 Hours

Calculate the energy use of your product's main subsystems, using estimations of power use and time spent running. Use whatever means you wish to estimate this (measuring with a watt meter, researching online, looking up spec sheets for components from your BOM, etc.). A few hints:

- Include components that transform energy (e.g., from electricity to heat, or from AC to DC electricity).

- Remember that whatever uses the most power may not use the most energy. For example, if a microwave oven uses 1,000 W of power to cook food, but only uses it for 6 minutes per day (thus causing an energy use of 100 Wh per day), but its clock and other systems use 5 W of power all 24 hours of the day (for an energy use of 120 Wh per day), then it uses more energy idling than cooking.

- Your product may cause other things to use energy—even more than your product itself. (For example, clothing uses no energy while being worn, but causes energy to be used by washing machines and perhaps dryers.)

- If your product does not use energy during its useful life and doesn't cause energy to be used by anything else, estimate what materials in the product cause the most "embodied energy" impacts. For example, aluminum has a high embodied energy because of the large amounts of processing energy required to refine it from ore. (If you did an LCA, you can quickly estimate embodied energy by looking at CO_2 emissions per material, as CO_2 is generally a good proxy for energy use in traditional material extraction and processing.)

Finally, be careful with your time. Digging into details like this will take as much time as you let it. This assignment is meant to only take a couple of hours, so don't worry about spending 30 hours making it perfect.

STEP 2: Estimate Energy Use Per Functional Unit
Time Estimate: 10–30 Minutes

Translate the above calculations into energy per unit of service. That is, take the calculations you made above and divide by the number of units of service the user gets from the product. For example:

- If a microwave oven averages 220 Wh per day (see calculations above), and its service is measured in minutes of cooking food, its energy intensiveness per unit of service is (220 Wh/day) / (24 hrs/day) / (60 min/hr) = 0.15 Wh per minute of service for average use. (A more sophisticated analysis would separate cooking energy and idling energy for different usage scenarios.)

- If it takes 52,000 Wh of energy to mine and refine all the steel, glass, electronics, etc. to make the microwave, and its useful life is 10 years, its embodied energy per unit of service is (52,000 Wh) / (10 yrs) / (365 days/yr) / (24 hrs/day) / (60 min/hr) =.01 Wh per minute of service.

STEP 3: Document Your Energy Priorities
Time Estimate: 10–30 Minutes

- List the top three users of energy in your product's whole system, in order (biggest energy user first).

- For each of these energy users, show your math and your data sources. (e.g. "#1. Motor: uses 10W for an average of 5 hrs/day, for 50 Wh/day average. Power data from motor spec sheet, http://company-x-motor-spec-sheet.com, daily usage data from our interviews.")

Checklist for Self-Assessment

To score your success on this exercise, see if you...

☐   *Calculated the energy use of your product's main subsystems.*

☐   *Translated your calculations into energy per unit of service.*

☐   *Listed the top three users of energy in your product's system in order.*

☐   *Provided your math and data sources for your calculations.*

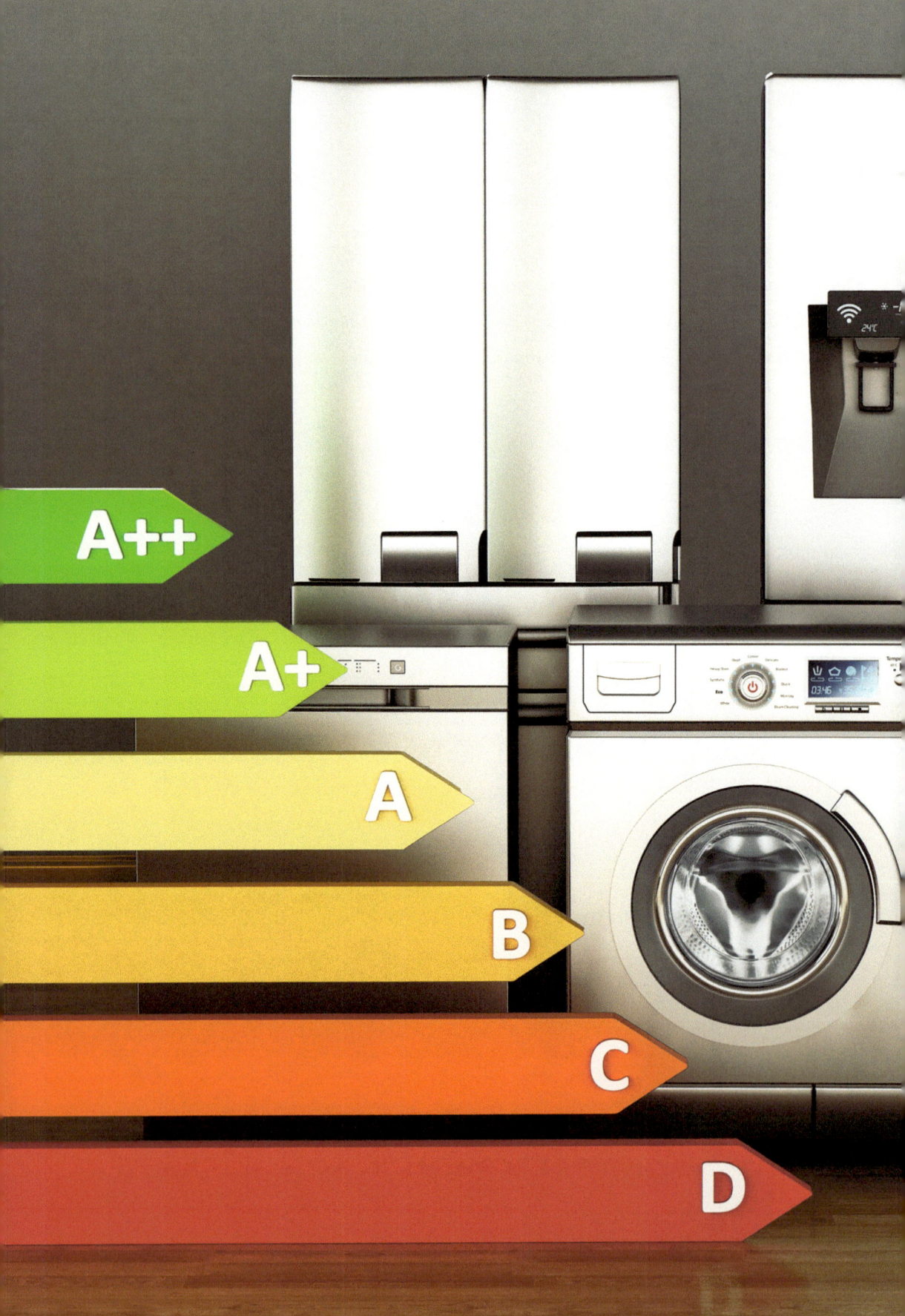

CHAPTER 18

Energy Efficiency

Jeremy Faludi,
Ruud Balkenende,
and Conny Bakker

Goals

- List the main causes of energy loss, and strategies of how to mitigate them

- Describe how to set priorities to improve product energy efficiency

- Evaluate solutions to improve product energy efficiency

DOI: 10.4324/9781003504672-20

Why It Matters

Because energy use causes such enormous environmental, social, and financial costs, even small improvements in energy efficiency at a global scale can cause large improvements to people and the environment. Understanding common causes of energy loss can help you find and fix them, and thus improve your design's energy efficiency.

Summary

- The most common physical inefficiencies are friction between moving parts, heat transfer, and fluid drag. You can improve them by choosing product geometry, materials, and components.

- The most common electrical inefficiencies are oversized motors, standby power, AC-DC converters, and lighting/display technologies. You can improve these by "right-sizing" motors, avoiding idle time/power, and choosing efficient components.

- Both physical and electrical system efficiencies can also be improved by automating operations with controls.

- One of the best ways to improve efficiency can be avoiding energy conversions: use heat as heat, motion as motion, etc.

18.1 Sources of Energy Losses

To make products more efficient, minimize energy losses. Once again, energy is never really created or destroyed, but when it's converted into heat or vibration or other forms we can't use, we call it "lost." The most common physical sources of energy inefficiency in products are friction, heat transfer, and fluid drag, all described below. (Electrical inefficiency is discussed later.)

18.1.1 Friction

Whenever one object rubs against another, some of their kinetic energy will be lost to unwanted heat, noise, deformation, and wear. This is friction. You can reduce friction in many ways (see Figure 18.1):

- Choose materials with low coefficients of friction together.

- Bushings (a layer of material between other materials).

- Lubrication (oil or powder).

- Wheels (trading gliding friction for rolling resistance).

- Vibration (reducing contact).

- Magnetic levitation (eliminating contact).

Note: When choosing materials for low coefficients of friction, you must choose pairs of materials together—it depends on both materials, as well as on how they are in contact (pressure, surface roughness, contact area). For example, steel on steel has a medium to high friction coefficient, as does polystyrene on polystyrene, but steel on polystyrene has a low coefficient. Table 18.1

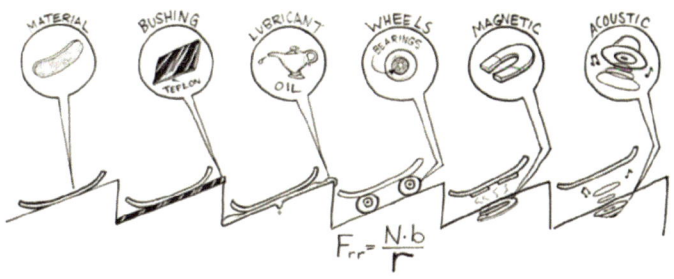

Figure 18.1 Ways to reduce friction

Source: based on Autodesk (2011).

Table 18.1 Coefficients of friction between materials

Material combinations			Frictional coefficient	
			Static μ_{static}	Kinetic (sliding) $\mu_{sliding}$
Aluminum	+	Aluminum	1.05–1.35	1.4
Aluminum	+	Mild Steel	0.61	0.47
Brass	+	Steel	0.51	0.44
Cast iron	+	Cast iron	1.1	0.15
Copper	+	Mild steel	0.53	0.36
Glass	+	Glass	0.9–1.0	0.4
Leather	+	Oak	0.61	0.52
Leather	+	Cast iron	0.6	0.56
Nylon	+	Nylon	0.15–0.25	0.06
Nylon	+	Steel	0.4	0.35
Oak	+	Oak (parallel grain)	0.62	0.48
Oak	+	Oak (cross-grain)	0.54	0.32
Polystyrene	+	Polystyrene	0.5	
Polystyrene	+	Steel	0.3–0.35	
Polyethylene	+	Polyethylene	0.2	0.2
Polyethylene	+	Steel	0.2	
PTFE (Teflon)	+	PTFE (Teflon)	0.04	0.04
PTFE (Teflon)	+	Steel	0.05–0.2	
Rubber	+	Rubber	1.16	
Silk	+	Silk	0.25	
Silver	+	Silver	1.4	
Skin	+	Metals	0.8–1.0	
Steel	+	Steel	0.5–0.8	0.42
Steel	+	Steel + Castor oil	0.15	0.081
Steel	+	Steel + Graphite	0.1–0.2	0.058

Source: Data from EngineeringToolbox.com and EngineeringLibrary.org.

shows some selected friction coefficients; other books and websites have much more extensive lists.

18.1.2 Heat Transfer

Heat transfer happens in all kinds of products that produce heat or cooling—refrigerators, coffee makers, vehicles, etc. Depending on the situation, efficiency may mean less heat transfer or more heat transfer. Heat generation is usually simple: generation from electrical resistance (as in toaster wires) and from burning gas (as in a gas stove) is usually 100% efficient. However, getting that heat where you want it (and getting it to stay there) are not. Many products unintentionally generate waste heat through inefficiencies; the best solution is to fix those inefficiencies, but the next-best solution is to dissipate that waste heat elsewhere.

As Figure 18.2 shows, heat transfer depends on conduction (direct heat transfer from one object touching another), convection (heat transfer through a moving fluid like air or water), and radiation (heat transfer through infrared light). The following list tells how these can be increased or decreased, and by how much. An "x" means doubling the variable doubles heat transfer; "x^2" means

doubling the variable quadruples heat transfer; "x^4" means doubling the variable multiplies heat transfer by 16.

Conduction increases with:

- Materials of higher thermal conductivity (x).
- Larger surface areas for the heat to flow through (area x, or length x^2).
- Thinner parts (x).

Convection increases with:

- Increased fluid flow (x).
- Larger surface area exposed to the fluid (area x, or length x^2).

Radiation increases with:

- Higher emissivity (x). Note: this is emissivity at heat-emitting wavelengths of light; surfaces that appear light and reflective (low emissivity) in visible light might be matte black (high emissivity) to heat radiation.
- Larger surface area exposed (area x, or length x^2).
- Larger temperature differences between the object and its surroundings (x^4).

18.1.3 Fluid Flow

Moving fluid through an object (like water through a pipe) or an object moving through a fluid (like a car through air) causes energy loss through fluid drag. Reduce this by streamlining moving objects and pipes—smoother bends, or less surface friction. An even more effective strategy is to reduce the velocity of the moving object or the fluid.

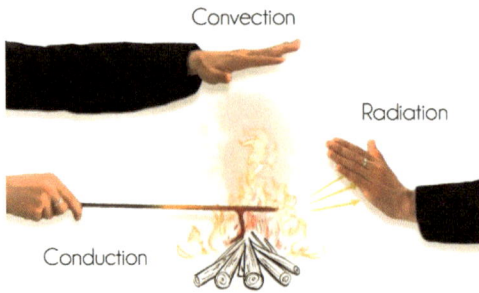

Figure 18.2 Different types of heat transfer
Source: based on Autodesk (2011).

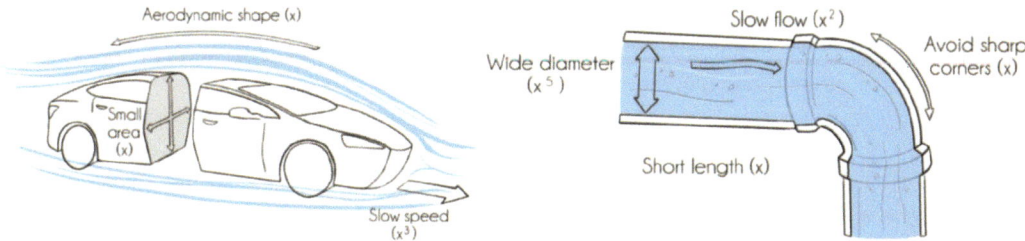

Figure 18.3 Drivers of fluid drag for objects in fluids and fluids in pipes, where x, x², etc. show proportions of change

Source: based on Autodesk (2011).

Figure 18.3 shows how fluid drag increases with relation to all major variables; as above, "x" means a linear change, "x²" means the variable squared, etc. Energy loss in pipes is reduced by the fifth power of diameter, so doubling diameter lowers drag by 32 times.

18.2 Electrical Components

Many products and systems are more electronic than mechanical, at least in terms of their energy use and losses. Therefore, efficiency of electronic components is often critical to system efficiency. Electronic inefficiencies also generate waste heat, which must be dissipated by heat transfer as described above.

18.2.1 Right-Sizing and Controlling Motors

Motors (and related components like pumps) should be sized and run for optimal efficiency in the most common circumstances, rather than grossly oversized for large factors of reliability and run at the same speed no matter the power needed. In developed

countries, motors consume around 45% of all electricity (ECN, 2016). Motors run everything from air conditioning and refrigerator compressors to water pumps to computer fans, they even make your mobile phone vibrate. Swapping inefficient motors for better motors already on the market would save 1350 terawatt-hours of electricity per year (7% of global electricity use), thus also saving hundreds of billions of Euros per year (ECN, 2016).

Why choose the wrong size or speed? For size, engineers often have to overbuild things with a "factor of safety" to account for uncertainty in how the product is used, or what conditions it is used in. For example, mechanical engineers build bicycles to withstand several times the rider's weight to withstand the force of hitting a curb at high speed. However, when oversized components are motors (and some other energy-using equipment, like lighting), this can cause serious inefficiencies. For speed, adding electronics for variable speed motors costs money. Today the cost is relatively small, but many motors are decades old, and even new systems are designed with old rules of thumb. The electricity saved during operational life is usually worth much more money than the initial cost of extra controls, but many people only think about purchase price. Even inexpensive motors can be 90–95% efficient when running in the optimal range of torque and speed.

When you purchase motors, choose the right size to make it efficient under normal operating conditions and change speed as needed, rather than greatly oversizing it. Every motor spec sheet shows torque versus speed as a graph or table of numbers, and lists a torque load rating for how hard the motor is designed to push. Often, people only choose motors based on the rated torque load, or the rated power (torque times speed). However, peak efficiency doesn't happen at peak torque or peak speed or peak power. Good manufacturer spec sheets also list the peak efficiency range for torque and speed. If it isn't listed in the spec sheet graphs or tables, ask the manufacturer. Choose a motor whose peak efficiency happens in the torque and speed range you need.

Figures 18.4 and 18.5 illustrate what you might see on a motor manufacturer's spec

sheet showing efficiency along with torque and speed. There are different kinds of electric motors—mainly DC permanent magnet motors and AC induction motors—and their efficiency curves are different. For DC motors, peak power happens around 50% of max torque, and at high speed; peak efficiency happens at around 10% of max torque, and lower speed; usually around 75% of max power. Efficiency falls off more slowly at higher torques than at lower torques, where it rapidly plummets. This means if you don't choose exactly the right motor, it's better to choose a smaller one (which will operate at higher torque and speed relative to its rated capacity) than a larger one (which will operate at a lower percent of its rated torque/speed). For AC motors, Figure 18.5 shows that efficiency is good across a wide range of higher torques, but it often plummets when below 20–30% of rated torque load. This is also true of high and low speeds, though they're not shown in the graph.

If you're driven to choose an oversized motor to handle large factors of safety just in case, consider more creative solutions to such situations. Can you use a variable speed motor, or add an auxiliary motor that only powers on when needed, or change the mechanics of the system to avoid such extremes, or other strategies?

18.2.2 Power Converters

Cheap transformers can be as low as 50% efficient, i.e., half the electricity becomes waste heat. Good transformers are 90+% efficient and are now becoming a market standard. Inverters that convert solar panel DC power to household AC power are usually 95% efficient. The EU's Ecodesign Directive is gradually outlawing inefficient transformers in Europe (requirements vary by size), and

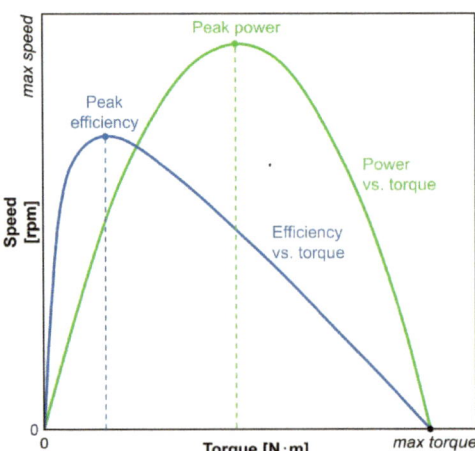

Figure 18.4 Generic efficiency curve for DC motors; note that its peak efficiency range is at a lower speed and lower torque than its peak power

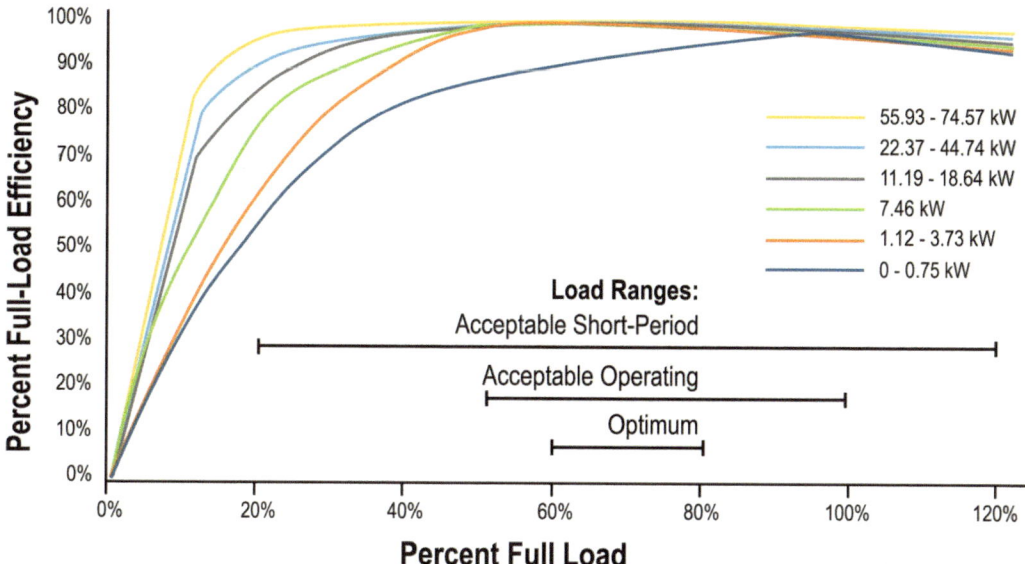

Figure 18.5 Efficiency curve for some AC induction motors; they have a wide range of high efficiency loading
Source: adapted from EERE (US DOE, 1997).

US "80 Plus" transformer certification is required for Energy Star product certification.

18.2.3 Lighting and Displays

Traditional incandescent lights only convert about 5% of incoming electricity to visible light; the rest becomes heat (MIT Technology Licensing Office, 2014). Today, LEDs are usually the most efficient lighting choice. The EU could save around €40 billion and 100 million tons of CO_2 a year by replacing remaining incandescent with LEDs (Verhaar, 2021).

Table 18.2 shows the efficiency of various lighting technologies, measured as "luminous efficacy" (lumens per watt). Remember, however, that not all lighting looks the same—most people like the warm yellow glow of candles, while many people dislike low pressure sodium's harsh yellow monochrome and the pinkish or bluish tones of some

fluorescents. LEDs can be many colors, and some are reasonable approximations of daylight. Such considerations are important to become popular replacements for inefficient technologies.

In addition to choosing efficient technology, you can improve lighting efficiency by separating task lighting from ambient lighting. Task lighting is located where the user needs it, ambient lighting fills the room. This matters because the brightness of a normal light source decreases with the square of the distance ($1/r^2$). For a user reading a book at their desk, a lamp on the ceiling or wall ten times farther away than a lamp on the desk will only shine 1/100th as much light on the book. There is no need to waste electricity over-lighting entire rooms so that people can work at their desks. Task lighting provides the user the desired brightness, while low ambient light levels provide comfort and save energy.

Table 18.2 Luminous efficacy of various lighting technologies

Category	Type	Luminous efficacy (lm/W)
Combustion	Candle	0.3
	Gas mantle	1–2
Incandescent	Tungsten	14–15
	Halogen	17–25
Fluorescent	Compact fluorescent	46–75
	T8 tube	80–100
	T5 tube	70–104
Gas discharge	Metal halide	65–115
	Low pressure sodium	100–200
LED	LED screw base lamp	80–150
	LED retrofit for T8 tube	75–180

Light-emitting displays are also important. They comprise up to 40% of electronics energy use, especially due to televisions (Urban et al., 2021). Some display technologies are better than others. Figure 18.6 shows power use and size of different display technologies. Better technology choice is not the only way to save display energy; displays can also be smaller or shut off when not in use, to avoid standby power.

Note that e-paper displays do not emit light, they require external lighting. However, this is usually present. Once an e-paper display shows an image, it uses no power to continue; it only uses energy to change the image. However, its slow refresh rate makes it unsuitable for video or fast interaction.

18.3 Standby Power

Standby power (or "vampire power") is electricity used by a device or component when idle, simply waiting for user interaction. Both the EU Energy Label and US Energy Star label require minimizing standby power. Even computer CPUs reduce

their power usage by powering down groups of transistors when not in use, nanosecond-by-nanosecond. Standby power happens in many consumer electronics—historically televisions, computers, game consoles, and chargers, but nowadays more often networked smart home devices, like smart lightbulbs, door locks, and even some kitchen appliances. See Figure 18.7.

Standby power can cause surprisingly large impacts: a smart LED lightbulb wirelessly controlled via an app (see Figure 18.8) may use 8 watts of power when turned on, but if only used for 2 hours/day (16 Wh per day), and it uses 0.75 W power for network connection all 24 hours of the day (18 Wh per day), then it uses more energy idling than shining light. Its idle efficiency becomes a higher priority than its light efficiency. Some smart bulbs use 2.5 W to idle—that's 60 Wh/day!

18.3.1 Minimizing Standby Power

Standby power of many products, such as laptops and TVs, has been cut by 80–90% over the past decades. This is a great success.

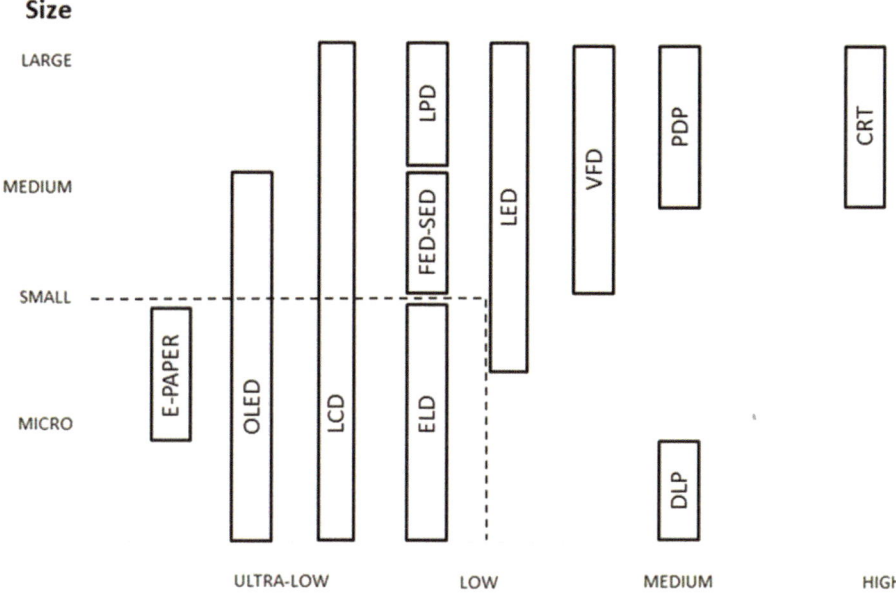

Power Consumption Density

E-PAPER: Electronic Paper
OLED: Organic Light-Emitting Display
LCD: Liquid Crystal Display
ELD: Electroluminescent Display
FED: Field Emission Display
SED: Surface-Conduction Electron-Emitter Display

LPD: Laser Phosphor Display
LED: Light-Emitting Diode
VFD: Vacuum Fluorescent Display
DLP: Digital Light Processing
PDP: Plasma Display Panel
CRT: Cathode Ray Tube

Figure 18.6 Power consumption of display technologies versus screen size, qualitatively

Source: Fernández et al. (2015).

The simplest way to minimize standby power is a traditional "hard" off switch that cuts power entirely. They are perfect for products with little to no startup lag time, like lamps, as long as no remote control or scheduling is required. For longer waits due to product startup time, such as TVs or stereos, you can help your user's patience with attractive graphics or other experience. Beware smart power outlets, as they consume some standby power. Will they save more in the products they shut off than they consume themselves waiting for a signal?

For products requiring instant response, identify which components use the most standby power. Common culprits include:

- Remote controls
- Continuously-on LEDs and displays (e.g., digital clocks)
- Touch screen keypads
- Battery chargers
- Network connectivity

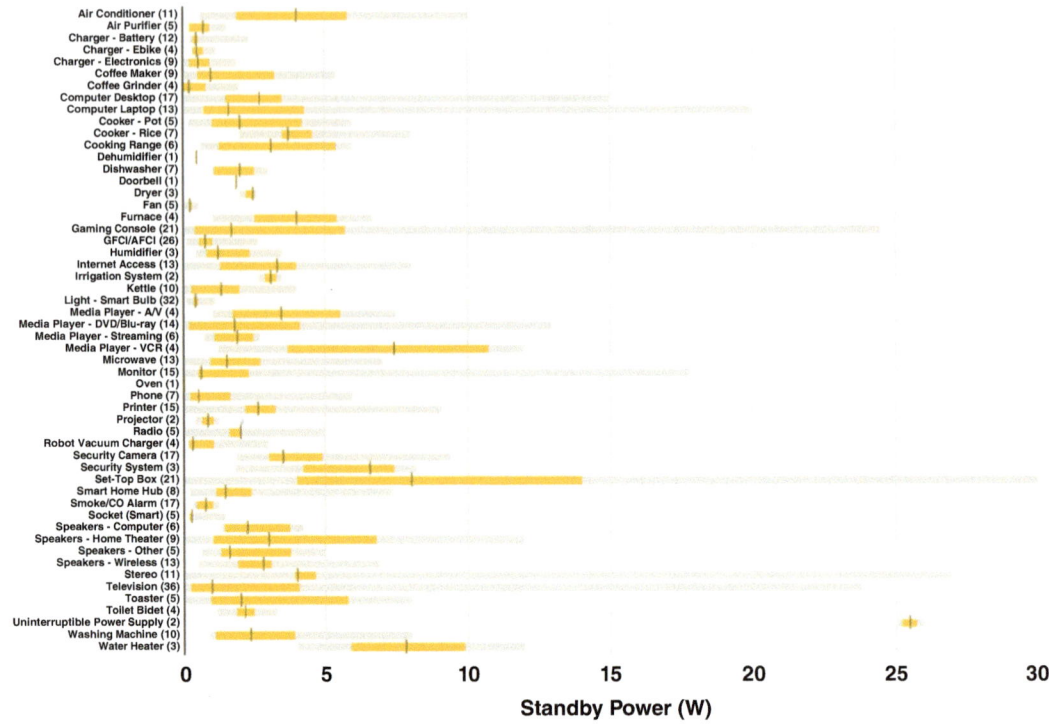

Figure 18.7 Standby power use of common consumer devices. Black lines are mean values, yellow bars are one standard deviation, gray bars are two standard deviations, and the number of data points are in parenthesis.

Source: Lawrence Berkeley National Lab (2023).

There are several ways to fix standby power:

- Choose better components. For example, network connectivity power demand can often be cut by 20–50% (Wang et al., 2020).

- Minimize the number of always-on components. For a mobile phone's idle connectivity, only the receiver circuitry waiting for calls must be active; transmission circuitry may be powered down, along with the screen, memory, CPU, etc.

- Choose different signaling technology. If the device only needs to respond to queries from another device, Wi-Fi

modules can be replaced with RFID, which requires no power supply, gathering energy from the broadcast signal triggering its response.

- Design idle components away. Does your kitchen really need the microwave, stove, and clock to all tell you the same time? Probably not.

18.4 Controls and Automation

Energy-saving lighting, laptops, etc. save as much or more of their energy through better controls than they do through fundamental

Figure 18.8 Smart lightbulb

changes in electronics technology. For example, office lighting is usually controlled manually, but automating it has shown energy savings of up to 38% (Williams et al., 2012). Vehicle driving is slowly being automated; it's expected to save up to 45% of fuel in an optimistic scenario (Chen et al., 2019), and better routing algorithms have already improved shipping efficiency by 10–20% without driving automation (Boggio-Marzet et al., 2021).

Some controls are passive—they don't use electronics, but instead happen naturally as a result of clever hardware design. For example, thermostats on radiators work by thermal expansion of a fluid, not electronic sensors. Diesel engines don't have spark plugs with timers to control fuel ignition, they cause ignition purely through compressing the fuel-air mixture to a critical pressure. Both are a matter of physics rather than timers or sensors, so they use no power and no electronics.

18.4.1 User Behavior

Not all controls are automated—sometimes the controls are user behavior, which is influenced by your design. For example, with car fuel efficiency, eco-style driving can cut fuel use and greenhouse emissions by 15–45% without increasing travel time in urban traffic (Fafoutellis et al., 2021). Some design interventions auto makers have used to encourage eco-style driving include live dashboard display of fuel economy, "eco-mode" buttons to limit acceleration, and more. See Chapter 20 for strategies you could try to influence user operation of your products or systems.

18.5 Avoiding Energy Conversions

The last way to save energy is to avoid conversions—use heat as heat, use motion as motion, etc. Here are some examples:

18.5.1 Kinetic

A bicycle or car transmission converts mechanical motion of one speed and torque to another speed and torque, but it remains mechanical motion, it isn't converted to electricity and back again. Such mechanical transmissions, even with many gear stages, are over 95% efficient. Mechanical energy can be stored in a flywheel or spring tension or lifting a weight in gravity. However, these are usually only useful for smaller amounts of energy, because large amounts of mechanical energy can become dangerous if mechanisms fail (e.g., heavy weights falling on users, or rapidly spinning flywheels cutting through enclosures like saws). See Chapter 19 for limits of energy storage.

18.5.2 Heat

Waste heat is common; some swimming pools are heated with waste heat from homes or nearby industry. Some buildings are too cold at the same time that other buildings are too warm, including different rooms in the same building. Stanford University restructured their campus heating and cooling system to be a network of hot and cold water pipes between buildings, reducing carbon emissions by 68% by moving heat from where it's unwanted to where it's wanted rather than using boilers and air conditioners (Bleavans and Winslade, 2022).

18.5.3 Light

All lighting ultimately uses power from the sun, converted several times before being converted back to light. (Even fossil fuels are sunlight converted to biomass, converted to petrochemicals). If a solar panel converts sunlight into electricity at 15% efficiency, then a battery converts that electrical energy to chemical energy and back at perhaps 98% efficiency, then an LED with 100 lumens per watt converts the electricity back to light at 15% efficiency, the total system efficiency is 2.2%. Other conversion paths are even less efficient. Some phosphorescent minerals or biological proteins absorb light, convert it to chemical energy for storage, and later convert it back to light (usually a different color), with higher total system efficiencies. Most phosphorescence is short-lived, low brightness, and only provides limited color choices; but even with these limitations, there are some practical applications like safety signage. Studio Roosegaarde's "Van Gogh path" outside of Eindhoven is a bicycle path with phosphorescent elements storing sunlight to glow in the dark at night (Figure 18.9). Its technology has been used for "smart highway" experiments elsewhere.

Figure 18.9 The phosphorescent "Van Gogh path" stores sunlight as chemical energy and re-emits it at night
Source: Daan Roosegaarde.

Resources and References

Resources for Further Study

- MacKay, D. J. C. (2009). Sustainable energy – without the hot air. Chapters 1 and2. Available at: https://www.withouthotair.com/

- Hawkens, P., Lovins, A., & Lovins, L. H. (1999). *Natural capitalism*. Little, Brown and Company.

References

Autodesk. (2011). Autodesk Sustainability Workshop. Autodesk Academy. Available at: https://sustainabilityworkshop. autodesk.com [offline as of May 2018].

Bleavans, L., & Winslade, A. (2022). Fact Sheet: Stanford Energy System Innovations (SESI) Project. Stanford University. Available at: https:// sustainable.stanford.edu/sites/default/files/SESI_Condensed_factsheet_2022.pdf

Boggio-Marzet, A., Monzón, A., Luque-Rodríguez, P., & Álvarez-Mántaras, D. (2021, September 22). Comparative analysis of the environmental performance of delivery routes in the city center and peri-urban area of Madrid. *Atmosphere*, 12(10), 1233.

Chen, Y., Gonder, J., Young, S., & Wood, E. (2019, April). Quantifying autonomous vehicles' national fuel consumption impacts: A data-rich approach. *Transportation Research Part A: Policy and Practice*, 122, 134–145.

ECN. (2016, November 1). More efficient electric motors could mean plans for 200 coal-fired power plants around the world can be scrapped. Available at: https://www.ecn.nl/news/item/more- efficient-electric-motors-could-mean-plans-for-200-coal-fired-power-plants-around-the-world-can-1/index.html (accessed September 22, 2022).

Fafoutellis, P., Mantouka, E. G., & Vlahogianni, E. I. (2021). Eco-driving and its impacts on fuel efficiency: An overview of technologies and data-driven methods. *Sustainability*, 13(1), Article 1.

Fernández, M. R., Casanova, E. Z., & Alonso, I. G. (2015, August 11). Review of display technologies focusing on power consumption. *Sustainability*, 7(8), 10854–10875.

Lawrence Berkeley National Lab (2023). Products that use standby power. Available at: https://standby.lbl.gov/standby-power-data-metering (accessed July 12, 2023).

MIT Technology Licensing Office. (2014). High efficiency incandescent lighting. Available at: https://tlo.mit.edu/technologies/high-efficiency-incandescent-lighting (accessed September 22, 2022).

Urban, B., Roth, K., & Olano, J. (2021). Energy consumption of consumer electronics in U.S. homes in 2020. Fraunhofer USA CMI Report to the Consumer Technology Association.

US DOE. (1997). Determining electric motor load and efficiency (No. DOE/GO-10097-517). US Department of Energy Motor Challenge. Available at: https://www.energy.gov/eere/amo/articles/determining-electric-motor-load-and-efficiency

Verhaar, H. (2021, July 20). The need for speed. The European Alliance to Save Energy. Available at: https://euase.net/tag/lighting/

Wang, W., Su, J., Hicks, Z., & Campbell, B. (2020). The standby energy of smart devices: Problems, progress, and potential. 2020 IEEE/ ACM Fifth International Conference on Internet-of-Things Design and Implementation (IoTDI).

Williams, A., Atkinson, B., Garbesi, K., Page, E., & Rubinstein, F. (2012). Lighting controls in commercial buildings. *LEUKOS*, 8(3), 161–180.

How to Apply #18: Energy Efficiency Brainstorm
Time Estimate: 2–5 Hours

In this exercise, you'll brainstorm to improve your product or system's energy impacts.

STEP 1: Estimate What Causes Your Product's Largest Energy Losses
Time Estimate: 5–15 Minutes

Think critically about where your product likely loses most energy: is it mechanical, electronic, thermal, or other? What components or processes likely cause it? If you can measure and quantify the energy use, even better. List your top priority loss(es) to fix.

STEP 2: Brainstorm Ways to Reduce Your Product's Energy Use by a Factor of Ten
Time Estimate: 0.5–1 Hour

Hold a brainstorm session, using the Rules of Brainstorming and whatever collaboration tools you prefer, to generate ideas for reducing your product's energy use to 1/10th the energy it uses today. Especially focus on your top priority losses. Because it's a brainstorm, don't limit yourself to 10x ideas, but also try to radically re-envision things for 50x or 100x improvement. What need is the product fulfilling, and how else could that need be fulfilled?

If your product does not use energy itself but causes energy to be used by something else in its system (like clothing and a washer/dryer), brainstorm ways your product could reduce that energy use. If your product does not use energy during its useful life, and doesn't cause energy to be used by something else in its system, then brainstorm ways in which its manufacturing or materials can use 1/10th as much embodied energy.

For the brainstorm:

- Start with the Whole System Map you created for your product, to keep in mind all the components of the system, and how they connect to each other, and the product's role in the larger system.

- Have a good number of ideas (30+) to fix top priority losses.

- Have at least one idea for every major component or step in the system.

- Have at least six ideas that eliminate a step or component of your system. Eliminating multiple steps/components is even better.

STEP 3: Narrow Down Your Brainstorm Options to Three or Four Finalists
Time Estimate: 0.5–1 Hour

Using dot voting or whatever tools you desire, narrow down to just three or four best ideas. In addition to judging them by energy impacts, use considerations from your Design Brief to rule out options that don't meet business criteria such as cost or usability.

STEP 4: Estimate the Total Energy Impacts for Each Winning Option, and Choose the Best Idea
Time Estimate: 10–30 Minutes

If your product uses energy during its life (or causes significant energy use in its system):

- Estimate the total lifetime energy use per functional unit of each of the finalist ideas you've chosen, from both brainstorms.

- Using LCA or other estimation tool, calculate the total lifetime impacts of that energy use per functional unit, for each idea. You do not need to do a full LCA, just record the impacts for energy use.

If your product does not use energy during its life:

- Use LCA or other tool to estimate the improvement in embodied CO_2 impacts per functional unit of your new ideas.

STEP 5: Choose One Final Idea to Move Forward with, and Illustrate It
Time Estimate: 5–30 Minutes

- Choose one winning idea (or combination of ideas), based on the estimated impacts and your other design brief priorities.

- Illustrate the final idea (rough sketch or fancy rendering) to clearly convey its energy improvement, and why it's a compelling design.

STEP 6: Document Your Decision and Brainstorms
Time Estimate: 10–30 Minutes

- Create a PDF with a short description and illustration of the winning design(s), and the reasons for your choice.

- Show your brainstorm, making it clear that you had at least one idea for every part of the system, and many ideas that skipped steps in the system.

- Show the illustration of the winning design(s).

- Succinctly describe the winning design(s), either as annotations to the illustration or as a stand-alone sentence or two. Describe why it is the best of all your new ideas.

- Show your estimates of total lifetime energy use per functional unit (or, for products that don't use energy, embodied CO_2 impacts per functional unit), both for the final winning idea and all the other ideas you ran numbers for. The ideas that didn't get chosen as the final winner don't need descriptions, but at least make their titles suggest what they are.

- Show your math for all the above estimates.

- State how much you expect the winning design to improve ecological impact compared to the original design.

Checklist for Self-Assessment

To score your success on this exercise, see if you…

- ☐ *Had 30+ ideas related to top energy losses.*

- ☐ *Had at least one idea for every major component or step in the system.*

- ☐ *Had at least six ideas that eliminate a step or component of the system.*

- ☐ *Chose 3–4 finalist ideas.*

- ☐ *Estimated improved impacts of all finalist ideas.*

- ☐ *Illustrated and described final winning idea(s).*

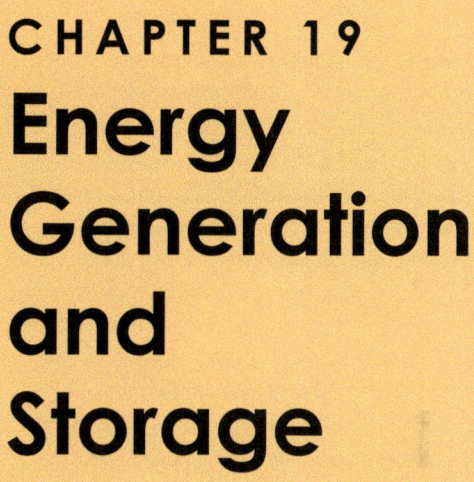

CHAPTER 19
Energy Generation and Storage

*Jeremy Faludi,
Ruud Balkenende,
and Conny Bakker*

Goals

- Identify opportunities and limitations for portable renewable energy sources
- Calculate the feasibility of portable solar power in devices
- Identify opportunities and limitations for energy storage options

DOI: 10.4324/9781003504672-21

Why It Matters

Is a solar panel on a desk lamp greenwashing or a real benefit? The global warming impacts of solar, wind, and other renewable power generation at grid scale are 1/10th to 1/100th that of fossil fuels. They also avoid the geopolitical problems of concentrated oil wealth, though building them requires mining materials, some of which have environmental and social concerns. Small portable energy systems on products are often expensive, and sometimes ineffective. Energy storage also has its own costs and limitations. Is it viable for your product to be self-powered, or not? You can calculate this by comparing your product's energy demand against its generation and storage potential.

Summary

- The world is full of vast clean renewable energy sources.

- Some of these can be viable at portable household product scales.

- Power and energy generation numbers are shown for solar photovoltaic, micro-wind, micro-hydro, human power, and waste heat scavenging.

- Viability of powering your product from on-board renewable energy can be calculated with energy balance equations. These are shown for PV power.

- Energy storage technologies are usually chosen by their energy density and/or power density; numbers are shown for batteries, fossil fuels, fuel cells, and others.

- Other considerations for energy storage are charge-discharge efficiency, financial cost, and environmental impacts.

19.1 Sourcing Clean Energy

Even if you can't change how much energy your system demands, its energy impacts can be greatly reduced by switching to clean renewable energy. Renewable energy is energy from sources in nature that are replenished faster than they are consumed. Sunlight, wind, and rivers are the main examples, but it also includes ocean waves, burning biomass that's grown faster than it's burned, and more. Technically, geothermal energy is not replenished, but the amount available is so large that human use isn't expected to deplete it, so it's considered clean energy. The main point of renewable energy is to be clean—to not emit CO_2 or other pollutants and not to deplete scarce resources like fossil fuels. Building the equipment to harvest renewable energy (solar panels, wind turbines, etc.) uses resources and causes pollution, but they last so many years that the impact per megajoule is tiny: LCAs show that compared to burning coal for electricity, solar panels and offshore wind turbines both emit roughly 7% as much CO_2/MJ, and large hydroelectric dams emit 0.5% as much CO_2/MJ (IDEMAT 2021).

There are many forms of clean energy, usually providing electricity, heat, or mechanical motion. Historic windmills and water wheels turned fluid motion into mechanical motion to grind grain, pump water, or do other work. Today, wind turbines and hydroelectric dams usually generate electricity. Heat is generated by solar thermal, geothermal, biomass burning, and more. Some are dual-purpose: cogeneration plants mentioned in Chapter 17 for heat and electricity can also run off of renewable carbon-neutral fuels.

Today, renewables provide a small percentage of total global electricity, but they could provide all we need and more. Current global electricity consumption is 65 EJ (exajoules, 10^{18} J) per year, expected to rise to 240–400 EJ/yr by 2070; but wind, solar, hydroelectric, and geothermal together can easily produce several times more than this by then, even in conservative scenarios (Deng et al., 2015). See Figure 19.1. Of course, building this much renewable power generation requires large financial and material investments, plus political will, but large-scale solar and wind are already the cheapest electricity on the planet.

Note that, because the sun doesn't shine all the time and the wind doesn't blow all the time, energy storage is required to handle their intermittency. This may mean batteries, running hydropower dams backwards to store energy as higher water levels, freezing water into ice reservoirs for later cold water use, electrolysis of water to form hydrogen for fuel cells, or many other creative solutions. Even growing trees for future burning as fuel

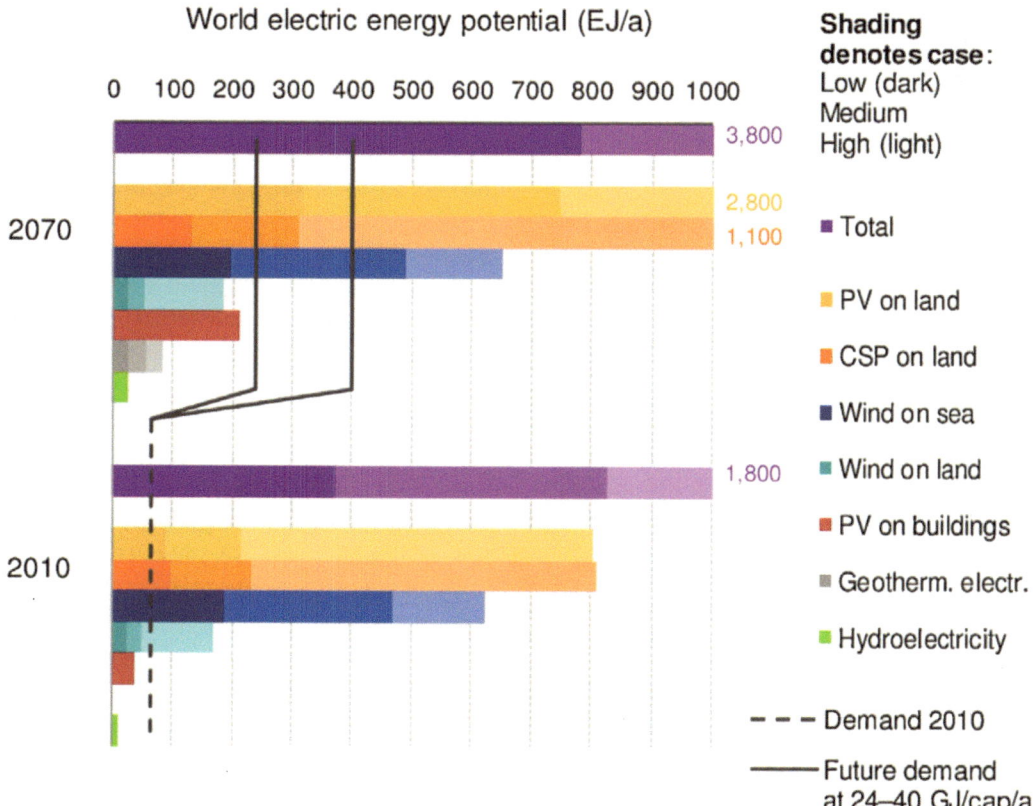

Figure 19.1 Global renewable power availability versus demand in 2010 and 2070.

Notes: EJ/a = exajoules per year; PV = photovoltaic; CSP = concentrated solar power; GJ/cap/a = gigajoules per year per person.

Source: Deng et al. (2015).

could be considered energy storage, though perhaps not a convenient one.

19.2 Portable Renewable Energy Generation

Most electricity requires grid connection, but some products can generate their own power portably. Power generation at the product scale is usually easiest with solar photovoltaics (PVs). Micro-scale wind or water power or biofuels can be viable for some niche applications. Human power, either from movement or body heat, can also be used for some products (such as a bicycle). For all renewable energy sourcing, minimizing product energy demand is a key enabler, so less energy needs to be generated (Flipsen, 2006). This is especially true for portable power, because energy generation and storage cost size and weight as well as money. Portable energy generation usually also requires energy storage; that's discussed in the next section.

19.2.1 Portable Solar

Solar photovoltaics (PVs) are the easiest form of portable renewable power generation. Solar calculators have existed since the 1980s; the components are affordable, small, robust, simple to design and operate and maintain, and have no moving parts and are silent. The only downside is requiring exposure to light, making them inconvenient for products like phones that live in users' pockets.

Available power: PV power generation depends on the area of PV panel, the brightness ("irradiance") of light exposure, and the PV conversion efficiency (which also changes with irradiance). See Table 19.1 and Equation 19.1.

$$P_{PV} = GA\eta \qquad (19.1)$$

Equation 19.1 *Power generation P (measured in W), where G = irradiance (W/m^2), A = PV area (m^2), and η = system efficiency (%).*

Different kinds of solar cells also have different efficiencies under different lighting conditions, so Table 19.1's numbers are rough rules of thumb. But it shows that roughly speaking, charging a AA battery with 1800 mAh at 1.2 V (2160 mWh of energy) from a 100 cm^2 solar panel would take 2 hours outdoors in direct sun, 20 hours outdoors in cloudy skies, and roughly 1,000 hours (that's multiple months!) in indoor lighting.

Available energy: To design for any portable power, solar or otherwise, you need to balance energy generation (E_{in}) against

Table 19.1 Different light conditions and respective solar panel efficiency

Place of light exposure	Irradiance, W/m^2	Solar panel efficiency (rough average) (%)
Direct sunlight	794	20
Bright indirect sun	79	20
The brightest office/factory	7.9	10
Most office/factory lighting	2.4–4.0	10
Hallways	0.79	5
Evening home living rooms	0.4	5

Source: Apostolou et al. (2016); Faludi (2011); Verwaal (2019).

energy demand (E_{out}). To calculate solar Ein, multiply irradiance by time, PV surface area illuminated, and PV efficiency under those lighting conditions (see Equation 19.2).

$$E_{in} = tGA\eta \qquad (19.2)$$

Equation 19.2 *PV energy generation (E_{in}), where t = light duration. If t is in seconds, E_{in} is in Joules; if t is in hours, E is in watt-hours.*

For the product's energy demand (E_{out}), measure or look up the power use of your product's different functions or components or operational modes (e.g., playing video versus sitting idle). Multiply each function's power (W) by its duration (h) for that function's energy demand in Wh. (Watch the units, and convert to MJ if needed.) Summing the energy demands of all functions provides your total energy use, see Equation 19.3.

$$E_{out} = \sum_i P_i t_i \qquad (19.3)$$

Equation 19.3 *Product energy demand E_{out} (in J or Wh), where Pi is power per function i (in W) and t_i is duration per function i (in seconds or hours), summed for all i.*

To decide whether PV will be feasible for your product, see Table 19.2, adapted from Verwaal (2019), for a quick rule-of-thumb judgment.

Table 19.2 Portable PV feasibility

$E_{in} / E_{out} > 10$	Feasible, but PV system too large: reduce it to save money and impacts.
$1 < E_{in} / E_{out} < 10$	Feasible
$0.1 < E_{in} / E_{out} < 1$	Marginal: PV system may be too small; try to enlarge it or reduce E_{out}.
$E_{in} / E_{out} < 0.1$	Not feasible

19.2.2 Micro Wind Power

Wind power is quite economical at large scales and high wind speeds for grid power, but very expensive at small scale with slow wind speeds. Wind speeds are generally low near the ground, because of obstacles, and faster high off the ground. Winds at 50 m heights often have double the power density of winds at 10 m, as shown in Figure 19.2. Wind turbines also don't begin generating power until their "cut in" wind speed is reached for the blades to start spinning— usually 5 m/s or higher. Thus, small-scale wind ("micro-wind") is almost never economically viable compared to solar, often costing five times as much per watt-hour of energy generated, even at architectural scales (Miller et al., 2009). It's even less viable for portable power. However, it can be useful for some specialized situations, like sailing boats.

Available power: The power available from wind is shown in Figure 19.2 and calculated in Equation 19.4:

$$P_W = \frac{1}{2}\rho A v^3 \eta \qquad (19.4)$$

Equation 19.4 *Wind power PW (in watts), where ρ = air density (kg/m^3), A = area (m^2), v = air speed (m/s), and η = system efficiency (%).*

The "area" is area of wind captured by turbine blades or scoops; the efficiency includes aerodynamics of the blades and all mechanical and electrical systems. Notice especially how wind speed is cubed—that means twice the speed generates eight times the power!

Available energy: As with all energy, wind energy is the integral of power over time.

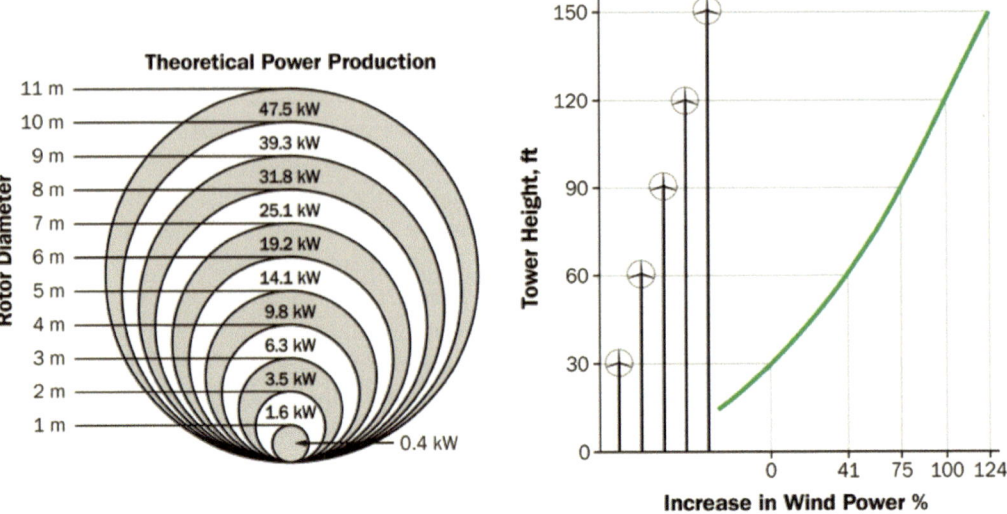

Figure 19.2 Theoretical power increases from increasing rotor size and height of wind turbines
Source: Clarke (2018), NREL (2007).

Unlike solar and some other energy, wind speeds often fluctuate greatly from minute to minute, and power varies with wind speed cubed, so multiplying average power by time may not be very accurate. However, some websites or databases let you look up average wind power density (W/m²), in which case, you can calculate energy by simply multiplying this by time, turbine area, and efficiency.

Average wind speeds and directions, and their variability, depend on your specific location (including height). You can look up annual average wind speeds and wind power densities at different heights on websites like https://globalwindatlas. info, or various government weather sites. However, because wind speeds vary so much by terrain, including surrounding trees and buildings, it's best to measure the winds at your location for some months to be sure. Store energy according to the energy balance equations

listed in the solar section, with Eout being the same and Ein being expected wind energy generated.

19.2.3 Micro Hydro Power

Like wind, hydropower at a large scale is some of the most affordable power on the planet, but small-scale systems are often expensive. Also, micro-hydro power is limited to locations with moving water. However, this actually includes flows of any fluid in any pipe. For example, flow sensors and water quality sensors are sometimes self-powered by the flow of fluids through pipes (Carminati et al., 2020).

Available power: Water power depends on your local water source. It tends to be higher efficiency than wind (50–80%) because it usually uses a pipe to force all moving water through the turbine. Also, it lacks wind's exponential scaling factors, so small-scale systems don't suffer from their size as much as wind.

Power from falling water is shown in Figure 19.3 and calculated in Equation 19.5:

$$P_H = \rho q g h \eta \qquad (19.5)$$

Equation 19.5 *Hydropower PH (in watts), where ρ = fluid density (kg/m³), q = volumetric flow rate (m³/s), g = gravitational acceleration (m/s²), h = height fallen (m), and η = system efficiency (%).*

Note the flow rate is volumetric flow, not speed. The height the liquid falls is called the "head" height. Even if the liquid is not falling from gravity but is being pumped through pipes, the pressure differential is usually written as gravitational acceleration and "head" height, by industry convention. As mentioned above, there are no exponents, everything is linear, so a flow of 1 liter per second (0.001 m³/s) of water, whose density is 1 kg/l (1000 kg/m³), falling a height of 1 m in normal gravity (9.81 m/s²) driving a 75% efficient generator will provide 7.36 W of power.

Available energy: Micro-hydro energy generation is easier to generate consistently

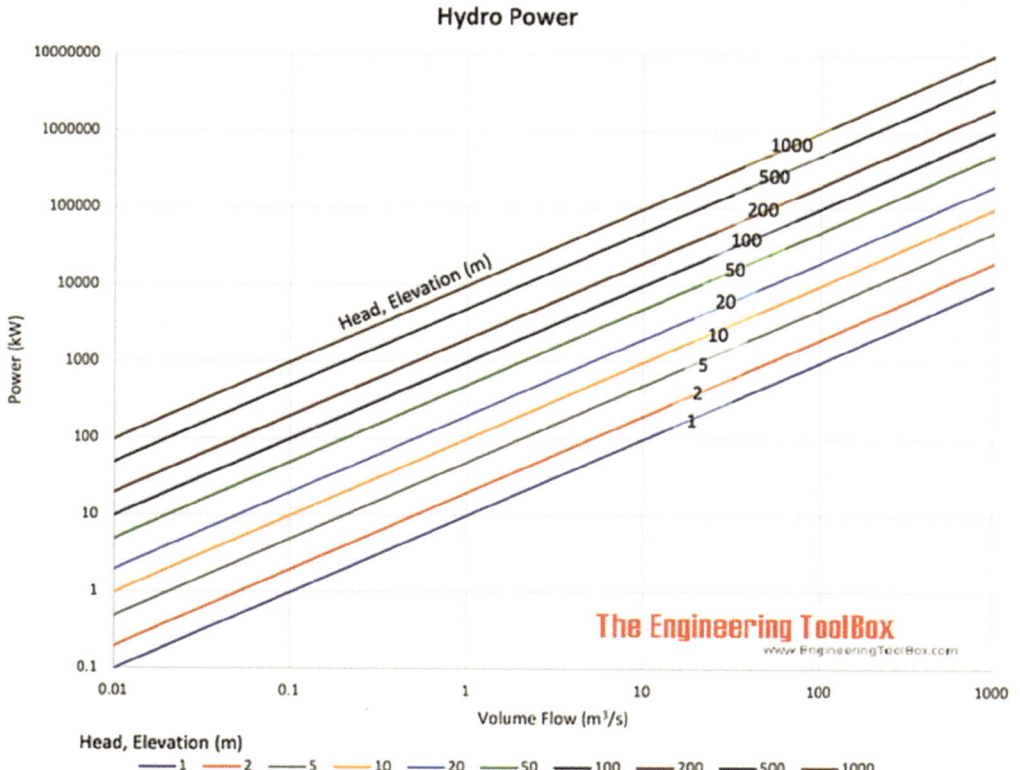

Figure 19.3 Hydropower theoretical power versus flow rate for several different heights of water fallen
Source: The Engineering Toolbox (2008).

than wind energy, so it's usually just power × time. This is because river flows are set by gravity and seasonal rainfall, not minute-by-minute wind fluctuations. River speed and volume can still vary greatly over time, but micro-hydro systems can be sized to only use a small percentage of river flow that is always reliably there, or a storage reservoir can save water during high-flow times for use in low-flow times. It can generate arbitrarily large amounts of energy, but at some point is no longer "micro."

19.2.4 Human Power

For portable electronics or other low-power applications, human power may be a good option. Sometimes the product can be powered by the very activity the user desires the product for, for example, the Octane Q47 elliptical cross-trainer machine was powered by the user's workout. They can also be powered by other activity or waste heat. Mechanical "self-winding" watches, powered by random motion, have existed since Hubert Sarton's in 1778, and perhaps before (see Figure 19.4). The Matrix Industries PowerWatch is an electronic smart watch powered by body heat. Experimental products have included pedal-powered blenders, shoes that charge phones, a medical breathing monitor powered by the act of breathing, and a remote control powered by piezoelectrics when the remote's buttons are pushed (Paradiso & Starner, 2004).

Available power: There are many kinds of human power, and each has different generation potential. Figure 19.5 from Paradiso and Starner (2004) shows how much can be expected from various activities or metabolic processes. Note the values for heat are waste heat from the entire surface of the body; small devices will recover much

Figure 19.4 Wrist watches powered by the body: (a) a 2020 electronic smart watch powered by body heat; (b) a 1778 Hubert Sarton self-winding watch powered by random movement

Source: the-gadgeteer.com and Jeremy Faludi.

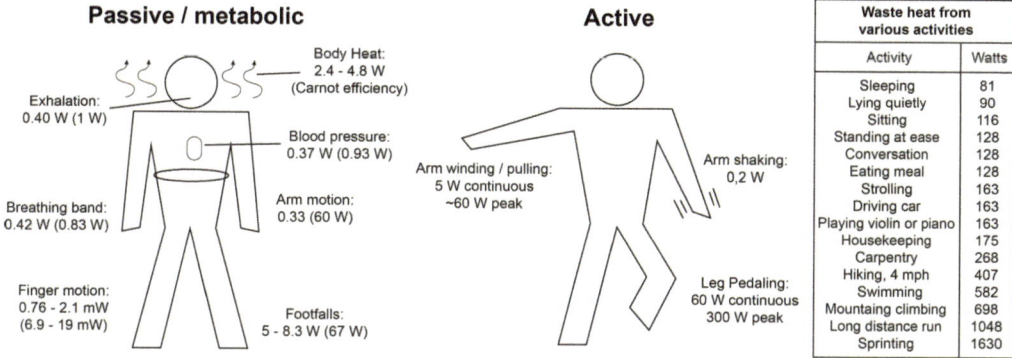

Figure 19.5 Typical amounts of power available from various body activities, both long-term (no parentheses) and short-term maximum values (in parentheses)

Source: Paradiso & Starner (2004).

less, and their efficiency will need to be accounted for. Paradiso and Starner note that a 0.5 cm² thermoelectric device might only generate 10 μW.

Available energy: Just as available power varies widely for different human power generation modes, so does the energy they can provide. Some activities are relatively constant over time (e.g., breathing), while others are not (e.g., sprinting). Energy, as always, is the integral of power over time, but you usually won't measure human power output from second to second—instead, guess an average power generation and multiply it by average time. Store energy according to the energy balance equations listed in the solar section, with E_{out} being the same and Ein being power from Figure 19.5 times time.

19.2.5 Recovering Waste Heat

Waste heat is everywhere, but little is usefully recoverable. Many technologies can recover waste heat: thermoelectric generators turn heat into electricity, heat engines turn heat into mechanical work

(e.g., Stirling engines, Rankine engines, or thermo-acoustic engines), and heat exchangers leave the heat as heat but move it from one place to another. Most of these are not very portable, though some can be hand-held. Thermoelectric devices ("Peltier devices") are very portable—they're solid-state plates that can be less than a square centimeter in area and one millimeter thick. Figure 19.6 shows a more typical sized one, roughly 6 cm².

Using waste heat as heat (simply moving the heat to where it's useful) has often been successful on industrial and architectural scales. Internet data centers that use water to cool their servers have worked with cities like Helsinki, Stockholm, and others to turn their waste heat into district heating for nearby buildings (Stockholm Data Parks, 2017). While datacenter waste heat is often not hot enough for district heating, it can be—the city of Mäntsälä, Finland, used this to cut its annual fossil fuel use and carbon emissions by 40%, heating 20,000 homes (DatacenterDynamics, 2016).

Available power: Power recovered from waste heat is simply equal to the power

Figure 19.6 A thermoelectric plate, which can generate microwatts of electricity from a temperature difference, or pump heat with electrical power input

of the waste heat times the efficiency of recovering it. Heat engines ("Carnot" engines) can generate arbitrarily large wattages, but are only efficient or cost-effective at large temperature differences between the heat input and ambient temperature: to even be 10% efficient, most heat engines need to be 500°C hotter than a surrounding 20°C room temperature (Snyder, 2008). Heat exchangers can be any size and can be anywhere from 40–90% efficient; small ones used for gaming computer CPUs can transfer tens of watts even as passive devices, and active heat pumps can transfer more.

Thermoelectric devices work at low temperature differences, but are usually 5% efficient or less. Some can generate hundreds of watts, but most can only generate a few watts maximum. However, theoretically they cool the object as they draw power away, which for a wearable product could provide comfort as well as energy. Their inefficiency often results in both sides of the device feeling similarly warm unless the hot side has a good heat exchanger to radiate the remaining waste heat away (which adds size and weight).

Waste heat is more useful at larger scales: many factories have high temperature waste heat from furnaces or other machinery, and even large residential buildings can have boilers big enough to provide useful amounts of waste heat. On average, factories and buildings might be able to recover 20% of their waste heat (Jouhara et al., 2018).

Available energy: As with power, this depends on the source. Some waste heat sources are relatively constant, so energy is simply power × time, but other sources vary.

19.3 Energy Storage

Energy generation usually (but not always) requires energy storage. Generally, the most efficient and affordable way is with batteries, often lithium ion for consumer electronics and are expanding to buildings and grid storage. However, airplanes use liquid fuels (ideally biofuels), because they store far more energy per unit mass than the best commercial batteries.

There are five main considerations when choosing energy storage: energy density, power density, charge-discharge efficiency, financial cost, and environmental cost.

19.3.1 Energy Density

Energy density is either the amount of energy per unit mass (MJ/kg) or the amount of energy per unit volume (MJ/l) that the battery stores. Vehicles usually optimize energy density by mass because moving more mass requires more energy, so a lighter battery is like a higher- capacity battery. Portable or wearable electronics often optimize energy density by volume, to easily fit in pockets and the like.

Figure 19.7 shows energy density by mass and volume. Note, however, that when designing a product you should consider the whole energy system, not just the storage. For example, hydrogen gas has one of the highest energy densities by mass, but to replace a battery with it, you also need the mass of a tank or other hardware to contain it, and a fuel cell to make electricity from it. Thus, hydrogen's 120 MJ/kg in the figure becomes 1.4 – 2.6 MJ/kg for a hydrogen fuel cell system.

19.3.2 Power Density

Power density, also measured per unit mass (W/kg) or per unit volume (W/l), see Figure 19.8, is the peak power that can be delivered, i.e., the speed at which you can release energy from the battery or fuel. A battery with low power density but high energy density might not be able to power the high-wattage lamp, but could power a low-wattage lamp for a long time. Similarly, a low power density battery might require a long time to charge.

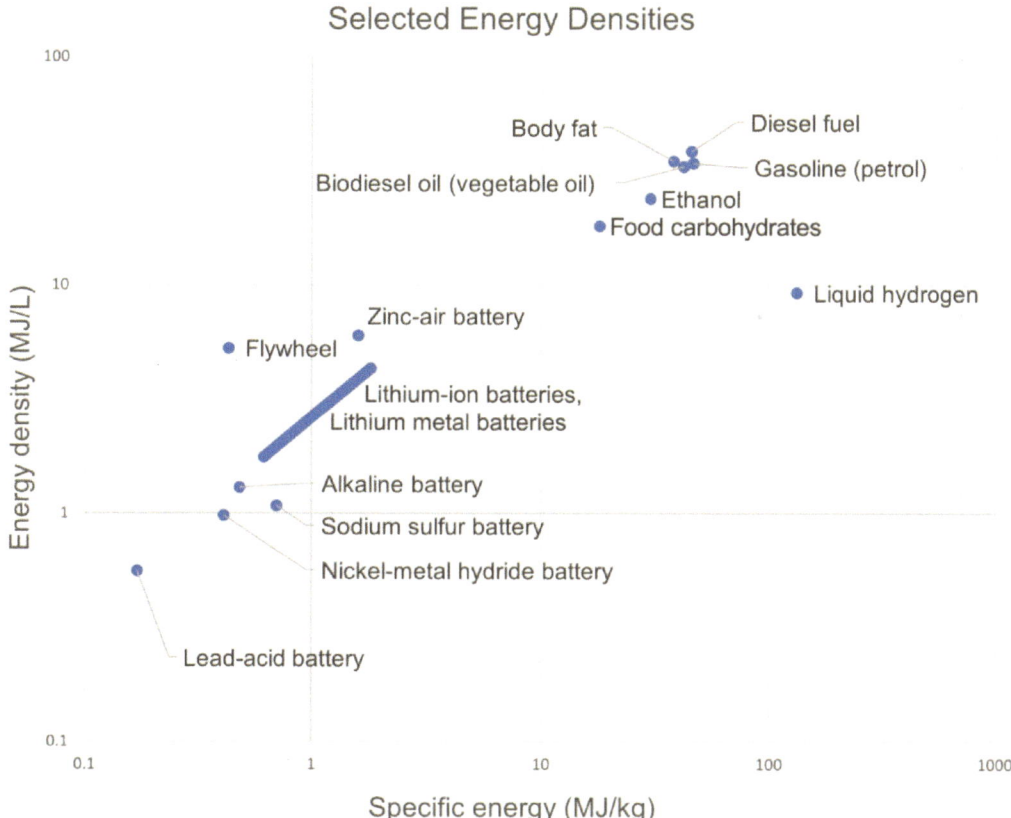

Figure 19.7 Energy density by mass and by volume for various energy storage media. Supercapacitors/ultracapacitors are too low to appear on the chart (<.02 MJ/L and <.05 MJ/kg). Lithium ion batteries appear as a line due to a wide range with rapidly increasing capacity.

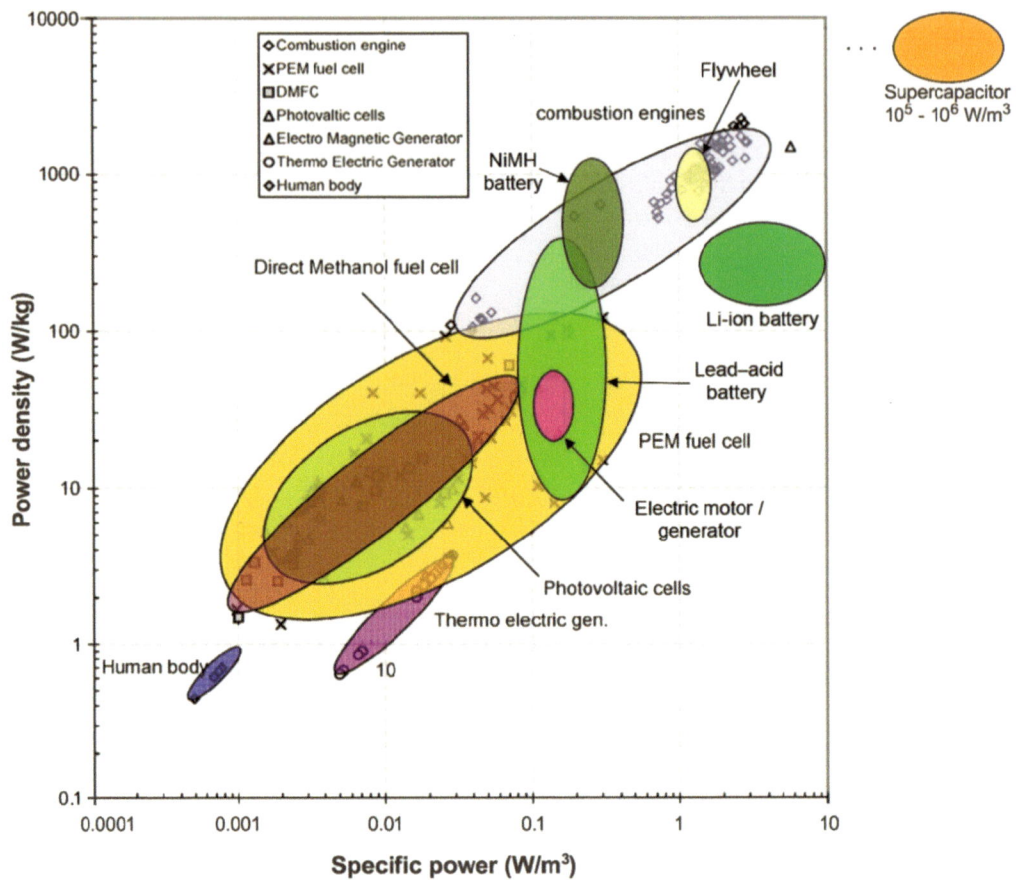

Figure 19.8 Power density by volume and mass
Source: Flipsen (2006) and Holm et al. (2002).

19.3.3 Charge-Discharge Efficiency

Charging efficiency measures how much of the electrical energy you put into a battery (or fuel cell, or flywheel, or other energy storage device) is successfully stored there (as chemical energy, rotational kinetic energy, or other), rather than being lost as heat or vibration or chemical corrosion, etc.

Discharge efficiency measures how much of that stored energy can be usefully recovered

as electricity. For example, lithium ion batteries are often 85–98% efficient (see Figure 19.9), so if they are used to store electricity from a solar panel, almost all the electricity used to charge them will be directly recovered. By contrast, most fuel cells are only ~50% efficient at turning hydrogen into electricity, and the process of extracting hydrogen from water by electrolysis is also usually only ~50% efficient (at best 80% efficient). So, storing electricity from a solar panel in a fuel cell system is usually (50%) • (50%) = 25% efficient, meaning it would

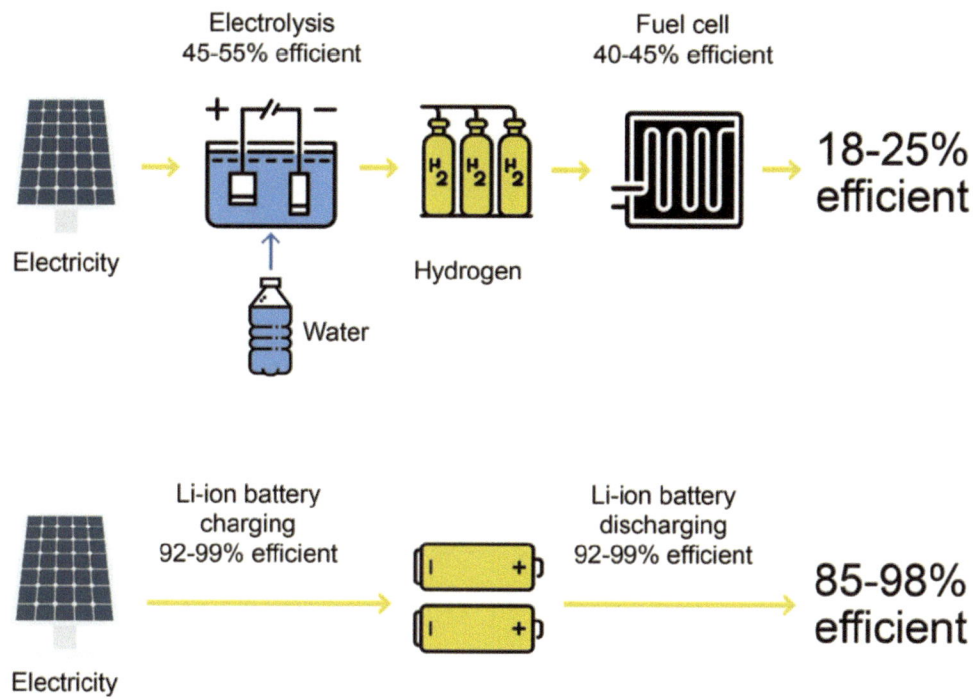

Figure 19.9 Charge-discharge efficiency for Li-ion batteries and hydrogen fuel cells with electrolysis of water

require roughly four times the area of solar panels to run as for lithium ion battery storage.

19.3.4 Financial Cost

Financial cost is an obvious limitation. The energy balance equations (E_{in} versus E_{out}) show how much energy needs to be stored, which lets you calculate the cost of generation and storage. A common misconception in renewable energy systems is that generation and storage efficiency must be maximized, but when the energy is free, such as sunlight, this is not true. Instead, the financial cost per unit energy is more often a limiting factor. For portable devices, size and weight are also limitations. For example, the portable solar energy balance numbers

assume a typical 20% maximum efficiency PV panel. Some solar cells are far more efficient but also more expensive, trading size for cost.

19.3.5 Environmental Cost

Environmental cost is another obvious limitation—sometimes the impact of producing batteries for portable renewable power is greater than the impacts they will avoid from fossil fuel-powered grid electricity. You can calculate the carbon payback time or more comprehensive ReCiPe points payback time with LCA, as shown in Chapter 17.

For batteries storing renewable energy to replace grid power, grid electricity impacts are simply calculated per megajoule of energy used on site. A battery's environmental costs are mainly from its raw materials, manufacturing, and end of life, which means its impacts per unit of service depend on how long the battery lives. For perfect accuracy, the lifetime is not how many years it lives, but how many charge-discharge cycles of how many MJ each, including its charge-discharge efficiency. However, practically speaking, the battery lifetime can be considered in years—either the same number of years as the overall product's life, or whenever the user would likely replace the battery on average. These numbers will not be precise, but even with large uncertainties, you can verify that the renewable energy system in your product will actually be an environmental win. Such calculations let you avoid greenwashing, and help ensure you're spending your time and money and creativity where it matters.

Resources and References

Resources for Further Study

- MacKay, D. J. C. (2009). Sustainable energy – without the hot air. Chapters 1 and 2. Available at: https://www.withouthotair.com/

- Verwaal, M. (2019). The small PV systems design guide: A short introduction in the design of small solar powered products. Handout for TU Delft course Design for Sustainability. TU Delft.

References

Apostolou, G., Reinders, A., & Verwaal, M. (2016). Comparison of the indoor performance of 12 commercial PV products by a simple model. *Energy Science & Engineering*, 4(1), 69–85.

Carminati, M., Turolla, A., Mezzera, L., Di Mauro, M., Tizzoni, M., Pani, G., Zanetto, F., Foschi, J., & Antonelli, M. (2020). A self-powered wireless water quality sensing network enabling smart monitoring of biological and chemical stability in supply systems. *Sensors*, 20(4), 1125.

Clarke, S. (2018). Electricity generation using small wind turbines for home or farm use [Fact sheet] (No. 18–005). Ontario Ministry of Agriculture, Food and Rural Affairs.

DatacenterDynamics. (2016). Yandex data center heats Finnish city. Data Center Dynamics. Available at: https://www.datacenterdynamics.com/en/news/yandex-data-center-heats-finnish-city/ (accessed September 24, 2022).

Deng, Y. et al. (2015). Quantifying a realistic, worldwide wind and solar electricity supply. *Global Environmental Change*, 31, 239–252.

Faludi, J. (2011). Measuring light levels. Autodesk Sustainability Workshop. Archived on VentureWell.org. Available at: https://sustainabilityworkshop.venturewell.org/buildings/measuring-light-levels.html (accessed September 24, 2022).

Flipsen, B. (2006). Power sources compared: The ultimate truth? *Journal of Power Sources*, 162(2), 927–934.

Holm, S. R., Polinder, H., Ferreira, J. A., Van Gelder, P., & Dill, R. (2002). A comparison of energy storage technologies as energy buffer in renewable energy sources with respect to power capability. IEEE Young Researchers Symposium in Electrical Power Engineering, s 1, 6.

IDEMAT (2021). IDEMAT database. Available at: www.ecocostsvalue.com (accessed July 5, 2022).

Jouhara, H., Khordehgah, N., Almahmoud, S., Delpech, B., Chauhan, A., & Tassou, S. A. (2018). Waste heat recovery technologies and applications. *Thermal Science and Engineering Progress*, 6, 268–289.

Miller, M., Faludi, J., Scheer, D., & Bellafiore, A. (2009). The science and technology of advanced structural, material and environmental systems in quick-to-deploy, high-performance green solutions. Office of Naval Research.

NREL (United States National Renewable Energy Laboratory). (2007). Small wind electric systems: A U.S. consumer's guide (DOE/ GO-102007-2465). NREL.

Paradiso, J., & Starner, T. (2004). Human-generated power for mobile electronics. *Low-Power Electronics Design*, 45, 1–35.

Snyder, G. J. (2008). Small thermoelectric generators. *The Electrochemical Society Interface*, 17(3), 54.

Stockholm Data Parks. (2017). Data center cooling with heat recovery: Cost efficient and sustainable data center cooling. Available at: stockholmdataparks.com/2019/11/12/three-new-data

The Engineering Toolbox. (2008). Hydropower. Available at: https://www.engineeringtoolbox.com/hydropower-d_1359.html (accessed September 24, 2022).

Verwaal, M. (2019). The small PV systems design guide: A short introduction in the design of small solar powered products. Handout for TU Delft course Design for Sustainability. TU Delft.

How to Apply #19: Brainstorm Energy Source Improvements
Time Estimate: 1–2.5 Hours

Brainstorm cleaner energy sources for your product, and estimate whether they're actual sustainability improvements.

STEP 1: Brainstorm Clean Energy Sources to Reduce the Impacts of Your Product's Energy Use by a Factor of Ten
Time Estimate: 0.5–1 Hour

Hold a brainstorm session, using the Rules of Brainstorming and whatever collaboration tools you prefer, to generate ideas for switching energy sources to reduce its environmental impacts to 1/10th the impacts it has today. Don't limit yourself to factor-ten ideas, there are no bad ideas in a brainstorm, but be bold about radically re- envisioning the product's energy generation and storage systems. Can it even generate more power than it uses, and feed free clean energy to other products for regenerative design? How could your product harvest free energy from heat, motion, fluid flow, light, or other phenomena happening around it?

If your product doesn't use energy itself but causes energy to be used by something else in its system (like clothing being washed), brainstorm ways you could switch that to clean energy. Ideally make ways your product could cause it, but if that's not possible, you can think up ideas unrelated to your product.

If your product does not use energy during its useful life, and doesn't cause energy to be used by something else in its system, then brainstorm ways in which its manufacturing or materials can switch to clean energy to have 1/10th the impact, or even positive impacts like sequestering carbon.

- Start with the Whole System Map you created for your product, to keep in mind all the components of the system, and how they connect to each other, and the product's role in the larger system.

- Have a good number of ideas (30+) to replace or change parts of the system that have high impacts.

- Have at least one idea to change/replace every major component or step in the product system.

- Have at least six ideas that eliminate a step or component of your system. Eliminating multiple steps/components is even better.

STEP 2: Narrow Down Your Brainstorm to Three or Four Finalist Ideas
Time Estimate: 5–30 Minutes

Using dot voting, common sense, or whatever tools you desire, narrow down to just three or four best ideas. In addition to judging them by energy impacts, use considerations from your Design Brief to rule out options that don't meet business criteria such as cost or usability.

STEP 3: Estimate the Total Energy Impacts for Each Finalist Idea, and Choose the Best Idea
Time Estimate: 10–30 Minutes

Using dot voting or whatever tools you desire, narrow down to just three or four best ideas. In addition to judging them by energy impacts, use considerations from your Design Brief to rule out options that don't meet business criteria such as cost or usability.

If your product uses energy during its life (or causes significant energy use in its system):

- Estimate the total lifetime energy use per functional unit of each of the winning ideas you've chosen, from both brainstorms.

- Using LCA or other tool, calculate the environmental impacts of that energy use per functional unit, for each idea. You do not need to do a full LCA, just estimate the impacts for energy use.

If your product does not use energy or cause it in the larger system:

- Use LCA or other tool to estimate the embodied CO_2 impacts per functional unit of your new ideas. Include the whole product in the estimated LCAs, as you'll likely be replacing several different materials.

STEP 4: Choose One Final Idea to Move Forward with, and Illustrate It
Time Estimate: 5–30 Minutes

- Choose one winning idea, based on the estimated impacts and your other design brief priorities. (You could also decide to pursue multiple winners, or combine ideas.)

- Illustrate the final idea (rough sketch or fancy rendering) to clearly convey its energy improvement, and why it's a compelling design.

STEP 5: Document Your Decision and Brainstorm
Time Estimate: 10–30 Minutes

- Create a PDF briefly describing and illustrating the winning design(s), and the reasons for your choice.

- Show your brainstorm, making it clear that you had at least one idea for every part of the system, and many ideas that skipped steps in the system.

- Show the illustration of the winning design(s).

- Succinctly describe the winning design, either as annotations to the illustration or as a stand-alone sentence or two. Describe why it is the best of all your new ideas.

- Show your estimates of total lifetime energy use per functional unit (or, for products that don't use energy, embodied CO_2 impacts per functional unit), both for the final winning idea and all the other ideas you ran numbers for. The ideas that didn't get chosen as the final winner don't need descriptions, but at least make their titles suggest what they are.

- Show your math for all the above estimates.

- State the estimated ecological impact improvement from original design to the winning design.

Checklist for Self-Assessment

To score your success on this exercise, see if you…

- ☐ *Had 30+ ideas for the energy source brainstorm.*

- ☐ *Had at least one idea for every major component or step in the system.*

- ☐ *Had at least six ideas that eliminated a step or component of the system.*

- ☐ *Chose 3–4 winning ideas.*

- ☐ *Calculated improved impact estimates for all finalist ideas.*

- ☐ *Illustrated final winning idea(s).*

- ☐ *Described final winning idea(s).*

CHAPTER 20
Behavior Change

Jeremy Faludi, Ruth Mugge, and Conny Bakker

Goals

- Explain and exemplify user ability, motivation, and prompts in Fogg's model of behavior change

- Explain and exemplify how attitude, subjective norms, and perceived behavioral control shape an individual's intentions and thus their behavior in Ajzen's theory of planned behavior

- List other approaches to change behavior from the Design With Intent tool

- Apply persuasive design tools to drive user/consumer behavior for sustainability

DOI: 10.4324/9781003504672-22

Why It Matters

All design influences user behavior. Many sustainability problems are problems of choices and lifestyles: overconsumption, the "throwaway society," and immediate gratification over long-term wellness. Changing people's behavior can guide them to reducing consumption, reusing and repairing products, traveling less far and less fast, etc. For some products, persuading users to behave more sustainably can have more positive impact than reengineering the product physically. Persuasive design interventions can be used instead of or in addition to better engineering.

Summary

- Fogg's model of behavior change says: motivation is whether people want to change, ability is whether people can change, and prompts are stimuli that provoke actual change.

- Ajzen's theory of planned behavior says: attitude is what a person thinks and feels, subjective norms are what that person thinks others believe, and perceived behavioral control is how easy or hard they think it is to change their behavior.

- Lockton's "Design with Intent" tool includes these and many more theories of change with 101 persuasion tactics grouped into eight theoretical "lenses."

- To encourage better behavior with your design, focus on the user experience, make it easy and compelling for the user to act better.

20.1 Designing Sustainable Lifestyles

Design for sustainable behavior includes anything that influences users to take concrete actions that are more responsible for the planet and society. Behavior design does not just happen in marketing materials, it suggests changing user behavior through interactions with the product or service itself (e.g., via features or interface).

Environmental consequences are usually abstract and psychologically distant to people—they will only happen in the future, or they affect the collective, not the individual. People are more likely to act on consequences that will happen immediately and that will bring them personal benefits, according to "construal level theory" by Trope and Liberman (2010). As a result, people who are environmentally concerned may still book an intercontinental flight to have exciting personal experiences.

Designing behavior also enables you to reach beyond your industry for bigger impacts. Arguably the most important design innovation in public transit in the last 50 years was not a redesign of trains or bus stops, but the Google Maps app on mobile phones adding transit trip planning (Figure 20.1). The app enables people to get transit information when and where they need it, even in cities they have never been to before, even in languages they don't speak. Digital products like this are often the most powerful persuaders, because they are so interactive.

Behavior design does not just happen in marketing materials, it suggests changing user behavior through interactions with the product or service itself (e.g., via features or interface). Digital products like the maps

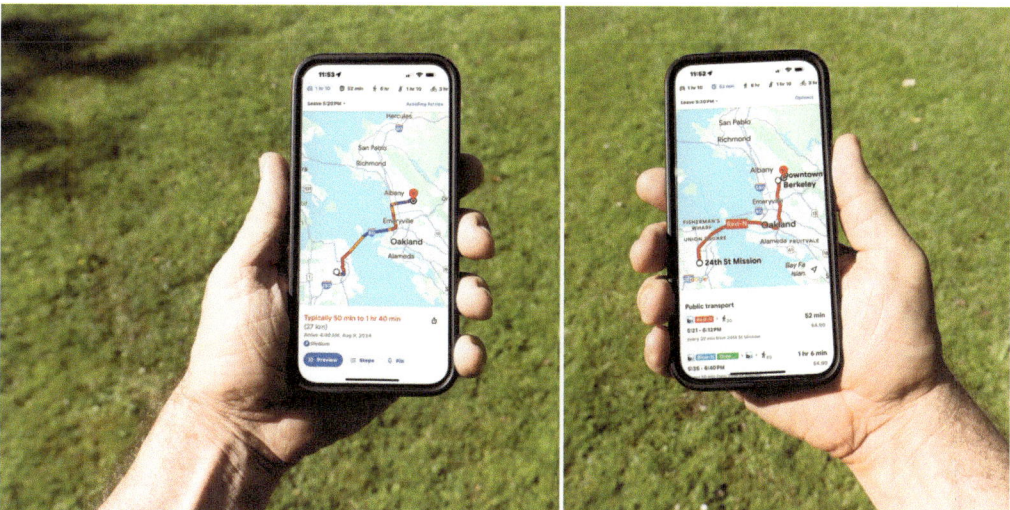

Figure 20.1 Google Maps driving directions and transit directions on a phone

app are often the most powerful persuaders, because they are so interactive.

There are many theories of how to encourage behavior change. This chapter mainly discusses models by B. J. Fogg (2009) and Icek Ajzen (1985) because they are the most widely used, have strong grounding in psychology research, and have been shown to predict behavioral outcomes well. The "Design with Intent" system by Dan Lockton includes these plus broader theories of change, with a very thorough list of actionable persuasion tactics. First, Fogg's model of behavior change generally results from the co-occurrence of three factors: motivation (if people want to do it), ability (if people can do it), and a prompt (a stimulus that provokes people to do it). He lists many tactics to build both ability and motivation. Second, Ajzen's "theory of planned behavior" assumes that people make intentional choices about their actions. These intentions are based on their individual attitude, subjective norms, and perceived behavioral control. In the end, though, people

are complex and unpredictable, and small details of implementation can have surprising effects. Be prepared to try and test several strategies before you find the best one.

20.2 Fogg's Model of Behavior Change

According to Fogg's (2009) model of behavior change (or "persuasive design"), motivation, ability, and a prompt are all needed for behavior change to take place. Figure 20.2's action line shows that if people want to change their behavior and are able to do so, a behavior change is relatively easy. However, in many situations, people are lacking either motivation or ability or both, and then a prompt is likely to fail. Motivation and ability can also compensate for each other: if a person feels highly motivated to act, they will be more likely to put in great effort.

Possible motivators are pleasure/rewards, money or time, hope, social deviance, or

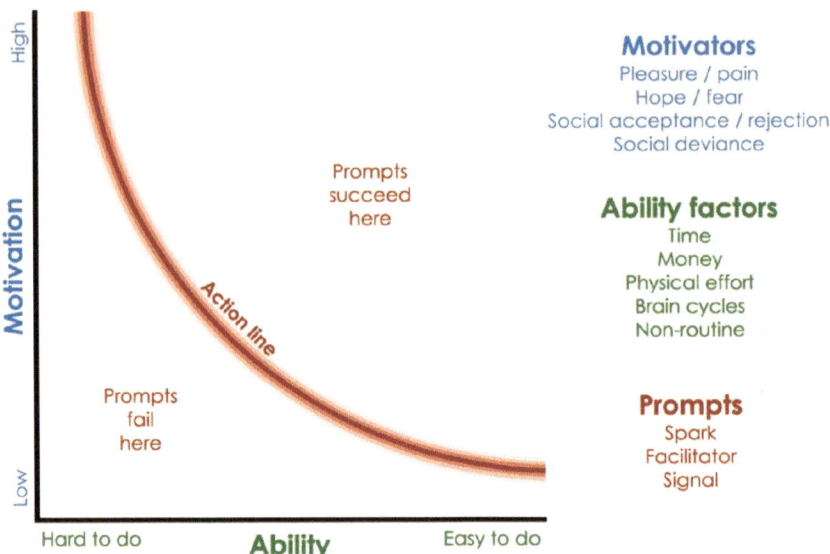

Figure 20.2 Fogg's model of behavior change

social acceptance. For example, Figure 20.1 shows that taking transit rather than driving during rush hour will save you time that day. For another example, if you're trying to persuade people to meet by videoconference rather than drive to physical meetings, you can make video calls more pleasurable by improving audio and video quality, or by providing features like transcripts or easy screen-sharing, thus increasing people's motivation to video call. Similarly, the desire to be socially accepted strongly influences people's behaviors. For example, if the social norm in an organization is to videoconference, individual employees will be more motivated to act accordingly. In addition, there may also be negative motivators if people strive to prevent pain, discomfort, fear, or social rejection. For example, people might be motivated to videoconference to avoid rush hour traffic, or avoid disapproving comments from coworkers.

Ability consists of five parts: reducing costs of money, time, physical effort, or brain

cycles required (including going outside one's routine). It's more demanding to change one's behavior if it takes a lot of time, money, or physical effort, resulting in a lower perceived ability. For example, someone with an old computer that doesn't have videoconferencing software has much less ability than someone with the latest software and fast computer. Furthermore, the amount of brain cycles or cognitive effort can reduce people's ability. If a task takes ten steps, even if all ten are easy by themselves, the accumulation feels difficult. Also, someone who's already multitasking between several things will usually find an added thing overwhelming. Finally, people's ability is relatively low for non-routine behaviors, people find it easier to perform behaviors that they do regularly, because habits reduce brain cycles. The assessment of ability depends on the individual.

The last factor that is needed for behavior change is a prompt, also known as a trigger.

A prompt is something that pushes people to perform a behavior. For example, a calendar event popping up inviting you to a video call. Three types of prompts exist: sparks, facilitators, and signals. "Sparks" increase the person's motivation, for example by evoking a feeling of pleasure. "Facilitators" enable a person to behave in the desired way and thereby increase their ability. "Signals" work if a person is already both motivated and has the ability, just needing a reminder or notification to do it now. A popup invitation to a video call might be a signal, but if it also lets you push a single button to join the call, it's also a facilitator.

Design interventions can increase both people's ability and motivation, moving them to different parts of Figure 20.2's graph. Even when both are high, the behavior still may not happen automatically if it's outside the user's current habits. Hence the need for prompts.

Because design for sustainable behavior is about changing user behavior through interaction with the product, or software app, or product-service system itself (not just marketing around the product), it is all about user experience design. By changing either the physical product design, digital interaction design, service design, or the communication related to it (e.g., packaging or user manual), it is possible to boost either motivation or people's ability.

Design for sustainable behavior can happen on macro or micro levels. In Google Maps, there's very little persuasion toward transit, but lots of people use the app. This is sometimes called "microsuasion." You can also make a product that exists entirely to change behavior; this is called "macrosuasion." Such products are often only used by a small specific target audience. For example, a car-sharing app allows users to get rid of their own car by letting them rent a car on demand, but that app would be used much less than Google Maps, which comes with the phone and handles all mapping. Both microsuasion and macrosuasion can nudge people's behavior in both the short term and the long term.

There is power in behavior change. That means there can also be unintended consequences of persuasion that you want to avoid. For example, if the car-sharing app were designed to reduce overall driving but ended up persuading users to drive more by swapping their previous bike trips with car trips, it would increase their environmental impact. This is called a "rebound effect." As a designer, it is important to think through whether your design interventions may bring about possible rebound effects.

20.3 Ajzen's Theory of Planned Behavior

The theory of planned behavior (TPB) says that a person's behavior comes from their intentions, and those intentions come from a combination of three components: attitude, subjective norms, and perceived behavioral control, shown in Figure 20.3 (Ajzen, 1985). Attitude is an individual's evaluative response to an object, such as a product, brand, organization, or behavior. It includes a cognitive component (beliefs, knowledge) and an affective component (feelings, emotions) and has strength and a sense of direction. For example, whereas one person can have a strong, positive attitude toward riding a bicycle, another can have a weak, but negative attitude toward the same behavior. According to the theory of planned

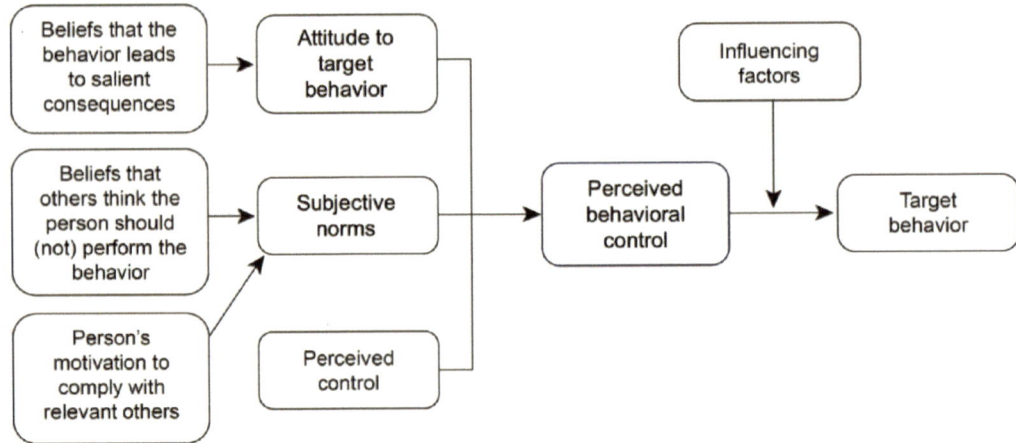

Figure 20.3 The theory of planned behavior
Source: Redrawn from Ajzen (1985).

behavior, people are more likely to change their behavior if they are positive about this behavior.

The subjective norms are what your user believes other people's attitudes are (e.g., friends, spouses, parents, coworkers, etc.). If people who are important to the user consider a certain behavior to be important, the user will be more inclined to change their behavior, too.

Perceived behavioral control is how easy or difficult your user thinks the behavior is for them to perform in their specific context. Obviously, the easier they think it is to do something they want to do, the more likely they are to do it, as with Fogg's model of ability and motivation. The difference is that here it is perceived ability.

The three components of intention also influence each other. If your user's friends are enthusiastic about the behavior, your user's attitude will likely become similar; if

their parents are enthusiastic, their attitude might become negative. Behaviors that seem difficult or inconvenient drive more negative attitudes, but an extremely positive attitude might lower perceived difficulty. Finally, of course, behavioral control has some direct relationship to behavior: even if your user has good intentions, they would not do the behavior if it is too hard. This component clearly overlaps with the ability factor of the Fogg model.

As a designer, you can focus on all components in the TPB to create new design interventions that will encourage people to change their behavior. Please be aware that even when people have the intention to perform a certain behavior, this will not necessarily result in executing this behavior.

There may be influencing factors that prevent the person from doing so, as shown in Figure 20.3. An example would be if someone wants to take the bicycle to travel but sees that it has a flat tire and decides to take the car instead. As a designer, it is important to think through potential influencing factors and explore possibilities to prevent these from inferring with the behavior change.

20.4 Design Tactics to Change Behavior

Both Fogg and Ajzen discuss the many specific tactics designers can use to influence user behavior. Their recommendations overlap greatly, so the list below includes both, but organizes them by Fogg's categories of ability and motivation.

20.4.1 Ability to Change

As mentioned above, people often encounter barriers to sustainable lifestyles. How can you give users more ability to do their desired behaviors?

20.4.2 Convenience

Making things easier or more convenient increases ability by saving time, physical effort, and/or brain cycles. For example, if you are trying to help your users reduce their environmental impacts from transportation, you can make it easier to use transit and bicycles instead of cars. Some examples are separating bike lanes from car traffic, and providing ways for bikes to get on transit (see Figure 20.4).

You can also make unsustainable behaviors less convenient. Some interventions in the domain of transit are to have less parking for cars, or by making it more expensive. Beware that people will like your company more if you

Figure 20.4 Easy transport of bicycles in public transport

only do positive things, not negative things; you might accompany any inconvenience with more convenience somewhere else. For example, less parking near desired destinations could be paired with more parking by transit stations.

20.4.3 Defaults

Setting defaults increases ability by solving the non-routine problem (a default becomes a new routine), and saving brain cycles, perhaps also time and money or physical effort. Most people don't bother to change the factory settings on a product, so why not make the most sustainable settings the default? An example is a washing machine that automatically activates the eco-mode as the default program to wash (see Figure 20.5).

20.4.4 Leading the Process

Leading the user through the behavior when the process is complicated improves ability by solving the barriers of brain cycles and non-routine. In this respect, it is worthwhile to

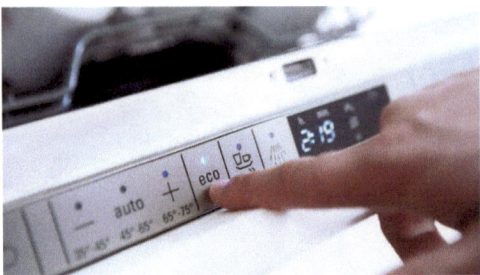

Figure 20.5 Eco-mode as the default mode

explore design interventions that use the rich possibilities of mobile devices, because they are almost always with the user. For example, a mapping app that shows public transit options and gives you turn-by-turn directions in real time is a great way to prompt more transit use.

20.4.5 Calculate/Simulate/ Measure

Helping the user understand both the consequences and the requirements for their action by calculating, simulating, or measuring increases ability by reducing their brain cycles; it may also save them time and money they would spend to calculate themselves, and can help overcome non-routine by showing benefits. For example, Google Maps not only tells users where to go on a map (a simulation of reality), it also calculates the time and cost, giving users the ability to plan and compare public transit options to driving or other modes.

20.4.6 Real-Time Feedback

Giving users feedback on the environmental or social impacts of product use right at the moment of action is much more

powerful than feedback at other times, because it eliminates the need to remember (brain cycles). It may also help overcome non-routine by showing impacts the user didn't know previously. For example, a car showing gas mileage in real time allows users to see the impact of driving faster than necessary, or coasting to a stop, or other behaviors in the moment when they can still do something about it. Only seeing their gas consumption the next time they fill up the gas tank a week later is both too late to act and eliminates the specificity of knowing what moments were good and bad.

Real-time feedback can also be more individually targeted, which makes it more effective. The car showing real-time gas mileage above shows it to the person driving now. Seeing consumption when filling the tank could include other drivers' behavior as well, making it less clear.

20.4.7 Motivation to Change

Even if your users have the ability to change their behavior, they need to want to. How can their interactions with your product or service build their motivation?

20.4.8 Rewards/Punishments

Rewards and punishments provide simple Pavlovian pleasure/pain. Rewarding the behavior you want to encourage will make it more likely for the user to do it again. For example, a fitness tracker that congratulates your user for biking today. Consider how to design these rewards, as they need to be sufficiently strong and relevant to motivate people to take action. An example of punishment is a car that beeps at your user

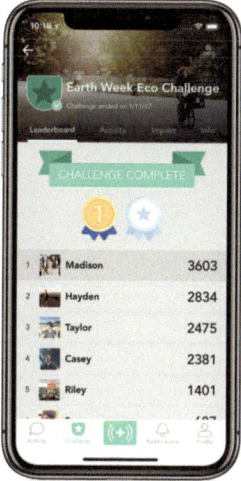

Figure 20.6 The JouleBug app's scoreboard and points
Source: John Williard, johnwilliard.com.

for leaving the lights on. It's usually best to minimize punishment, though—your users will like you more for positive reinforcement.

Sophisticated versions of rewards and punishments can "gamify" your user experience, prompting users to "level up" into higher and higher levels of green behavior. This can be great for turning short-term actions into long-term goals. For example, the "JouleBug" phone app encourages users to recycle and save household energy by giving points for different actions, challenges to accomplish, and scoreboards ranking players against each other (see Figure 20.6).

20.4.9 Attractive/Likeable

Attractive or likeable products stimulate pleasure or social acceptance feelings. Even though we all know products are inanimate objects, we often treat them in semi-social ways. Your user's feelings toward the product

matter—can you design in an emotional connection? This motivates people to use it and keep it in service longer, like a beloved old car or favorite mug.

20.4.10 Modeling Behavior

Showing your user someone else behaving a certain way can inspire your user to do the same thing, by stimulating feelings of social acceptance and hope; this is "modeling" behavior. It's especially effective when the other person is someone that your user aspires to be, or likes, or is attracted to. For example, on Instructables.com, many people have shown how they have modified their bikes to carry cargo so they do not need a car (see Figure 20.7).

20.4.11 Pride

Conversely, people like looking good to other people—pride stimulates feelings of social

Figure 20.7 Modeling bicycle usage can inspire your users to use bicycles rather than cars
Source: based on the Autodesk Sustainability Workshop.

acceptance and hope and pleasure. When sexy electric sports cars appeared, even people who didn't care about sustainability started switching to electric.

20.4.12 Social Interaction

Humans are social creatures, so one of the strongest motivators is social acceptance, which happens in social interactions. Letting your users connect with others through the desired behavior builds confidence, comfort, and competence. It might provide peer pressure. It also associates the feelings of the relationship with the behavior. These relationships can be cooperation, like your fitness tracker helping you find other cyclists to ride with, or it can be competition, like winning a prize for riding the most. Social interactions can also include rejection and fear and pain, but it's usually best to drive positive experiences for your users.

20.4.13 Prompting Change

Remember, ability and motivation are often still not enough, people may need a prompt at the right time and place to act. Most of the time, a simple notification such as a blinking light or sound or message can prompt a behavior, as described above. Sometimes, however, it's not enough. You can increase the insistence of the signal

Figure 20.8 Nagging to motivate users

by nagging (reminding your user frequently until they do it). For example, a car's "Check engine" light keeps reminding users until they deal with it (see Figure 20.8). Unlike pleasure/pain motivators, it's a gentle but persistent request. However, if it's over-used, it becomes annoying.

20.5 Combining Multiple Theories into One Tool

The Design with Intent (DwI) toolkit by Dan Lockton, available at http://designwithintent. co.uk, is intended to "describe systems (products, services, interfaces, environments) that have been strategically designed with the intent to influence how people use them" (Lockton et al., 2010). It collects many different theories of change, not just one

like Fogg or Ajzen. Thus, it also includes a broader range of persuasion tactics: 101 cards or "patterns" organized into eight theoretical "lenses" (see Figure 20.9). The eight lenses are a way of grouping the 101 persuasion tactics by disciplinary worldviews or fields of research. The descriptions given here are based on Lockton's introduction to the DwI toolkit.

The **architectural lens** takes the physical environment as starting point and uses insights from environmental design (in architecture, urban planning, and traffic management) to influence user behavior. For example, the labyrinth-like interiors in casinos and department stores (like IKEA) force you to walk past endless rows of slot machines or products before you reach the exit or cash register. This increases the chance that you will engage by gambling or buying something.

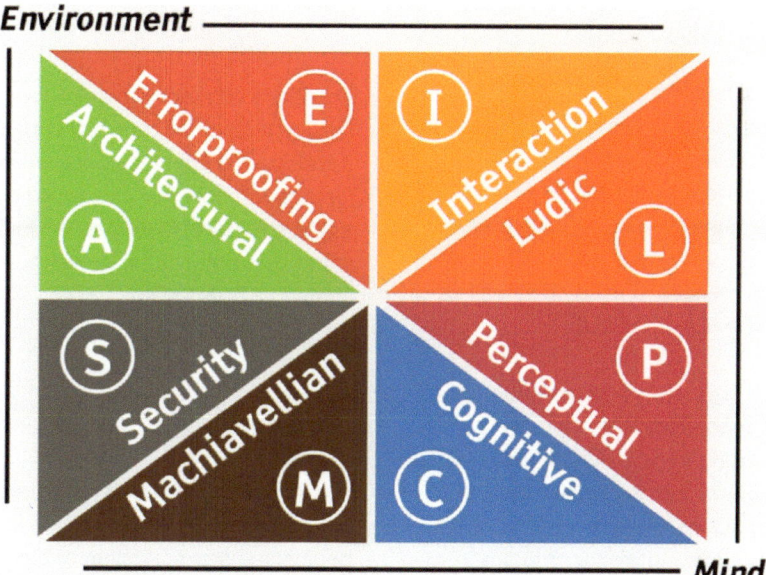

Figure 20.9 Lenses of the toolkit
Source: Lockton et al. (2010).

The **errorproofing** lens tries to help users avoid "errors" (deviations from the target behavior), either by making it easier for users to work without making errors, or by making errors impossible in the first place. It's often found in ergonomics and health and safety-related design. Errorproofing doesn't care whether the user's attitude changes, as long as the target behavior is met.

The **interaction lens** looks at users' interactions with the system and how it influences their behavior. Much used strategies are feedback, progress bars, and previews (used in human-computer interaction), but the lens also includes patterns from persuasive technology, building on B.J. Fogg's work.

The **ludic lens** is derived from games and play, ranging from basic social psychology mechanisms such as goal-setting via challenges and targets, to common game elements such as scores, levels, and collections.

The **perceptual lens** combines ideas from product semantics, semiotics, ecological psychology and Gestalt psychology, addressing how users perceive patterns and meanings as they interact with the systems around them. These are mostly visual, but they need not be: sounds, smells, textures, and so on can all be used, individually or in combination.

The **cognitive lens** draws on research in behavioral economics and cognitive psychology looking at how people make decisions, and how this is affected by heuristics and biases. For instance: "give your system a personality," "provoke empathy," "use reciprocation." These tactics draw particularly heavily on the work of Robert Cialdini, Dan Ariely, Richard Thaler, and Cass Sunstein.

The **Machiavellian lens** comprises design patterns which all embody an "end justifies the means" approach of the kind associated with Niccolò Machiavelli. These will often be considered unethical, but nevertheless are commonly used to control and influence consumers through pricing structures, planned obsolescence, lock-ins, and so on, and are central to work by authors such as Vance Packard and Douglas Rushkoff.

The **security lens** represents a "security" worldview, i.e., that undesired user behavior is something to deter and/or prevent though countermeasures designed into products, systems and environments, both physically and online, with examples such as surveillance (watching from above in a power hierarchy), "sousveillance" (watching from below), and "peerveillance" (watching peers).

20.6 How to Use the Toolkit

There are different ways you can use the Design with Intent toolkit. It was originally developed to help inspire brainstorming and idea generation, but it can also be used to analyze existing examples of design that influences behavior. See examples of two cards in Figure 20.10.

20.6.1 Can You...? What If...? How Could You...?

Each pattern is phrased as a question—a provocation to invite discussion about the behavior change question or brief you're considering. You could go through all the cards and quickly decide the patterns' relevance to your brief based on whether the answer is "Yes," "No," "Good," "Bad," "Not sure," etc.

Figure 20.10 Two patterns from the Architectural Lens

20.6.2 Going Through Lens by Lens

Lay out all the cards, grouped by lens, and go through each lens seeing whether the questions inspire any concepts for addressing your problem. In groups (e.g., 4 or 8 people) it often works well for one or two people to take a lens and become "mini-experts" for a few minutes before reporting back to everyone else. A group discussion can then proceed to refine the ideas.

20.6.3 Cards or Worksheets

The worksheets, available at http://designwithintent.co.uk/downloads/, are good for group work or where you want an overview of each lens, while the cards enable more detailed deliberations over each pattern—or looking at sets of a few patterns rather than all of them.

20.7 Testing and Ethics

Because people are complicated and unpredictable, behavioral design interventions should not be based on theory alone. Several alternatives should be tested, in as close to real-world target user contexts as possible. Testing can be done using any standard tools for testing user interface design. The most common is "A/B testing," where users get randomly assigned interface A or interface B, and their actions are measured to see which interface produces more of the desired behavior.

Ethics must also be considered when influencing user behavior. Users can be manipulated to act against their best interests. There is a book and website called "Evil by Design: Interaction Design to Lead Us Into Temptation" that's both a good tutorial on behavior design and a warning to not misuse it. When you design persuasion into your products or services, take time to think about unintended consequences that might harm your users, others, or the environment. For example, instead of guiding people to better decisions, you can rob them of choices that would benefit them. Or you could cause "rebound effects" where your users accidentally increase their environmental impacts—you might make shared cars to reduce overall car usage, only to find you accidentally motivated people to switch to shared cars from public transport.

Resources and References

Resources for Further Study

- Fogg, B. J. (2002). *Persuasive technology: using computers to change what we think and do* (1st edition). Morgan Kaufmann.

- Ajzen, I. (1985). From intentions to actions: A theory of planned behavior. In J. Kuhl & J. Beckmann (Eds.), *Action-control: From cognition to behavior* (pp. 11–39). Springer.

- Lockton, D., Harrison, D., & Stanton, N. (2010). The Design Intent Tool. ISBN 978-0-9565421-0-6. downloadable PDF at https://designwithintent.co.uk.

References

Ackermann, L., Mugge, R., & Schoormans, J. (2018). Consumers' perspective on product care: An exploratory study of motivators, ability factors, and triggers. *Journal of Cleaner Production*, 183, 380–391.

Ajzen, I. (1985). From intentions to actions: A theory of planned behavior. In J. Kuhl & J. Beckmann (Eds.), *Action-control: From cognition to behavior* (pp. 11–39). Springer.

Fogg, B. J. (2009, April). A behavior model for persuasive design. In *Proceedings of the 4th International Conference on Persuasive Technology* (pp. 1–7).

Lockton, D., Harrison, D., & Stanton, N. (2010). The Design Intent Tool. ISBN 978-0-9565421-0-6. downloadable PDF at https://designwithintent.co.uk.

Trope, Y. & Liberman, N. (2010). Construal-level theory of psychological distance. *Psychological Review*, 117(2), 440–463.

How to Apply #20: Behavior Change Brainstorm
Time Estimate: 2–3 Hours

Goal: Brainstorm persuasive design strategies to solve one of your sustainability priorities.

STEP 1: Brainstorm Designs to Change User Behavior
Time Estimate: 20–40 Minutes

Hold a brainstorm session to generate ideas for reducing your product's environmental impacts by changing the user's behavior. Choose one of your Design Brief priorities to focus on.

Brainstorm off of specific tactics, or general theories, of Fogg/Ajzen/Lockton. Your brainstorm should:

- Have 30+ ideas. Brainstorm off of multiple persuasion tactics or theories, with at least ten ideas for each one. For example, ten ideas that increase ability, ten ideas increasing motivation, and ten ideas for prompts (Fogg's behavior model). Or ten ideas for Lockton's "angles" pattern, ten for the "converging & diverging" pattern, etc.

- Bonus: Have some "microsuasion" ideas, where behavior change is prompted via a small feature in a larger product (e.g., in Google Maps, there are buttons for transit or biking directions, but the app is used for much more), and some "macrosuasion" ideas, where the sole purpose of the product is to persuade you (e.g., Quitnet.com, a website dedicated to helping people stop smoking).

- Focus on the design of your product system and the user experience. It can include revised product design, or an auxiliary product or service, such as user interactions with packaging, customer support, retail environment, additional devices, etc. Don't be afraid to add new functionality to your product!

STEP 2: Narrow Down Your Brainstorm Options to Three or Four Best Ideas
Time Estimate: 15–40 Minutes

Think of the relevant evaluation criteria (using different stakeholder perspectives) based on your Design Brief to evaluate your ideas. Use these criteria to rate your ideas and narrow down your brainstorm into the selection of three or four best ideas.

STEP 3: Estimate Idea Impacts, and Choose One Final Idea
Time Estimate: 20 Minutes–2 Hours

Estimate the improvement in impacts of your best ideas using LCA, an eco-label scorecard, or other quantitative analysis. Use the results to decide on a final idea to continue with. (You could also decide to pursue multiple winners, or combine ideas.)

Illustrate the final design intervention with a sketch, storyboard, customer journey or rendering to clearly convey the idea, its benefits and the user experience.

STEP 4: Document Your Design Process
Time Estimate: 15–40 Minutes

Create a PDF showing your brainstorm, the best design intervention(s), and the reasons for your choice. Describe why it is the best of all your new ideas.

Checklist for Self-Assessment

To score your success on this exercise, see if you…

☐ *Brainstormed 30+ ideas, including ideas about ability, motivation, and prompts (Fogg's behavior model) or multiple Theory of Planned Behavior factors, or multiple Design with Intent patterns.*

☐ *Had 5+ microsuasion ideas and 5+ macrosuasion ideas.*

☐ *Showed the analysis to choose the 3–4 best ideas.*

☐ *Showed sketches of the 3–4 best ideas.*

☐ *Showed the LCA or other analysis you did to choose the final design.*

☐ *Illustrated and explained the final design intervention(s).*

☐ *State how much you expect the winning design to improve ecological impact.*

Pure Product
Customer owns

Product Service System
Ownership varies, value comes from combining produc

Product Oriented

Use Oriented

Customer Ownership
Company manufactures, customer buys

Product Related Service
Sell product with related service (e.g. maintenance contract)

Product Lease
Exclusive use, but without owning product*

Th

Legend

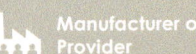
Manufacturer or Provider

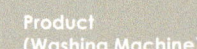
Product (Washing Machine)

Value Transaction

Product User

Service Provider

Advice & Consulting
Sell product & advise on use (e.g. training)

Product Renting / Sharing
Non-exclusive use; Renting: provider owns. Sharing: a customer owns.

P
Use

Product Related Software
Sell software running on product

  10101

Product Pooling
Simultaneous use

Servi
re

Software can be its own product, service, or mix.

Pure Service
Company owns

ted

g
oduct,
ce

Service Providing
Service performed with no
product, e.g. hand washing

Service

e Unit
service

ervice

Service System Likely Environmental Benefits:

- Reduce manufacturing material & energy demand
- Reduce use-phase material & energy demand
- Incentivize extended producer responsibility
- Incentivize durable products
- Incentivize product upgrade
- Professionalizing maintenance extends product life
- Efficiency/economy of scale
- Ease collection at end of life

esult
ers result,
ct type

rvice

*Note: Leasing sometimes shortens product life due to user carelessness.

Source: Jeremy Faludi.

CHAPTER 21

Business Models

Product Service Systems

*Jeremy Faludi,
Conny Bakker,
and Ruud Balkenende*

Goals

- Apply Product Service Systems (PSSs) to make circular design strategies economically viable

- Illustrate examples of a pure product, a product service system, and a pure service

- Explain the differences between a pure product, a product service system, and a pure service

- Analyze existing products/services, and ideate new ones, using the Product Service System Landscape

DOI: 10.4324/9781003504672-23

Why It Matters

Business models are great enablers of the circular economy—without a supporting business model, many circular product designs fail in the long run. The business model of selling products to a mass audience often drives premature obsolescence, because long-lived products reduce new product sales. Designing for durability, repair, etc., often increases the initial cost, while purchasers usually prioritize low initial cost over lifetime cost, making long-lived products less popular. Thus, unless governments mandate design for long life, you need to find a way for your company to make money from longer product lifetimes and reduced manufacturing. This might be done with branding, but a deeper system intervention is changing from a business model of selling products to a product service system.

Business models are important for designers and engineers to understand, even if you never run a business, because your product design will not be viable if it clashes with your company's business model. "Aligning the incentives" of economics and sustainability means your company makes more money when your product lasts longer. Your goal doesn't need to be maximizing profit, you could aim for sufficiency, but you need to show economic viability, and a PSS can do that.

Summary

- Product service systems (PSSs) are revenue models where customers pay the company for the service the product provides, not (only) for the physical product itself.
- Designing the revenue model along with the product can align economic incentives with environmental impacts, so the company has an incentive to design products for longer lives.
- Both the company and the customer can benefit financially from longer-lived products billed as services.
- Not all product service systems are sustainable environmentally or socially, but if properly designed, they can greatly enable the circular economy.

21.1 Product Service System Concept

A product service system (PSS) combines some kind of service with a physical product. Not all PSSs are sustainable, but the goal of a sustainable PSS is to align company revenue streams with resource efficiency by providing services that supplement or replace physical product manufacturing. In other words, decoupling company revenue from physical stuff, so that the stuff becomes a cost rather than a revenue stream. The company providing the service does not have to be the same company manufacturing the physical product. For example, most car rental companies are not car manufacturers, and they are the ones capturing the increased revenue, not the manufacturers. To create a sustainable PSS, be careful to align the incentives of profit and circularity.

There are countless different PSS business models—you can be as creative about them as you are in your physical product design. According to Sakao and Lindahl (2009), a successful PSS must address three things: (1) the user (or customer), (2) the offering, and (3) the service provider. This is because PSSs enable the provider and the product manufacturer to be different companies. The "offering" is whatever combination of physical product and service. Software can be a product or can enable a service—it depends on the revenue model. Tukker (2004) was the first to create a taxonomy of different possible variations of user, manufacturer, product, service provider, and offering; this chapter will detail it later for both brainstorming new PSSs and analyzing existing ones.

One of the earliest examples of a PSS is jet engines for aircraft: in the 1990s, Rolls-Royce went from selling engines to selling "power by the hour" (Figure 21.1). Airline accountants were delighted because it eliminated the large cost spikes of buying a new engine and decommissioning an old engine, turning it into a constant easy-to-budget monthly cost. Rolls-Royce was incentivized to make more durable, maintainable, repairable, and upgradable engines, because materials and manufacturing cost them money, while their revenue stream was constant. And not only did Rolls-Royce make more profit from their longer-lasting engines, but the airlines also saved money, because Rolls-Royce kept some of the savings and passed some of the savings on to its customers (Neely, 2008). Even if customers don't save money as they did here, reducing financial risk is often valuable to them.

Figure 21.1 Rolls-Royce jet engines sold as "power by the hour"

Source: Darren Koch, Wikimedia Commons.

A PSS is often easier to start with business customers like this than private customers, because professional buyers are often more methodical about planning for future costs. PSSs are also more attractive to customers when the product is expensive, as customers would already want to spread out payment. Also, when you're designing a PSS, you can afford to spend much more time and effort designing for one large-volume customer than thousands of small customers. Note that to realize sustainability improvements like design for repair and remanufacturing, if the manufacturer and service provider are different companies, they need to collaborate closely for the right knowledge to inform the design.

Product service systems can be especially good for startups. Some products are easy for competitors to imitate, and larger companies can manufacture with greater economies of scale. A PSS can be harder to copy, because the service may establish a closer and longer-lasting relationship with the customer, even beyond any technical or logistical advantages in your service operations. Because these closer and longer-term relationships can provide better customer feedback, PSSs can enable the company to respond more rapidly and easily to a changing market or changing user needs.

21.2 Kinds of Service Systems

There are many different kinds of PSS business models, with several levels between pure product and pure service. Figure 21.2 shows a taxonomy of PSS, based on Tukker (2004) with some modifications.

On the left of Figure 21.2 is the traditional revenue model, the "pure product." For example, the box shows the manufacturer selling a washing machine to the user. As mentioned above, there could be a circular end of life such as the manufacturer providing a takeback system, but the revenue model does not drive it.

The second column has "product-oriented" revenue models: you still mostly sell a product that the user owns, but you also make money from "product-related services" such as maintenance and repair, or "advice and consulting" such as teaching customers how to use it and what they can do with it. If designed correctly, these can let you bring in recurring revenue when the product lasts longer. For a washing machine, you might sell it plus delivery and installation, a warranty for repair, repair services, or even upgrade. Another service could be software (discussed later).

The third column has "use-oriented" models: the user no longer owns the physical product, they pay for using it. But they do still operate it. This includes some of the sharing economy, like an Interface carpet being leased by a building owner, or a communally owned washing machine in an apartment building, used by one person at a time ("product sharing"), or multiple people doing their laundry together in one machine at the same time ("product pooling"). Use-oriented models are often used for aircraft engines as mentioned above; office printers have used product pooling since the 1990s, including Xerox, Canon, and others. Printer driver software enables print jobs from many people to queue in one shared printer; this concept can apply to other products, too.

The next column has "result-oriented" models: the user not only doesn't own the physical product, but they often don't even operate it. They just pay for the result. For example, in the "functional result" model, you might run a cleaning service with some washing machines, some dry cleaners, and some other devices; people drop off their clothes, you wash them, and they pick them up again. In "outsourcing," you hire a third party contractor to do the washing because it's not your core business. An especially powerful strategy is "pay per service unit," as it often aligns incentives better. For example, since 2014, Signify (formerly Philips Lighting) has sold lighting to some customers, including Schipol Airport, as "pay per lux" (Philips, 2014). That means Signify does not sell them bulbs and fixtures, but pays for all hardware, maintenance, repair, etc.; they bill the customers per lux of light emitted over time. Because of this cost and revenue structure, Signify has an economic incentive to make the lighting energy-efficient with long-lived hardware. It's similar to Rolls-Royce's "power by the hour" billing, or Ricoh photocopier's pay-per-copy service system. These models are often called Product As A Service (PaaS) because the emphasis is no longer on the product.

These business models (and others) can also overlap: the ride share companies Lyft and Uber use "outsourcing" because they don't manufacture the cars and don't provide drivers standard employee jobs. They also use "pay per service unit" because customers pay only for the distance/time driven, not a monthly subscription or other fees. They also use "functional result" because a user could be picked up in any car, from a Mercedes sedan to a Toyota truck. Sometimes they even use "product pooling" when users share part of their ride with others.

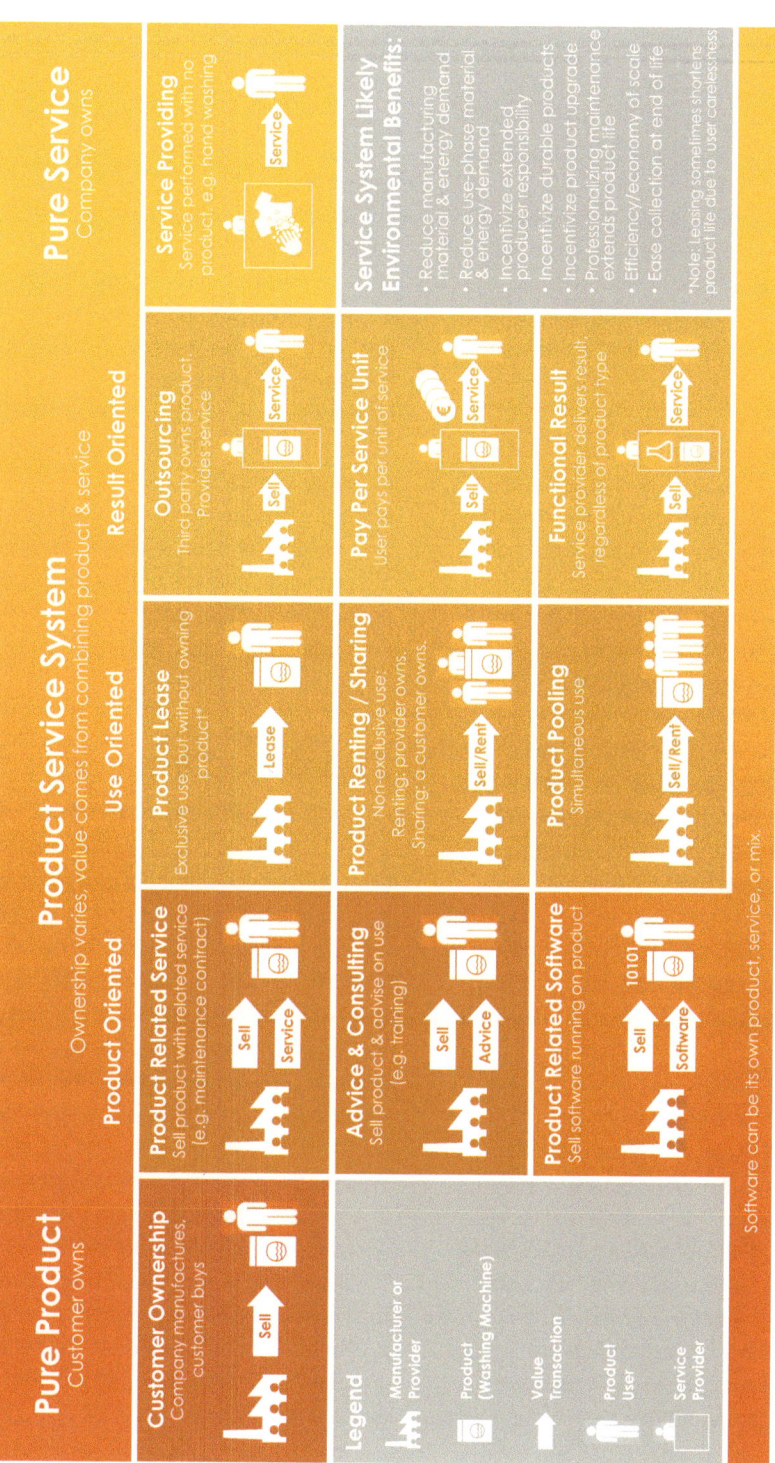

Figure 21.2 Product service system landscape

Source: Adapted from Tukker (2004) and Plan C.

And, finally, Figure 21.2's last column is "pure service." There is no physical product. In the case of washing clothes, this is not 100% literal—even if your business hand washes clothes, you'll still use soap and water—but it's close enough that hand washing is considered pure service. Most "pure service" businesses are like this—they usually need some back-end infrastructure to perform the service, but the user never sees it, and likely would not be able to operate it without training.

Besides these models from Tukker (2004), another business model is selling software running on the physical product. This means selling one piece of hardware that gets used for a long time and making most (or even all) of your money from recurring software sales. For example, videogame consoles like Nintendos are sold at very low margins because most of the money is made from game software sales. This hasn't been applied to washing machines yet (apps that do exist, like starting the washing machine remotely, are free, not a revenue stream). But who knows what the future holds? Be careful with this strategy: products making most of their money from software may become obsolete sooner, not later, if growing software demands outstrip hardware capabilities.

Software can, itself, be a digital product or service in any of these categories. Most web applications are software-as-a-service (SaaS), as well as some desktop software. In addition, software can enable any or all of the different product service systems described above, whether the software runs on the product itself or on a separate platform, whether or not the software itself generates revenue. For example, mobile phone services are a PSS using phone hardware that runs software to make calls and text messages, etc. Many of these apps are free, but keep users paying for the service. By contrast, car share services like Zipcar, GreenWheels, etc. use normal cars with minor modifications; the software where users make reservations and check availability runs on a website or phone app separate from the car. But such software is often critical for such product sharing and pooling services—car sharing can't give thousands of users each thousands of metal keys for thousands of vehicles in thousands of locations, even if they could manage thousands of paper calendars and maps for them. But an online database can easily give millions of people access to millions of vehicles in millions of locations, with the ability to schedule, track, sort, and filter them for a convenient reservation.

21.3 PSS, Circularity, and Sustainability

Circular business models can keep materials and even whole products circulating for long periods of time and/or many users. Maximizing customer use per physical product maximizes resource efficiency. The Value Hill, shown in Chapter 4, illustrates how this resource efficiency also creates financial value. Not all product service systems are circular business models, and not all circular business models are a PSS. A recycling plant is a circular business model, receiving waste and separating it into pure metals to sell to smelters, that is a circular business model that is not a PSS. A car rental agency is not a circular economy business model unless it makes explicit measures to extend its cars' lives, which most do not.

Companies can be circular and sustainable without using a PSS. Companies like Patagonia and Fairphone already make durable, repairable, or otherwise long-lived products with the standard business model of selling new products. They achieve this through brand recognition that, while their products may cost more initially, they will provide more user value over the long term. However, these are a small percentage of the market, both because consumers mostly prioritize low first cost over lifetime cost or other values, and because the business model of selling new products means faster product turnover increases profits. Companies with long-lasting products need very strong brand value with customers, and strong purpose within the company, to succeed.

Not all PSSs are circular, and not all circular businesses are sustainable. The peer-to-peer ride share company Uber provides a non-sustainable PSS. While its service does reduce the number of cars that need to be manufactured by increasing the usage of each car, the primary environmental impact of cars comes not from their manufacture but from burning fuel to drive them. Uber and similar services may cause 83% more driving than people owning their own vehicles (Henao & Marshall, 2019), because they drive "empty" so far to pick up passengers. Uber also isn't considered socially sustainable, because even though its system financially benefits both the company and the customers more than traditional taxi services, it often pays drivers poorly, with no job security.

Sometimes even PSSs designed for sustainability may cause "rebound effects," where they unintentionally cause more resource use rather than less. Uber's additional kilometers driven is one example of a rebound effect. Another could be mobile phone services where the service contract includes a phone. While this business model could incentivize the phone company to keep phones in service longer, the company might do the reverse by promising its customers they can always have the latest phone by paying for premium service. Rebound effects are especially likely if the products are not designed with the service in mind, or vice-versa. For example, shared products are often abused, due to users not feeling a sense of ownership. Rental cars often have shorter lifetimes because of this, since the cars are designed like any other car, not more repairable or more durable, usually just selected by rental companies for affordability or style. Bicycle share companies, by contrast, usually provide both the service and the bicycles, so they design the bikes far more robustly to enable longer lives with rougher use.

One way to drive sustainability without PSS or circular business models is for government regulations to outlaw or heavily tax short-lived products, using standard metrics for durability and repairability. However, legislation is slow in coming, especially on the global scale, and many sustainability professionals prefer to motivate companies through rewards rather than penalties. Rather than waiting for legislation, companies can motivate themselves by using product service systems to align profits with circularity.

For example, if you design for repair and long life, but make your money by selling new products, the incentives do not align unless you have exceptional brand value built around that specific reputation. You must work against the incentives, as mentioned above for Patagonia and Fairphone. If, instead, your company leases a product to customers monthly, then replacing the product hurts your profits compared to leaving the same product in place longer. You have an incentive to make the product durable, maintainable, and repairable. Likewise, if your company pays for the repairs, you have an incentive to make the product faster and easier to repair. Both the company and the customer can save money this way, as with Rolls-Royce and airlines.

Finally, another potential PSS benefit is that the continued relationship between customer and company throughout the life of the product can provide better communication, logistics, and trust, which can help facilitate product take-back for repair, remanufacturing, or recycling. The customer relationship can also help the company encourage customer maintenance or product care. This lowers the barriers to reuse, remanufacturing, dedicated recycling, or other end of life improvements.

21.4 PSS Barriers and Caveats

There are barriers to designing product service systems for the circular economy, like inertia and the need for new infrastructure. If you are a product manufacturer trying to go from pure product to a PSS where your company is also the service provider, you need to operate differently from a "normal" product manufacturer. Figure 21.3 shows a landscape of barriers to PSS and how to overcome them, from the European Commission's (2008) report, "Promoting Innovative Business Models with Environmental Benefits."

In Figure 21.3, magenta circles are barriers to any innovation, and green circles are barriers specific to PSS (and some other sustainable design strategies). The gray circles are ways to help address the barriers. For example, lack of trust can be a barrier to adoption of any new product, sustainable or not, because users are unfamiliar with it. You can overcome this in many ways, but the two ways shown here are to offer a guarantee (e.g., if customers aren't satisfied, they can get their money back), or to retain legal responsibility for the product (e.g., if it malfunctions and damages something, people will sue you, not your customer). As Figure 21.3 shows, the barriers to PSS are mostly the barriers to any innovation.

Some barriers are PSS-specific, though. For example, new infrastructure is often needed, such as a car share service's many cars, parking locations, and the software to schedule, locate, and otherwise manage car pick-up and drop-off. Figure 21.3 shows this barrier can be overcome by starting with a smaller market, or building on existing hardware. For example, the company Swapfiets (started by TU Delft students) provides bicycles as a service for a monthly fee, and broken bikes are swapped out for different ones while the original is repaired and then given to a different user. This requires an extensive network of repair centers (infrastructure), so Swapfiets started with a very limited geographic service area and expanded over time.

The gray circles are not the only solutions to these barriers. For example, lack of trust can be overcome by advertising or communication. Strong collaboration between your company and your customers (especially large institutional customers you can build long-term relationships with) can build trust,

Figure 21.3 Barriers to PSS and how to overcome them
Source: EU Commission (2008).

manage risks, create a common vision, overcome inertia, lower transaction costs, find ways to reduce infrastructure needs, build knowledge, and clarify incentives, all at once. See Chapter 26 for more details.

An important caveat, mentioned earlier, is that not all PSSs are sustainable. Uber and AirBnB have especially been criticized, and even outlawed in many cities around the world. Their outsourcing models have generated profits not by greatly increasing physical resource efficiency but by increasing cost efficiency through avoidance of standard employee contracts, hotel taxes, and the like.

21.5 Choosing a PSS Business Model

How do you apply these ideas to choose a business model that supports your material recovery or other sustainability goals? Figure 21.4 shows an overview.

- In Figure 21.4, **tightest circular loops** means considering what drives the most material and value recovery for your product type: durability, upgradability, product sharing, or something else? Different products get thrown away for different reasons.

- **Customer needs and values** means considering what your customers care most about, or what problem you can solve for them—your value proposition. For example, if initial cost is a problem, a "pay per service" model might be best. If the main problem is long-term cost, or convenience, or reliability, other PSS models might be better. And remember you can combine PSS models to meet multiple needs.

- **Community needs and values** means considering what's best for the people and local economies where the company manufactures, sells, and otherwise operates. Do the people there need more jobs? If so, more "service-oriented" models might be best. Or for poor communities that lack material goods, "product pooling" might be beneficial. Engage with the local communities and consider UN Sustainable Development Goals to find what your business can support.

- **Company strengths and values** means considering what best fits the company skills and resources, and how you can generate the most revenue. Do you currently sell to individuals or institutions? Do you already have a recovery strategy, like take-back programs or maintenance contracts? For example, if you're a car manufacturer that already has an extensive financing business, it's easy for you to expand into leasing, rental, and product pooling. Or if you're a software company starting a new laptop business, it's easy for you to make your laptops a platform for your software sales. You can build on your company strengths with a learning trajectory to overcome the barriers you encounter, such as Figure 21.3's barriers and suggested solutions to try out.

Together, these considerations of tight circularity, customer value, community value, and company strengths help you reach all three points of the Triple Bottom Line. In the end, the benefit of a well-designed circular PSS is that it aligns profits with long product life, repair, and recovery of materials for a circular economy. They can apply to many product categories, from jet engines to lighting to clothing to building materials. As mentioned earlier, they're often easier to set up when targeting institutional customers, but

Tightest circular loops

Customer needs & values

Community needs & values

Company strengths & values

Figure 21.4 Considerations when choosing a PSS

individual people use PSSs like Swapfiets, Rent the Runway, and more. So, what's a good business model for your product and company?

Figure 21.5 shows examples of this: washing machines mostly get thrown away when they break, not because fashion is driving everyone to have this season's hippest model. By contrast, clothing mostly gets thrown out because people want a new fresh look, not because the old clothing was broken. For products that are thrown out when they break, your business might use the "product-related service" model to sell maintenance and repair services, or might use "Product Lease" to capture value from a product designed for longer life. For products that are thrown out for obsolescence, you might use a "Functional Result" model, designing your product for upgrade and customizability, then capturing value from mixing and matching the same parts in new ways.

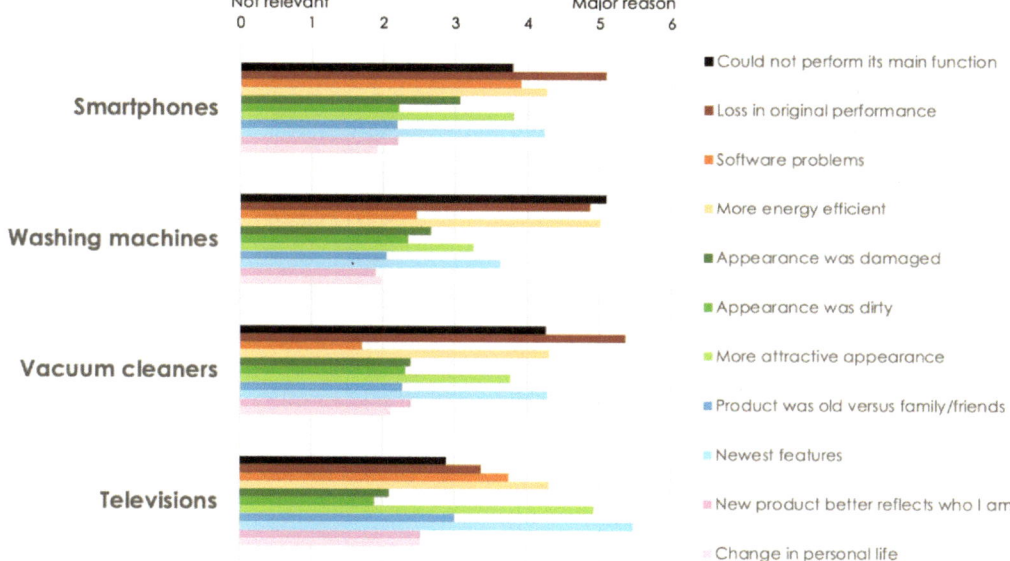

Figure 21.5 Why we throw things away

Source: Magnier & Mugge (2022).

Resources and References

Resources for Further Study

* EU Commission (2008). Promoting innovative business models with environmental benefits. Final Report. COWI. Available at: https://ec.europa.eu/environment/enveco/innovation_technology/pdf/nbm_report.pdf. Especially see Section 5.2, "Business Rationale" (pp. 29–31) and Section 7, "Realizing PSS" (pp. 39–46).

References

EU Commission. (2008). Promoting innovative business models with environmental benefits. Final Report. COWI.

Henao, A., & Marshall, W. E. (2019). The impact of ride-hailing on vehicle miles traveled. *Transportation*, 46(6), 2173–2194. https:// doi.org/10.1007/s11116-018-9923-2

Magnier, L., & Mugge, R. (2022). Replaced too soon? An exploration of Western European consumers' replacement of electronic products. *Resources, Conservation and Recycling*, 185, 106448.

Neely, A. (2008). Exploring the financial consequences of the servitization of manufacturing. *Operations Management Research*, 1(2), 103–118.

Philips. (2014). Case Study: National Union of Students. Philips Lighting.

Sakao, T., & Lindahl, M. (2009). *Introduction to product/service-system design*. Springer.

Tukker, A. (2004). Eight types of product–service system: Eight ways to sustainability? Experiences from SusProNet. *Business Strategy and the Environment*, pp, 246–260.

How to Apply #21: Brainstorm Turning Your Product into a PSS
Time Estimate: 1–3 Hours

Goal: Brainstorm product service systems to turn your product into a service for longer lifetime and/or sharing.

STEP 1: Brainstorm Revenue Models to Turn Your Product into a Circular PSS
Time Estimate: 30–60 Minutes

Hold a brainstorm session, based on the PSS landscape (Figure 21.2) and, using the Rules of Brainstorming, to generate ideas for changing the revenue model from a one-time purchase to a service for lifetime and sharing, where your company somehow stays involved in the product's life cycle.

- Ideally have 5–10 ideas for every box in the PSS landscape. (Ideally 60+ total ideas, but a minimum of 30 ideas.) Have more ideas for the boxes more promising for your product/service.

- You don't need to work through your Whole System Map, but you might find it helpful, especially to see how services connect to other parts of the system.

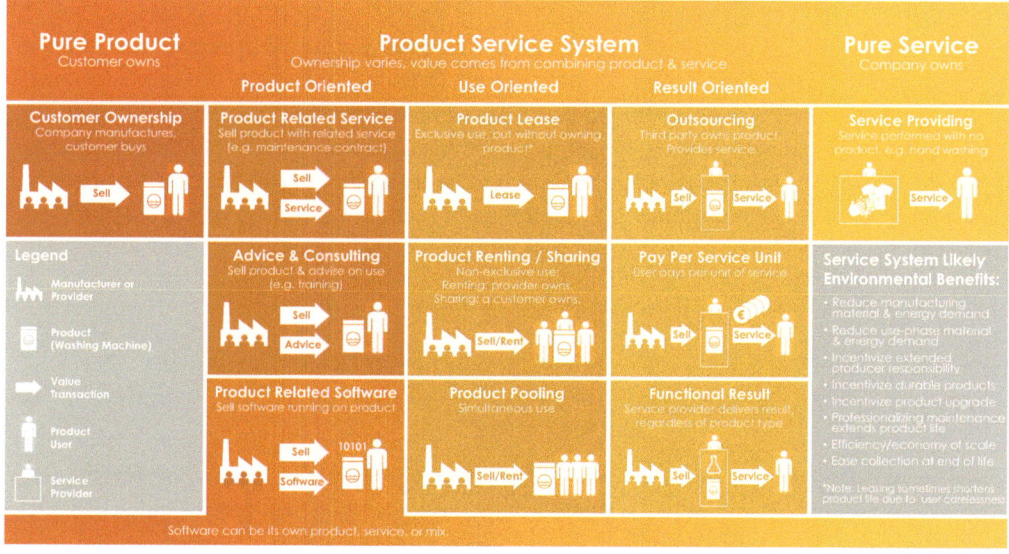

- For each service idea, list the added value for the customer compared with just purchasing the product. Does it save the user money, either that one time or in the long run? Does it help the user avoid financial or physical or other risk? Does it provide the user more convenience? Something else? Don't write long notes, just a few words like "more convenient" or "lower one-time cost."

- For each service idea, list the added value for you (the manufacturer and/or service provider), compared with just selling the product.

- If desired, you can also mention if there is value added or lost for other parties in the value chain (e.g., component suppliers losing money from fewer new products produced).

STEP 2: Narrow Down Your Brainstorm Options to 3–4 Winning Ideas
Time Estimate: 5–15 Minutes

Use dot voting, decision matrices, or whatever tools you desire, to narrow down your brainstorm to just 3–4 winning candidates. In addition to judging each idea by material reduction per functional unit of service, use considerations from your Design Brief to rule out options that don't meet business criteria, such as cost or usability.

STEP 3: For Each Winning Idea, Sketch and Estimate the Material Reduction of That Option
Time Estimate: 5–30 Minutes

For each of your winning ideas, draw a quick sketch or storyboard of how the customer uses the product and service. Then estimate each idea's percent reduction in material use per functional unit of service compared to the original product. For example, if one idea shares the product between two people but only lasts half as long, it does not save any material per functional unit. Something shared between two people and lasting twice as long would use just 1/4 the material per functional unit, a huge improvement. Show the math for how you got the estimated percent reduction for each idea.

STEP 4: Choose One Winning Idea and Illustrate It
Time Estimate: 20–40 Minutes

Choose one winning idea (or combination of ideas), based on the results of the percent reduction in materials intensity, and your other Design Brief priorities. Create a high-quality image of the winning idea, either by hand or digitally, to clearly convey how the idea is different from the current product, and why it's a compelling design.

STEP 5: Document Your Decision and Brainstorm
Time Estimate: 30–60 Minutes

Create a PDF with the winning redesign (or the top few) and the reasons why it's the best option.

Checklist for Self-Assessment

To score your success on this exercise, see if you…

- ☐ *Listed all the new ideas.*

- ☐ *Listed each new idea's value to the customer.*

- ☐ *Listed each new idea's value to you.*

- ☐ *Listed (or show labeled sketches of) your top 3–4 ideas.*

- ☐ *Listed the percent reduction in material use for each of these designs and show the math.*

- ☐ *Showed the illustration of the winning design.*

- ☐ *Succinctly described the winning design, either as annotations to the illustration or as a stand-alone sentence or two.*

- ☐ *Briefly described a convincing business case for the final design choice.*

CHAPTER 22
Business Models
Presidio Booster

*Jeremy Faludi,
Conny Bakker, and
Ruud Balkenende*

Goals

- Recognize sustainable business opportunities and barriers for companies

- Explain why sustainable business models are needed

- Illustrate examples of a business model, a sustainable business model, and a circular business model

- Analyze existing businesses using Presidio's Sustainability Booster tool

- Ideate new businesses ideas using Presidio's Sustainability Booster tool

DOI: 10.4324/9781003504672-24

Why It Matters

Sustainable business models go far beyond product service systems. In addition, some circular PSS and sustainable design strategies rely on other business factors like marketing and partners. For example, sustainable material choices require suppliers who sell those materials at affordable rates; designing affordable products must not compromise factory worker pay; etc. Finally, sustainable business models can help address social sustainability. Thus, it's important to look at all aspects of business models to see where social and environmental improvements can be made. This is critical for entrepreneurs, and even if you "only" do product engineering, it's still valuable to understand how your work depends on these factors, so you can push for the business practices you need to support your designs. You can even help the business managers discover what's possible.

Summary

- The Business Model Canvas (BMC) is a fundamental tool for both understanding and creating or changing businesses. It diagrams costs, revenues, your customers, suppliers, and more, so you can make sure they all align.

- Normal BMCs only consider economic profit, but the Presidio Sustainability Booster adds environmental and social sustainability considerations for every part of the BMC.

- The Presidio Sustainability Booster is a qualitative and creative tool; it does not quantify whether one business model is better than another.

- The Presidio Sustainability Booster can help you assess an existing business, and/or help you generate ideas for improving a business.

22.1 Business Model Canvas

A business model is not a product, or service, or company, or industry; it is a way for a business to create and capture value. One company can have many business models for different physical products, digital systems, personal services, or anything else. The models describe how the company intends to make money from each offering. The most widely used tool for understanding existing business models, communicating how they work, and designing new ones is the business model canvas (BMC). As described by Osterwalder and Pigneur (2010), the BMC lays out all the major parts of a business, with their locations on the canvas implying the relationships between them. See Figure 22.1.

In Figure 22.1, the center of a business model is the "value proposition"—usually a product or service that people will pay money for. You make your value proposition using "key resources." like raw materials and a factory, and by doing "key activities," like manufacturing. These resources and activities depend on "key partners," your supply chain. You pay "costs" to those partners for their services, products, or materials, shown in the bottom left corner. Your users and/or the people paying you are your "customers." You get your product or service to the user through distribution "channels" (e.g., retail stores or online shops). Your customer learns about your value proposition through "customer relationships" (e.g., marketing and advertising). The money customers pay you in exchange for the value you provide them is "revenue," shown in the bottom right corner. If your revenue exceeds your costs, you profit. That's the canvas in a nutshell.

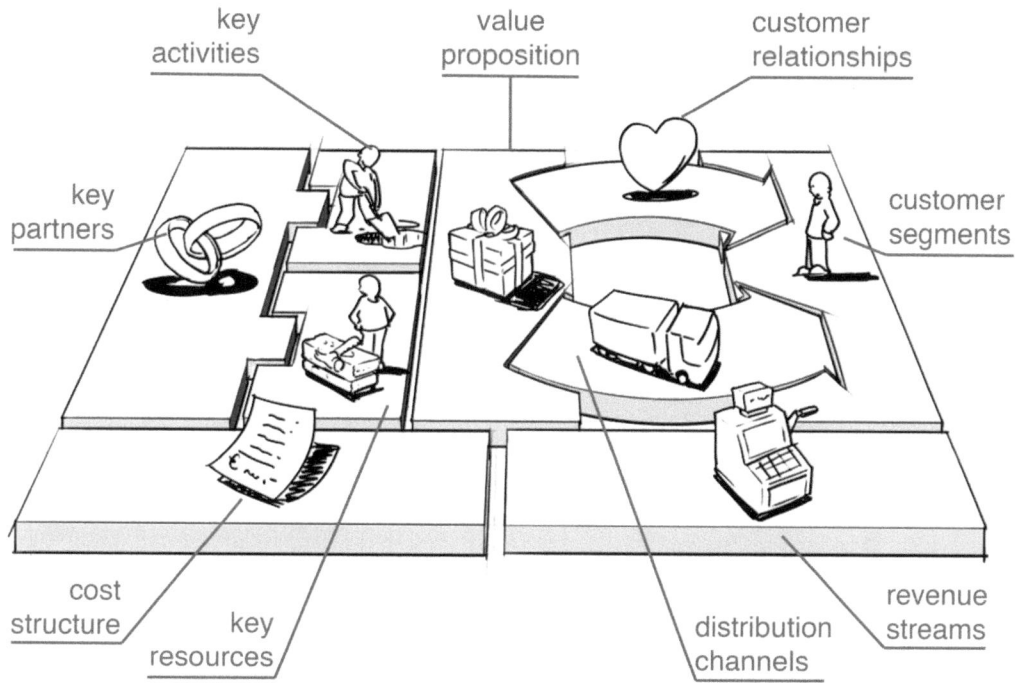

Figure 22.1 Business Model Canvas overview
Source: Osterwalder & Pigneur (2010).

Once the business is running, time flows from left to right on the canvas. But if we apply human-centered design to improve a business model or invent a new one, we start with your user/customer. You then decide what value proposition you can offer them, and fill out the rest of the canvas from there.

You don't need to start a new company to have a new business model. You can have a new model for any product or service in an existing company, and you don't even need a for-profit company, these tools can also be used for nonprofit organizations. You can even pair nonprofits and for-profits, like the Biomimicry Institute (an educational nonprofit) and Biomimicry 3.8 (their for-profit consulting arm). As shown above, the BMC simply lets you plan how costs and revenues balance, and all the processes or connections that support the costs and revenues to offer the value proposition to target users.

The BMC also illustrates how your company connects with outside actors through "customers" and "key partners." These are not just users and suppliers of parts or materials, but include shipping, retail, and could include repairers, recyclers, reverse logistics to collect products for remanufacturing, etc. If your product or service is digital, they include the other hardware and software your app runs on (e.g., internet browsers and cellular data networks). Partners can also include government regulators or industry consortia that could be opportunities for activism to drive sustainability standards.

22.2 Sustainable Business Models

Business models are important for designers to understand because they should support your product or service design rather than undermine it, as described in Chapter 21. They can help your company transition away from selling more stuff while remaining profitable. Research has shown that the two main factors that internally drive companies to sustainability are leadership and the business case (Lozano, 2015).

Filling out a BMC can help you discover where sustainability must be integrated into daily operations (e.g., supply chain management, labor practices, etc.) or company governance structures that drive long-term benefit to many stakeholders rather than only short-term profit to shareholders. Other stakeholders include workers in the company and supply chain, communities around factories and mines and waste disposal areas, plants and animals in those areas, and more. "Selection of stakeholders must consider moral justice for potential human and nonhuman stakeholders" (Upward & Jones, 2015).

The BMC also highlights external factors that can help drive sustainability, such as which materials are available as key resources, or which government agencies or communities could be key partners. Business models can also address other aspects of sustainability that physical and digital product design can't, such as social justice in worker wages, community support, or system-level interventions like partnerships across industries.

There are several BMC overlays for sustainability, including the "Flourishing Enterprise Innovation Toolkit" (Upward, 2016), the "Triple Layer Business Model Canvas" (Joyce & Paquin, 2016), and others. They

all help companies recognize business opportunities and barriers in environmental and social responsibility. They are not about maximizing profits at any cost, they are about harmonizing profits with your company's impacts on people and the planet.

22.3 Limitations of Business Models

Critics of the Circular Economy, Green Growth, and other movements for sustainable business have argued that business models cannot solve everything. As degrowth proponents argue, "economic growth without the destruction of nature is an illusion" (Schmelzer et al., 2022). Sustainability systems must also go beyond for-profit businesses to include government, nonprofit, and other community programs.

However, even in these cases, sustainable BMC tools may help non-business entities plan services in economically viable ways. All institutions have costs, value offerings, users, partners, etc. BMC tools can help understand where these align or misalign, where costs and revenues come from and why, and can help innovate all of these.

22.4 The Presidio Sustainability Booster

The Presidio Business School's "Business Sustainability Booster" (Willard et al., 2017), available online at https://www.presidio.edu/blog/business-sustainability-booster, is an especially good BMC overlay because it is relatively fast and simple, built directly on the BMC for ease of adoption, but still enables thorough examination of the whole business model and creates fertile ground for innovative positive solutions. See Figure 22.2.

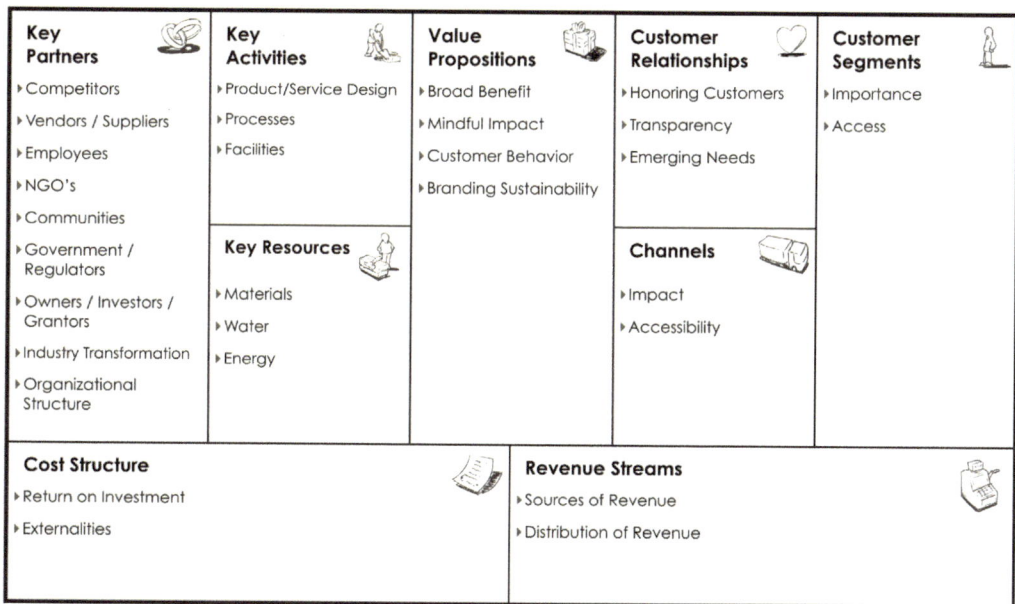

Figure 22.2 The Presidio Sustainability Booster layout

Source: Willard et al. (2017).

As Figure 22.2 shows, the Sustainability Booster is the normal BMC with added questions in each BMC box. It doesn't tell you what to do, it just adds the prompts to help you innovate. Because each box of the BMC is different, there are different questions in each box. Each prompt has a one-page "card" describing why it's important and listing specific questions to brainstorm, as shown in Figure 22.3.

This tool is exploratory—it won't tell you what's right or wrong or better or worse.

To make such judgments, combine the sustainability booster with other tools or metrics such as the UN SDG targets, LCA, or sustainability certifications. But it can help you thoroughly assess an existing business using your own judgment, or it can help you brainstorm many ideas to boost the sustainability of your business, both environmentally and socially aligning the company operations with your ethical values. Such explorations are crucial to making products or services that change the world.

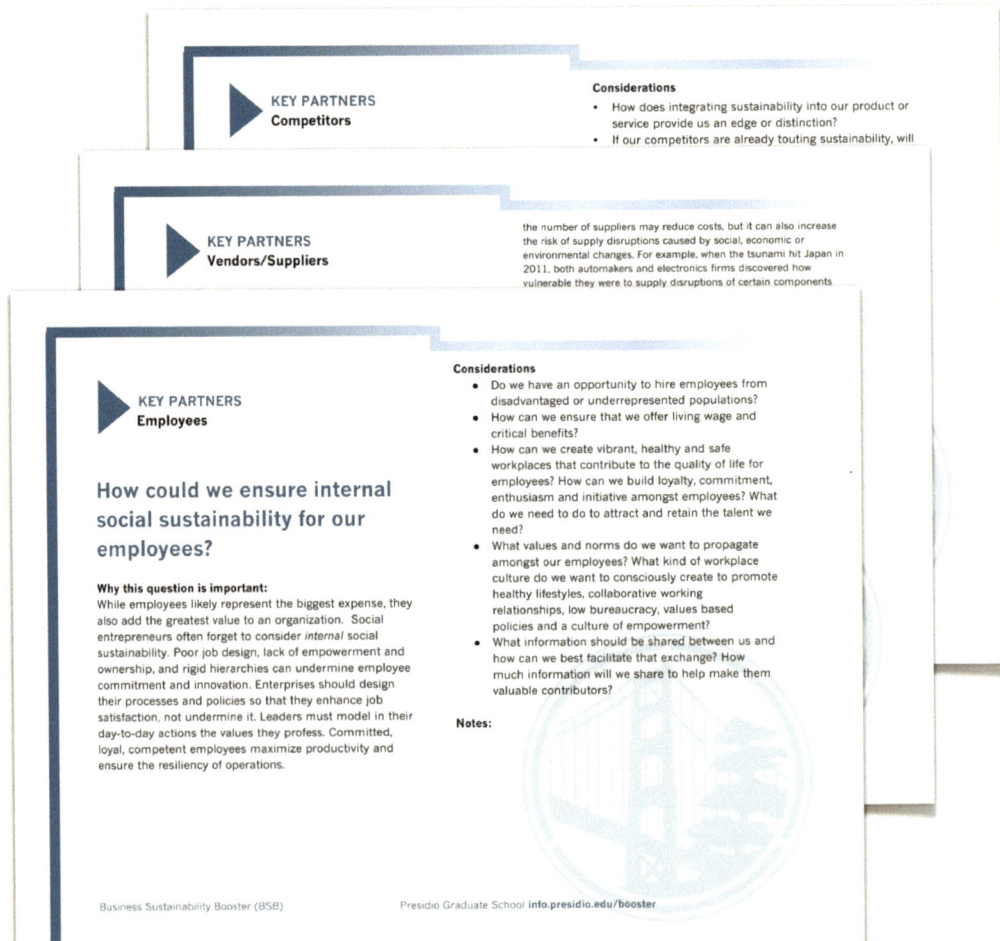

Figure 22.3 Prompting question "cards" in the Sustainability Booster

Source: Willard et al. (2017).

The Presidio Booster recommends circular economy considerations, but does not recommend specific product service systems. It offers abstract prompts that can operate on the higher-level system. For example, for circular material sourcing, it provides obvious suggestions like the "materials" prompt in the "Key Resources" box suggesting material reuse and low-impact materials, but it goes beyond that to create the business conditions to drive such material availability. In the "Key Partners" box, the "Vendors/Suppliers" and "Industry Transformation" prompts suggest working with supply chain vendors to increase the availability of better materials or components, and improve their production practices, share knowledge to co-create better options, and more. These and the "Governments/Regulators" and "Non-governmental Organizations" prompts suggest driving better policy across the whole industry. The "Industry Transformation" card even suggests working with competitors to drive such material availability or policy. Such larger systemic interventions don't replace product-level tools and strategies for circular design, but make them more viable by driving the availability of circular-sourced components.

22.5 Using the Sustainability Booster

For an example of how this works to analyze an existing business or start a new business, imagine your company makes a cargo-carrying kick scooter that fits on trains or buses.

In the Sustainability Booster, we'll start with the customer. What are your user's needs, pain points, and desires?

We have a profile on your user, see Figure 22.4 for details: she drives all the way

A 40-mile drive is caused by a one-mile gap at each end.

Target User: "Sally"

• Busy urban professional
• Carries laptop, briefcase, other items
• Commutes 30+ minutes each way
• Lives & works within 1-2 mi of transit
• Spends $2,500/yr on gas, $1,000+ on maintenance, repair, insurance
• Wastes 63 workdays per year driving

User Needs

• Convenient
• Easy
• Fit with transit
• Carry things
• Fit with lifestyle

The Problem

• Buses are slow, transfers are hassle
• Bikes, scooters, etc. don't meet user needs
• Thus drives, wasting time and money

Figure 22.4 User profile
Source: Freepik.com.

from San Francisco to Silicon Valley because the train stations aren't conveniently located within a kilometer of her home or workplace, and there is no easy connection from those stations to her trip's beginning and end. Thus, a 60-kilometer drive is caused by a 1-kilometer gap at both ends.

In the business model canvas "Customer Segments" box, the Presidio Booster has two questions: "Importance" and "Access" (see Figure 22.2). What are they? If you look up each page in the Presidio Booster PDF (Willard et al., 2017), the page on "Importance" says it's how much sustainability matters to your users. It's important because you need to know how to market to them. What can you brainstorm here? Let's assume your users don't care about sustainability, so that won't sell; that means you brainstorm on whatever value would resonate best for them. Maybe they value convenience: if so, show how much productive time they can recover by working on the train versus driving, see Figure 22.5.

The other Presidio Booster question for "Customer Segments" is "Access." The page on Access says it's how you can provide more people with access to your product. It's important because some people lack the money or other resources (e.g., language or technology) for durable high-quality goods. What can you brainstorm here? For your scooter, you might decide Access is already good, no further action needed, since a scooter is much less expensive than driving a car (see Figure 22.6). However, the initial purchase price may still be expensive for some people, especially those in disadvantaged communities (see Chapter 25). So, you could brainstorm ways to be accessible to even more people.

A change could be moving from selling scooters to running a scooter-share network (Figure 22.7), costing only a dollar or two per trip. This might radically expand who can afford your product, and earn you more revenue at the same time.

For the BMC "Value Proposition" box in Figure 22.2, there are four prompts. The

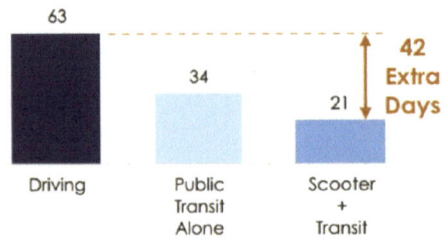

Figure 22.5 Data on how much time is recovered

Figure 22.6 Commute costs per year depending on the transport

Figure 22.7 Scooter rental service

Figure 22.8 Irresponsible scooter handling

"Mindful Impact" one is about considering the unintended negative consequences of your business model. For example, if you change from selling scooters to a scooter share service, what happens when people leave their scooters scattered irresponsibly (Figure 22.8)? Many scooter share companies have had problems like this, with neighborhood residents feeling like their streets are overrun with outsiders' vehicles. How can you redesign your system so users don't do that?

Finally, in the BMC "Revenue Streams" box, there are two prompts; let's look at "Distribution of Revenue." You could choose to structure your company as a worker-owned cooperative, where profits are distributed back to the workers because they own the company, and the largest ratio of pay between the lowest paid worker and the highest paid executive is set at a low ratio like 9:1. This might sound untenable, but in the Basque region of Spain, there are over 250 companies employing over 75,000 people organized in this way through the Mondragón corporation (Whyte & Whyte, 2014). They are the world's largest cooperative business.

All of the BMC boxes can be sources of sustainable innovation, but the "key partners" box also deserves special attention. These may be for your individual product's sustainability, for example., finding repairers, recyclers, or reverse logistics to bring products back to you for remanufacturing. They can also be for activism driving sustainability beyond just your company

to your whole industry, such as creating government regulations, or standards set by industry consortia. Sometimes external action is even necessary for sustainability action within your company, for example, when a greener process is more expensive and your company will not do it unless regulators mandate it across your whole industry.

Key partners can also form collective purchasing agreements. That's where one company does not buy enough of a material to make it worthwhile for a supplier to switch their factory over to making a greener material, but several companies coming together do, so they formally guarantee the supplier they'll collectively purchase a certain amount.

To recap how the Presidio Sustainability Booster works: you keep going through the rest of the BMC as described above, to

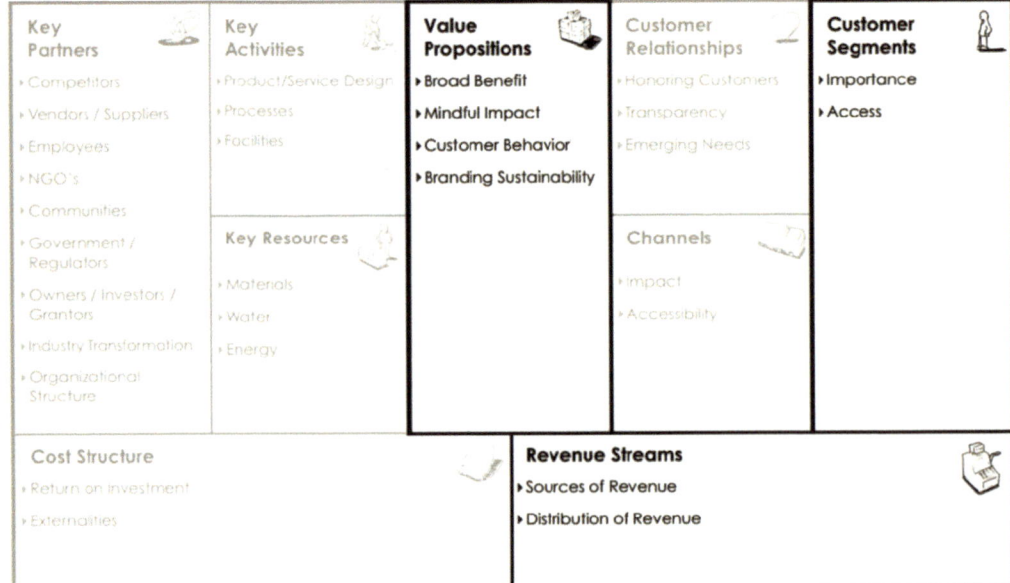

Figure 22.9 Highlighted focus areas for designers

either assess how sustainable an existing business is, or brainstorm on how your new business could be. When brainstorming, you can choose what prompts to brainstorm solutions to, versus what to ignore or consider already done. Focus especially on Customer Segments, Revenue Streams, and Value Propositions, see the highlighted fields in Figure 22.9. Those are the places where product and digital design can contribute the most, and those boxes can lead you to circular economy product service systems or more creative and novel business models. Keep in mind that changes anywhere on the canvas can ripple out to the rest of the canvas.

Resources and References

Resources for Further Study

- Willard, M., et al. (2017). Business Sustainability Booster. Presidio Graduate School. Available at: https://www.presidio.edu/blog/business-sustainability-booster.

- Joyce, A., & Paquin, R. L. (2016). The triple layered business model canvas: A tool to design more sustainable business models. *Journal of Cleaner Production*, 135, 1474–1486.

- Upward, A. (2016). Flourishing enterprise innovation toolkit – tools for the strongly sustainable revolution. Ontario College of Art and Design. Available at: http://flourishingbusiness.org/

References

Joyce, A., & Paquin, R. L. (2016). The triple layered business model canvas: A tool to design more sustainable business models. *Journal of Cleaner Production*, 135, 1474–1486.

Osterwalder, A., & Pigneur, Y. (2010). *Business model generation: A handbook for visionaries, game changers, and challengers*. Wiley.

Lozano, R. (2015). A holistic perspective on corporate sustainability drivers. *Corporate Social Responsibility and Environmental Management*, 22(1), 32–44.

Schmelzer, M., Vetter, A., & Vansintjan, A. (2022). *The future is degrowth: A guide to a world beyond capitalism*. Verso Books.

Upward, A. (2016). Flourishing enterprise innovation toolkit – tools for the strongly sustainable revolution. Ontario College of Art and Design. Available at: http://flourishingbusiness.org/

Upward, A., & Jones, P. (2015). An ontology for strongly sustainable business models. *Organization & Environment*, 29(1), 97.

Whyte, W. F., & Whyte, K. K. (2014). *Making Mondragón: The growth and dynamics of the worker cooperative complex*. Cornell University Press.

Willard, M., et al. (2017). Business Sustainability Booster. Presidio Graduate School.

How to Apply #22: Apply Presidio Sustainability Booster to a Business Model Redesign
Time Estimate: 2–4 Hours

In this exercise, you will brainstorm ways to improve sustainability strategies of a product/service and its business model, using the Presidio business model canvas sustainability booster.

You will need the Presidio Business Sustainability Booster document. The tool is available for free download in the official Presidio Graduate School website as a PDF. Take a first look at introduction and instructions on pages 1–6.

STEP 0 (Optional): Analyze Current Product/Service and Your Company's Business Model
Time Estimate: 10–40 Minutes

Analyze your current product/service and company, according to the normal Business Model Canvas (BMC):

- Above the BMC table, name the chosen product or service, and include a picture of it.

- Fill out every box in the BMC table with a few words on the current business model: Who is the main user (the "customer segment")? What is the main value proposition? What is the main revenue stream? Etc. If your product addresses a highly segmented market, you may divide up the boxes for different target users, or make multiple copies of the table.

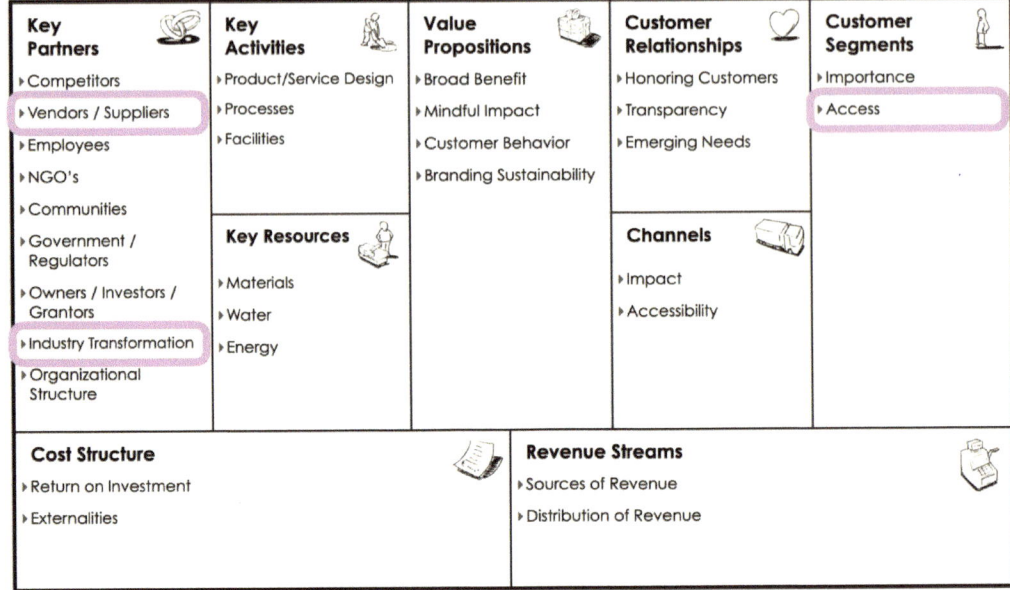

STEP 1: Select Presidio Prompts to Brainstorm On
Time Estimate: 1–10 Minutes

Look at the Presidio Sustainability Booster's business model canvas table, and read the prompts for brainstorming listed in each box. If there are any you don't understand, flip to the page describing that prompt to quickly familiarize yourself with it.

Select three prompts to brainstorm on. For example, you might consider the following prompts below:

- Customer Segments > Access

- Key Partners > Vendors/Suppliers

- Key Partners > Industry Transformation

STEP 2: Brainstorm on a Prompt
Time Estimate: 10–30 Minutes

In the Presidio Sustainability Booster text, flip to the page describing the "card" for the prompt you plan to brainstorm on. For example, the Customer Segments > Access card, prompting you to think about increasing accessibility for customers. Read through the card to understand the topic and its important considerations.

Brainstorm a minimum of 10+ ideas for this prompt (20 or 30 is better). And remember the brainstorming rules! Don't judge ideas, but also stay focused on your topic, be visual, have many ideas, etc. (Being specific is one easy way to have many ideas—listing many possible implementation details.)

Be sure to brainstorm several ideas for each "consideration" bullet point on the card. For example, with Customer Segments > Access:

- Minimizing specific barriers for buying (e.g., language, culture, technology)
- Changing revenue models to fit those most in need
- Diverting costs from end users
- Using product service systems to make it more affordable

STEP 3: Brainstorm on Other Prompts
Time Estimate: 10–30 Minutes

Repeat the previous step for all the prompts you want to brainstorm, always remembering the rules of brainstorming. For a thorough exercise (several hours), brainstorm every prompt in every box in the table. For a minimal exercise, brainstorm at least three prompts from three different boxes, to break out of your normal domain of consideration and get creative.

STEP 4: Choose Winning Idea(s) from All Three Brainstorms
Time Estimate: 10–40 Minutes

Take your brainstormed lists of all ideas for every prompt, and narrow them down to the best idea(s) to move forward with. To do this, balance your sustainability and business priorities (if you have your Whole System Mapping priorities, use those). Also consider what would be most easily implemented, and best fit with company/brand culture. To decide, you might use dot voting, a decision matrix, or just discussion. You might combine ideas, or have multiple winners, but be realistic about the limitations of how much you can implement. You might narrow it down to one winner per prompt before choosing an overall winner (or few winners). Although changes anywhere on the canvas influence the rest of the canvas so try to think how your winning idea(s) affects the others (which ideas reinforce others to boost the overall sustainability strategy further).

Finish with one idea (or set of ideas) to move forward with.

STEP 5: Communicate the Value of the Winning Idea(s)
Time Estimate: 10–40 Minutes

Write a concise description of the winning idea(s), and sketch an illustration of it. The description should include which BMC box(es) and Presidio prompt(s) are involved, and what the new solution is, in an engaging way that would inspire adoption by business executives. The sketch should illustrate the solution and its effect on the user, company, or larger system.

Checklist for Self-Assessment

To score your success on this exercise, see if you…

☐   *Chose three or more Presidio prompts to brainstorm on.*

☐   *Generated 10+ new ideas for each prompt (ideally 30+).*

☐   *Chose winning idea(s) to move forward with, based on sustainability and business priorities.*

☐   *Described the winning idea(s), including the BMC box and prompt, to sell it to business executives.*

☐   *Sketched an illustration of the winning idea(s) to support the description.*

CHAPTER 23
Biomimicry Basics and Mentors

Jeremy Faludi

Goals

- Ideate sustainability solutions using inspiration from nature
- Find inspiration from biological mentors

DOI: 10.4324/9781003504672-25

Why It Matters

Many sustainability problems require creative out-of-the-box thinking. Nature is a great source of such ideas, because (1) it designs very differently from modern industry, (2) it contains a vast variety of design strategies, and (3) it is the very world we are trying to emulate and harmonize with when we practice sustainable design. Thus, biomimicry can be a powerful ideation tool, can inspire us, and can change our relationship with nature.

Summary

- To design products and systems in harmony with nature, we can look to nature's designs.

- For inspiration from mentors:

  - Define the problem biologically.

  - Find mentor organisms and their strategies, either through direct observation or online/literature sources.

  - Translate biological strategies into buildable things.

23.1 Biomimicry: Inspiration from Nature

Biomimicry (also called bionics or bio-inspired design) is a form of design-by-analogy to find innovation inspiration from nature. It isn't automatically sustainable—historically, it is most often used for robotics and medical devices—but it can be excellent for out-of-the-box thinking, and can remind us we are part of nature, responsible for healthy contributions to the ecosystems we live in. It can especially inspire us to have a relationship with nature as mentor and family, rather than simply as resource extraction site and waste dump. To redesign products and systems in harmony with nature, shouldn't we understand how to design like nature? Of course, nature does not intentionally design anything—all life is evolved. But just as physicists create equations to describe "laws" of nature, we describe "design strategies" of nature to translate them into our language and bring analogous strategies into our designs. That is the heart of biomimicry.

Biomimicry can work at any scale, from the microscopic structure of waterproof fabric inspired by lotus leaves to a ten-story building's passive ventilation cooling inspired by a termite mound.

Figure 23.1 shows Amphico's waterproof yet breathable clothing inspired by arthropod exoskeletons. Rather than using multiple layers of different materials, some of which are toxic PFAS chemicals, it uses a single polymer which is non-polar and woven into threads with nanoscale and microscale pillars, giving water very little surface area to stick to. The combination of material and geometry makes their fabrics

Figure 23.1 Amphitex waterproof and recyclable mono-material fabric (left) and the lotus leaves that inspired them (right)

Source: Amphico.com and Unsplash.

superhydrophobic, and as water beads up and runs off, it also carries away most dirt, which saves washing energy and chemicals. Because the material is all one polymer, rather than layers of different ones, it can be easily recycled.

Zimbabwe's Eastgate shopping center (Figure 23.2) imitated a termite mound's

ventilation tunnels for passive cooling. It uses 50% less air conditioning energy (3% less total energy), and even reduced air conditioning installation costs by 90% because so little mechanical equipment was required. It's also more resilient to power outages, because it works passively. Interestingly, scientists later found that termite mounds do not work the way they were

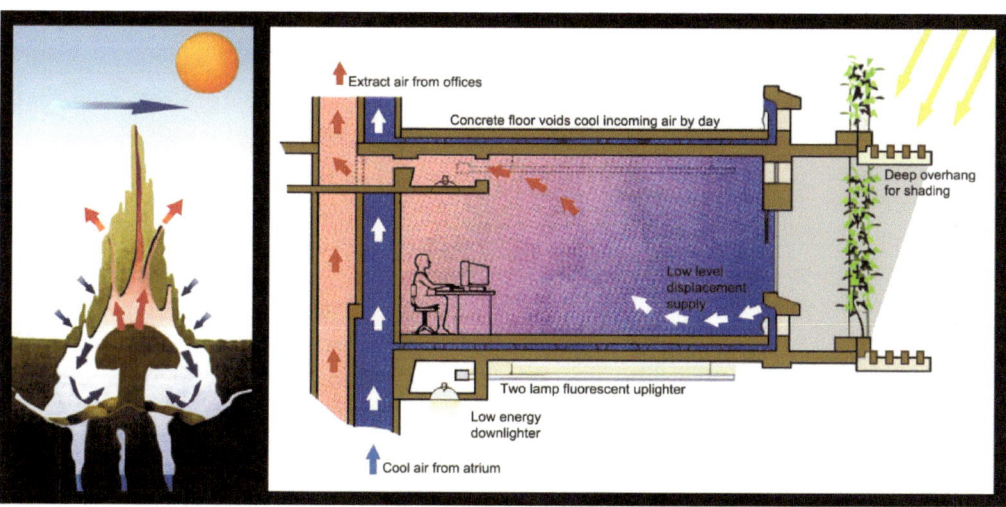

Figure 23.2 Eastgate Centre in Harare, Zimbabwe, was inspired by a termite mound's passive ventilation
Source: Mick Pearce.

Figure 23.3 Aesthetics and slavish imitation are not meaningful biomimicry

Source: Wikimedia commons.

believed to when Eastgate was designed, but the building design is still effective. This illustrates how biomimicry is not slavish imitation, it does not need to be completely accurate to provide inspiration for innovation.

Sustainable biomimicry isn't about aesthetics or slavish imitation of nature, it's about using nature's functional principles. A building that looks like a duck might be fun (see Figure 23.3), but if there's no useful duck-inspired function that measurably improves environmental impacts, the inspiration is shallow. The fabric in Figure 23.1 looks nothing like arthropod exoskeletons, nor does Figure 23.2 look like a termite mound. But they imitate their functions to provide sustainability benefits.

23.2 Biomimicry Levels and Roles

Janine Benyus's (1997) book, *Biomimicry*, explains that bio-inspired design can be applied at different levels: **form**, **process**, or **system**. For instance, Lotusan paint is biomimetic on the level of form, because it imitates the physical form of the lotus leaf (on the micro-scale). But it's not biomimetic on the level of process because the paint isn't manufactured by planting seeds and waiting for them to grow. Lotusan paint is also not biomimetic at the system level—at the end of its life, the paint is not recovered for use on other buildings. In nature, all waste becomes food for some other organism—that is what inspired Cradle to Cradle's concepts of technical and biological nutrients. A product doesn't have to be biomimetic on all three levels to be more sustainable than normal products; a great improvement on one level can be enough. Such improvement is best measured by LCA or other quantitative scorecards.

For example, Cypris Materials paint is biomimetic in both form and process (Figure 23.4). In form, its colors come not from pigment, but from structural color, like Morpho butterfly wings or the irises of blue-eyed and green-eyed people. The paint is a clear polymer with nanoscale ridges; different sizes reflect different wavelengths of light, for different colors. This eliminates the wide variety of chemicals and heavy metals traditionally used for paints. In process, it's

biomimetic because rather than using heat, energy, or nanoscale machinery to build it, it uses chemistry to self-assemble at ambient temperature, like butterfly wings.

Note the difference between biomimicry and biomaterials. A table built of wood is "bioutilization" but might not be biomimetic. Its form, manufacturing process, and system might have nothing to do with nature. But combining biomimicry and biomaterials is often a great strategy. For example, Ecovative packaging materials (Figure 23.5) replace styrofoam with mycelium, which means they use biomaterials, plus their manufacturing process is mycelium growth produced using agricultural waste feedstock, and they can be composted at end of life. Thus their process and system are biomimetic, in fact literally being a biological cycle.

Benyus's book also points out that biomimicry can play multiple roles. "Nature as model" means nature's role is a source of inspiration, but it also can also play the roles of "nature as measure" and "nature as mentor." Nature as measure means changing our metrics for success from weak goals like 10% less energy or 20% recycled material

Figure 23.4 Cypris Materials paints eliminate many of the toxic chemicals in paints by replacing pigments with structural color, inspired by butterfly wings

Source: Cypris Materials, Ryan Pearson.

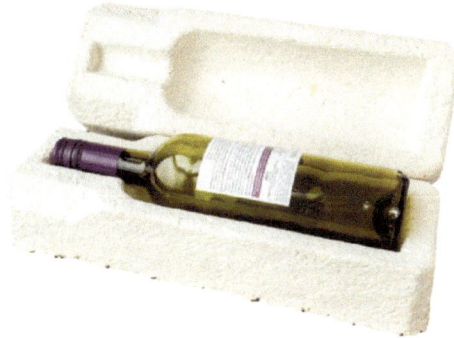

Figure 23.5 Mycelium packaging is both biomimetic and bioutilization

Source: Grown Bio.

to truly sustainable goals, performing like an organism in a healthy ecosystem. That means 100% circular material flows running on 100% clean renewable energy and contributing positively to the surrounding web of life with rich diversity. How would your product and production have to change to accomplish that? Finally, nature as mentor means changing our relationship with nature, as mentioned before, to treat the Earth and all its organisms as teachers to respect, learn from, and give back to.

23.3 Limitations of Biomimicry

Biomimicry's strengths are the inspiration it gives for creative novel solutions and connection with nature but like all tools and paradigms, it has its limits. First, it does not provide quantitative metrics for environmental or social impact; its principles for how nature works can provide inspiration for creating such metrics, but those must be validated on their own. Its creative solutions may not be sustainability improvements. Thus, biomimicry is not as useful for setting design priorities, benchmarks, or success thresholds as LCA or eco-certifications. In addition, biomimicry from mentors has the difficulty of bridging the gap between biological strategies and what current industry can build today. Biomimicry from principles usually circumvents this, but is often less inspiring because it lacks direct connection with natural mentors. However, these limitations can often be overcome by combining biomimicry with other sustainable design and engineering methods.

23.4 Where to Use Biomimicry

Combining biomimicry with other methods makes it more powerful and reliable. As mentioned above, it is an excellent ideation tool, but needs guidance to ensure its ideas are both practical and sustainable. Thus, it should be used after other design methods have established goals and metrics for success, and before gatekeeping steps to compare winning ideas for further development. For example, in Whole System Mapping, biomimicry should be used in the brainstorm step, after the system map is drawn and LCA has set design priorities, but before LCA is used again to evaluate new ideas. For another example, in using Cradle to Cradle, biomimicry is already the source of the Waste = Food principle, and can help suggest more specific target visions such as the Mannahatta Project's list of ecosystem services New York City should provide (Sanderson, 2013). By using biomimicry well, combined with other methods, we can find countless creative ideas and listen better to nature.

23.5 Doing Biomimicry

There are many ways to find inspiration in nature. Like most design, some methods push a new technology, while others are demand-driven. For biomimicry, the technology push starts with a well-understood organism and a lab that has successfully built something imitating the organism's strategy for some function, like the surface structure of a lotus flower. Then designers search for product applications for a good cause, like self-cleaning surfaces that avoid cleaning chemicals and water overconsumption. Its advantage is that it eliminates the main obstacle of other biomimicry methods: translating biologically-inspired strategies into buildable things. Its disadvantage is that it's limited to pre-existing solutions.

Most sustainable designers use demand-driven biomimicry: they identify a problem to solve, then seek inspiration from the natural world. The main hurdle here is translating nature's strategies to things you can build today in industry—it took the inventor of Velcro ten years to work out viable manufacturing processes and materials. Today's manufacturing and chemical engineering are advanced in many ways, but they are sorely lacking compared to nature's ability to build from the atomic level upwards, using only sunlight and nontoxic organic chemistry at ambient temperatures and pressures. Nevertheless, this approach is successful enough to be the default for most practitioners, such as the Biomimicry Institute.

The Biomimicry Institute's method is the "DesignLens" (Baumeister, 2013; Biomimicry 3.8, 2013), with eight steps shown in Figure 23.6 (details available online at Biomimicry.net). Other demand-driven methods include Nature Inspired Design (Tempelman et al., 2015), the method by Reap and Bras (Reap & Bras, 2014), and the Biomimicry Institute's earlier Design Spiral.

Here, we'll use a streamlined modular version of the DesignLens that fits better into the limited time of most design practice (Faludi, 2018). This modular version has two to five steps, depending on whether you find biological mentors in real life and learn their

Figure 23.6 The Biomimicry Design Lens method
Source: Biomimicry 3.8 (2013).

strategies yourself, use pre-analyzed data to find biological strategies directly, or use pre-analyzed and pre-abstracted principles of natural design to brainstorm from. See Figure 23.7.

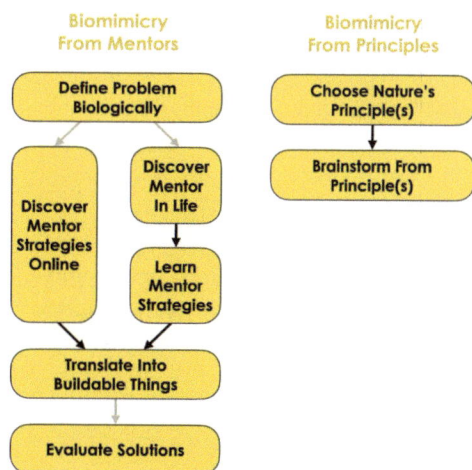

Figure 23.7 The streamlined modular biomimicry method; gray arrows are optional connections, black arrows are necessary
Source: based on Faludi (2018).

Figure 23.7 shows two main ways to do biomimicry: finding inspiration from mentors (left column) or from general principles (right column). The latter will be discussed in Chapter 24; it's fast and easy but doesn't connect you with nature as well. Biomimicry from mentors will be described in this chapter, below. It involves finding specific organisms in the wild or in literature; this is usually more time-intensive and difficult, but more inspiring.

23.6 Doing Biomimicry from Mentors

23.6.1 Define Problem Biologically

To find inspiration from natural mentors, as shown in Figure 23.7, first define your problem biologically. For example, to improve a refrigerator, should your goal be to keep food cool, or to prevent food from decaying by fungus or bacteria, or other? The more concrete you can make your goal, the better; if you feel that narrows your scope too much, you can repeat the process multiple times with other problem definitions.

Once you have a clear function, translate it into a biological analogy. This gives you starting places of where to look in nature. For example, how do animals stay cool in the heat of the desert? How do plants prevent decay by fungus or bacteria in jungle heat and humidity? The problem definitions suggest different answers. Also include the context your product or service is in: indoors or outdoors, wet or dry, large or small, rich or poor, etc. These can help refine your search.

23.6.2 Find Mentors and Strategies

Second, find mentors and their strategies. This is the inspiring part. You can find inspiration by direct observation (usually outside) and/or scientific media (usually online). For direct observation, go and explore nature, whether it's your own backyard or exotic locales. Examine the plants, animals, and other organisms you find, in search of things related to your problem definition. This is usually more inspiring than searching online media, but it requires an extra step: once you've found a mentor, you need to understand what strategy it's using, and how the physics or chemistry works. This is easiest to do with an expert on the local flora and fauna, or expert on the physics or chemistry of the organism, who knows these strategies and how they work.

It can sometimes take large amounts of time and effort (even scientific research) to understand these strategies. For example, the nose of Japan's Shinkansen bullet train (Figure 23.8) imitates the beak of a kingfisher bird to reduce noise from pressure waves in tunnels; separately from that, it also imitates owl feather structures to reduce noise in the rails pulling electricity from overhead wires. It cost Shinkansen engineers many months of both theoretical and empirical research to understand how the shapes of these bird beaks and wings performed, and how to imitate their principles on a train. However, the investment paid off: the redesigned trains not only improved user experience and reduced neighbor complaints, they also reduced energy use by 15% while enabling the train to travel 10% faster (see Figure 23.8).

Figure 23.8 Shinkansen 500 series train, kingfisher's beak, and closeup of owl feathers to show the sound-damping leading edge

Source: Wikimedia Commons by ↄ, Joefrei, & Mdf.

Instead of direct observation, or in addition to it, you can **find mentor strategies online**. The simplest option is to search the AskNature.org biomimicry database: you can do a keyword search for the function you're looking for, or browse through the taxonomy of functions. AskNature.org not only shows the mentors, but explains how their strategies work in language designers, engineers, and architects understand, so that's no longer an extra step in your process. When a product exists using that strategy, there's also an "innovation: industry" page on that; academic inventions are in "innovation: academia" pages. See Figure 23.9.

If you have more technical biology knowledge or are not afraid of jargon, you can search for mentor strategies in scientific literature using databases like Science Direct or PubMed. You could also search for examples in physical books; even some non-technical books with quality illustrations can provide good inspiration. See the Resources list at the end of the chapter.

Find as many mentors and strategies as you want, but 5–15 of them give you a good number of options without being overwhelming. When gathering your biological strategies, write them on sticky notes so you can group them or move them around. You'll often find related or redundant ones. That's okay, it may imply the strategy is particularly good. But make sure you have several different strategies, to give yourself choices. When you write or sketch each strategy, also list its mentor organism. If you

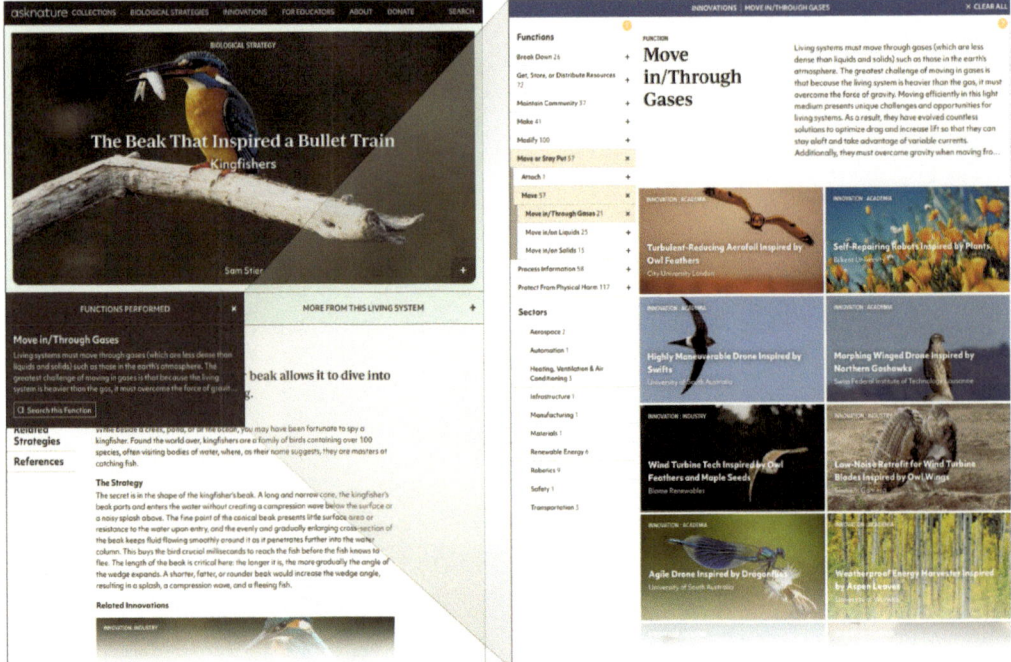

Figure 23.9 The AskNature online database of biomimicry, showing a biological strategy page (left) and a function taxonomy browsing page (right), with links to pages on academic and industry innovations

Source: Unsplash and Wikimedia Commons, Stuart Price & harum.koh.

want, include pictures and notes. It's best to keep a citation or URL for each organism and strategy, in case you need more detail later.

23.6.3 Translate to Buildable Things

Next, translate strategies into buildable things. For every natural strategy in the last step, list one or more ways to produce it with existing technology that's commercially viable for you. Contract manufacturers, organic chemists, and others can help you brainstorm on manufacturability, if needed. Line up all your sticky notes from the last step, and use a new color of sticky note to brainstorm buildable ideas for each strategy, one buildable idea per sticky note, as shown

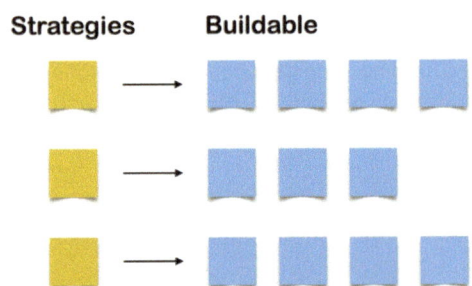

Figure 23.10 Brainstorming how to translate biological strategies into buildable things

in Figure 23.10. Each buildable idea will have limitations, and that's okay. To work around this, have several ideas for each biological strategy, especially ideas relevant to your product's context (e.g., used outside in high wind speeds).

Translating to buildable things is an important step, because nature's best strategies may not be feasible for you to build with today's technology. Therefore, the best strategy for you may not be the best strategy in nature; as long as it's better than what you have, it's still an improvement. Biomimicry is design-by-analogy, and these analogies are usually the hardest part. Velcro is a biomimetic product, imitating how burrs stick to a dog's fur; the principle is very simple, but it took ten years to commercialize, because a new manufacturing method had to be invented. And it's still not biomimetic on the process level, because nobody knows yet how to grow Velcro at ambient temperature with water-based non-toxic chemistry like fur or burrs grow. So don't choose a favorite biological strategy, choose a favorite way to build a strategy.

23.6.4 (Optional) Check Versus Principles

You could finish there, going back into your normal product development cycle. As Figure 23.7 shows, the next step is optional. But you could finish by using biological principles as quality control. You'll need to do some form of narrowing down to a winning idea (or a few ideas) to move forward with. To narrow down using biological principles, you evaluate your new buildable ideas by how well they align with principles of nature, discussed in Chapter 24. Go through your new buildable ideas one by one, and for each idea, see how many principles it follows well, especially ones relevant to your application. For example, with a bullet train, "use low energy processes" in Figure 24.1 in Chapter 24 is very relevant. This can help you choose your favorite buildable idea(s) to move forward with. If you notice a principle that your new ideas don't follow but should, you might stop and redo previous steps to find new mentors or new buildable ideas. Or you might choose to brainstorm from the natural principle.

This step is optional because LCA, eco-label checklists, and other sustainability measurements are more objective and relevant for design decision-making, but using biological principles can be a nice way of considering insights from previous biomimicry researchers in your design process.

Resources and References

Resources for Further Study

- Observing nature in your own back yard, a nature reserve, park, or botanical garden, etc.

- *Biomimicry*, by Janine Benyus (the book that started the modern biomimicry movement). 1997. William Morrow.

- *Nature Inspired Design*, by Erik Templeman, Bram van der Grinten, Ernst-Jan Mul, Ingrid de Pauw (a practical guide). 2015. Boekengilde.

- *The Way Nature Works*, edited by Robin Rees (a picture book for ideas). 1997. Wiley.

- *Cats' Paws and Catapults*, by Steven Vogel (an excellent source of mechanical engineering-related biomimicry). 2000. W.W. Norton & Company.

- *On Growth and Form*, by D'arcy Thompson (a large tome of theory and a sourcebook of images). 2018. Forgotten Books.

- *Out of Control*, by Kevin Kelly (an inspiring book of systems design theory). 1995. Basic Books.

- *Structural Biomaterials*, by Julian Vincent (a materials science analysis of natural materials). 2012. Princeton University Press.

- *Biomimicry for Optimization, Control and Automation*, by Kevin Passino (a computer science textbook on genetic algorithms, neural networks, etc.). 2005. Springer.

- *Zygote Quarterly*, by many authors (a magazine of bio-inspired case studies and interviews). 2012–present. Available at: https://zqjournal.org

References

Baumeister, D., Tocke, R., Dwyer, J., Ritter, S., & Benyus, J. (2013). *Biomimicry Resource Handbook: A seed bank of best practices*. Biomimicry 3.8.

Benyus, J. (1997). *Biomimicry*. HarperCollins.

Biomimicry 3.8. (2013). Biomimicry DesignLens: A toolkit of best practices. Biomimicry 3.8.

Faludi, J. (2018). Recommending sustainable design methods by characterizing activities and mindsets. *International Journal of Sustainable Design*, 3(2), 100–136.

Reap, J., & Bras, B. (2014). A method of finding biologically inspired guidelines for environmentally benign design and manufacturing. *Journal of Mechanical Design*, 136(11).

Sanderson, E. W. (2013). *Mannahatta: A natural history of New York City*. Abrams.

Tempelman, E., Grinten, B. van der, Mul, E.-J., & Pauw, I. de. (2015). *Nature inspired design: A practical guide towards positive impact Products*. Boekengilde.

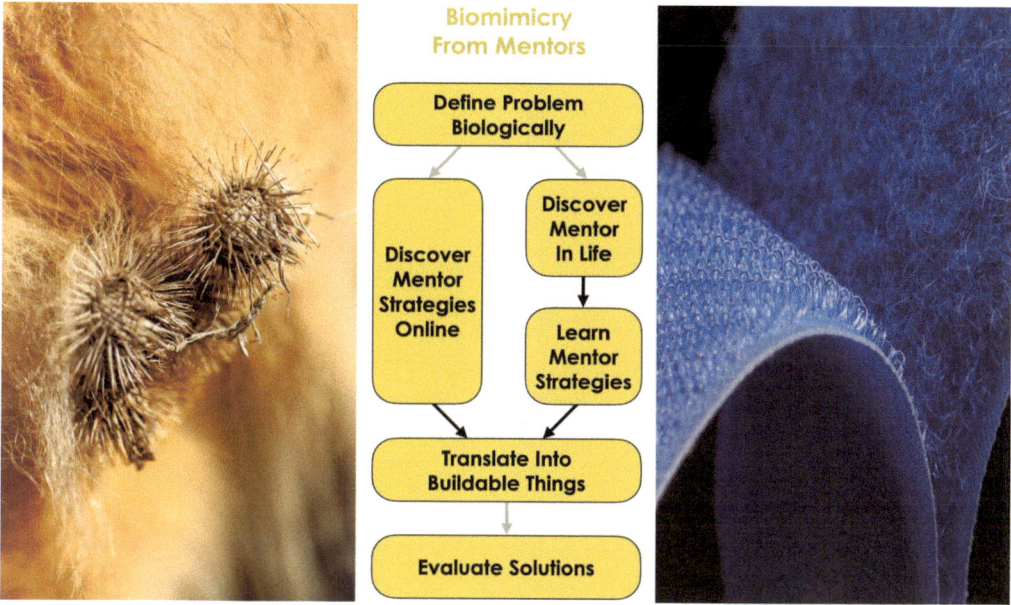

Biomimicry From Mentors

Define Problem Biologically

Discover Mentor In Life

Discover Mentor Strategies Online

Learn Mentor Strategies

Translate Into Buildable Things

Evaluate Solutions

Velcro was inspired by burrs sticking to fur. It took ten years to go from inspiration to manufacturable product. *Source*: Shutterstock and Velcro.com

How to Apply #23: Biomimicry from Mentors
Time Estimate: 70–130 Minutes

Goal: Use inspiration from specific organisms ("mentors") to ideate solutions to a problem.

STEP 1: Define the Problem Biologically
Time Estimate: 10–20 Minutes

Choose a problem you'd like to solve about your product. Usually it's best to choose specific mechanism-level problems, but you can also try larger-scale abstract problems. If you're using biomimicry in your Whole System Mapping brainstorm, the problem can be your system-level priorities (e.g., for a refrigerator, reducing energy impacts without sacrificing cost or user experience) or a more specific problem uncovered in that brainstorm (e.g., prevent heat transfer into the refrigerator when you remove food).

Translate your problem into a biological analogy, as described in the chapter (e.g., for a refrigerator, how do organisms keep cool, or how do they prevent rotting?). Ideally also include the sustainability concern (e.g., not using electricity) or usage context (e.g., in a hot humid environment). This helps you find where to look for mentors: animals in the desert? Plants in the jungle? Others?

STEP 2: Find Mentor Organisms and Their Strategies
Time Estimate: 30–60 Minutes

Find biological mentors and see what strategies they use to solve your problem. Three main sources are direct observation of natural organisms (usually outside), the AskNature.org website, and literature (e.g., books or online scientific journals).

- **Direct observation**: Go outside, to your own backyard or exotic locales, and directly see, hear, touch, etc. When you find an inspiring and relevant aspect of an organism (e.g., the shape of a leaf or the activities of ants), write it down or sketch it. Then try to understand what the physics or chemistry or other basic principle of the strategy is, as described in the chapter, and write or sketch it. You'll often need to search literature for this. Reminder: this is easiest with expert assistance, like a biologist.

- **Online**: Go to the AskNature.org website, and search for your function, or browse through the taxonomy of nature's strategies. It not only shows the mentors, but explains how the strategies work. Write down both the organisms and sketch or write how their strategies work. It's best to keep the URL for each organism and strategy, in case you need more detail later.

- **Literature**: browse the books in the chapter's Resources list, or search academic databases like https://www.sciencedirect.com or https://pubmed.ncbi.nlm.nih.gov if you have strong technical expertise. Again, write both the organisms and sketch or write how their strategies work, and keep citations or URLs, in case you need more detail later.

Write or sketch the strategies on sticky notes, so you can group them or move them around. Find at least 5–15 strategies. Have at least two or more from AskNature, and one or more from somewhere else, to diversify.

Once you're done collecting strategies, group them by strategy—you'll usually find several overlaps. If several organisms use the same strategy, this is a hint it's a very effective strategy. However, make sure you also have several different strategies, to give yourself choices.

STEP 3: Translate Strategies into Buildable Things
Time Estimate: 20–40 Minutes

Line up all your sticky notes from Step 2 in one column, and use a new color of sticky note to brainstorm rows of buildable ideas for each strategy, as in Figure 23.10. For every natural strategy from Step 2, think of several ways to commercially produce it. Especially think of ideas relevant to your product context. Experts such as contract manufacturers, organic chemists, and others can help you with the ideation.

Draw or briefly describe each buildable idea on its own sticky note. Each idea will have limitations, as discussed in the chapter; having many ideas gives you many options.

STEP 4 (Optional): Using Principles as a Quality Control
Time Estimate: 10–20 Minutes

If you want, you can evaluate how biomimetic your new ideas are, by comparing them to principles from nature, such as Biomimicry 3.8's "Life's Principles" or others listed in the chapter. This can help you choose your favorite idea(s) to move forward with.

Go through your new buildable ideas one by one, and for each idea, see what principles it follows better than the original product did. If you notice a principle that your new ideas don't follow but should, you might stop and redo previous steps to find new mentors or new buildable ideas. Or you might choose to do the exercise on biomimicry from principles.

This step is optional, as most people find LCA, a Whole System Mapping decision matrix, eco-label checklists, or other sustainability measurements to be more objective and targeted, but this can be another way of considering natural principles in your design process. You'll need to do some form of narrowing down to a winning idea (or a few ideas) to move forward with.

Checklist for Self-Assessment

To score your success on this exercise, see if you…

☐   *Formulated a clear problem statement.*

☐   *Rephrased the problem biologically.*

☐   *Found 5+ strategies (2+ strategies from AskNature, 1+ strategies from elsewhere).*

☐   *Included mentor organism names (ideally also citations/URLs).*

☐   *Sketched or described multiple buildable version(s) of each strategy.*

☐   *(Optional) Evaluated winning strategies based on principles of nature.*

CHAPTER 24
Biomimicry from Principles

Jeremy Faludi

Goals

- Recognize principles of designing like nature
- Ideate sustainability solutions using inspiration from biomimetic principles

DOI: 10.4324/9781003504672-26

Why It Matters

While biomimicry is a great source of inspiration and can help us reconnect with the natural world in professional practice, designing based on specific mentor organisms is often difficult and time-consuming, requiring much basic research. Designing by principles that other researchers have already identified is much faster and easier.

Summary

- Using design principles of nature is much faster and easier than researching mentor organisms, though it does not provide the deeper connection to nature.

- To ideate from principles, find a list of biological design principles, such as Biomimicry 3.8's "Life's Principles." Steven Vogel's list in *Cats' Paws and Catapults*, or this chapter. Select one or more relevant principle. Brainstorm buildable ideas from that principle. Repeat as necessary.

- Principles can also be used to assess how biomimetic a design idea is, though this is highly subjective.

24.1 How Nature Designs

As mentioned in Chapter 23, nature does not design, it evolves. But there are many commonalities in how organisms and ecosystems work. People interpret these as design strategies, so we can apply the ones we find relevant and helpful to our own designs. For example, Cradle to Cradle's "Waste = Food" is a fundamental principle of natural ecosystems, which is used as a design principle to eliminate the concept of waste: make all materials in one product recoverable for useful lives in other products.

When designing, it's more inspiring to look to specific mentor organisms, but it's faster and easier to brainstorm from biomimetic design strategies that other scientists, designers, and engineers have identified. Especially when brainstorming buildable ideas. This is doing biomimicry from principles.

24.2 Doing Biomimicry from Principles

To design based on natural principles, simply find a list of biological design principles, such as those listed below. Select one or more principle relevant to your design problem. Then brainstorm buildable ideas following that principle. Repeat as necessary.

Many people have made lists of nature's principles. The most famous is Biomimicry 3.8's "Life's Principles," in Figure 24.1. For more details on it and how Biomimicry 3.8 uses it in design, see Biomimicry 3.8 (2013). A few are explained below, as examples:

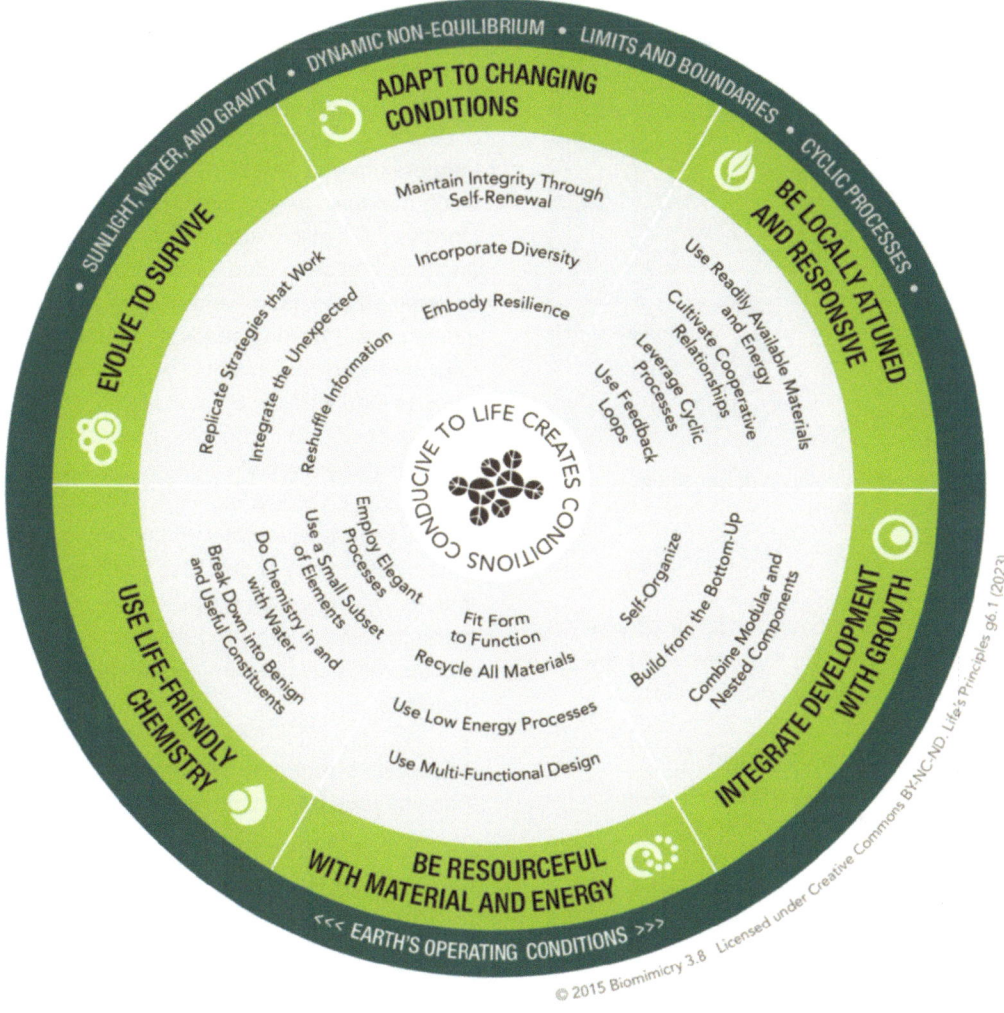

Figure 24.1 Biomimicry 3.8's "Life's Principles"

Source: Biomimicry 3.8 (2013).

Note: For explanations, see link in Resources for Further Study at the end of the chapter.

Evolve to survive: Biology does not design, as previously mentioned—it simply tries many random variants and the strategies that work best survive and propagate; repeating this countless times finds optimized solutions for the conditions at hand. Genetic algorithms imitate this in design—you feed parameters of variation and success criteria into a software optimization tool, and it simulates survival of the fittest for many variations across many virtual generations. The NASA spacecraft antenna in Figure 24.2 owes its odd shape to genetic algorithms, which optimized it in ways a human engineer would never have thought of. This means designers are no longer authors, but gardeners,

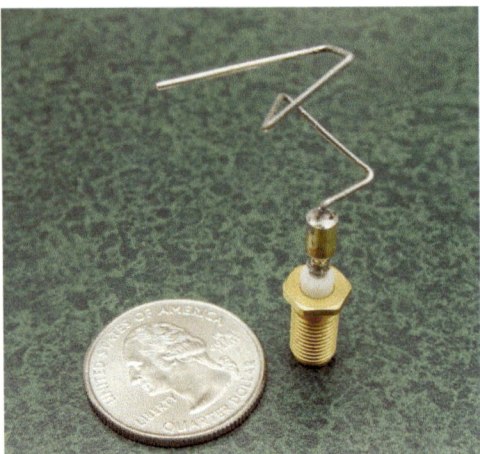

Figure 24.2 NASA antenna designed by evolutionary algorithms

Source: NASA.

creating the right conditions for great designs to emerge from. Kevin Kelly described it as "letting go, with dignity" (Kelly, 1995).

Do chemistry in water: Most industrial chemistry is petroleum-based (non-renewable), happens at high temperatures and/or pressures (energy-intensive), and frequently involves toxic solvents or other chemicals that must be carefully managed to avoid harm. Biological chemistry generally happens at ambient temperature and pressure, is water-based, using non-toxic renewable organic chemicals. However, it is also extremely complicated, and thus has been hard for industry to understand and use. Chemical and biotech industries are making headway, and design engineers can help invent new materials, such as Figure 24.3's Spintex fabric, which imitates spider silk. Producing it uses 1/1000th the energy of synthetic plastic fibers, uses no hazardous ingredients, and its only byproduct is water. In the long run, such biochemistry should also be cheaper than today's industrial chemistry because of the reduced energy and need for safety procedures, plus the plentiful (renewable) materials.

Be locally attuned and responsive: Trees change from summer to winter; many

Figure 24.3 Spintex Engineering's bio-based and biodegradable silk fibers are made at room temperature in non-toxic, water-based chemistry

Source: Spintex Engineering.

animals hibernate for the winter, or change color and behavior. Products can also adjust to their user or current circumstances. For example, Figure 24.4 shows a child's high chair that adjusts to grow as the child grows, thus greatly extending the product's life. For another example, many computers save power by having their CPUs and other chips selectively shut down areas not being used, constantly adapting to usage intensity as needed.

Figure 24.4 The Goodevas Growing Chair
Source: modernnursery.com

24.3 Other General Principles

Other principles of nature not included in the wheel above include the following.

Cooperate *and* compete simultaneously: For the first hundred years of biology, most people described Darwinism as a world of pitiless competition; today's biologists emphasize the cooperative interdependencies of creatures in their ecosystems. Both are true, and both are useful in their own ways. Dee Hock, the creator of the VISA credit card system, coined the word "chaordic" to describe a partly-chaotic, partly-ordered system where the relationships can be both cooperative and competitive at the same time, as they are in natural ecosystems. He structured the VISA system like this, so that many banks could cooperatively agree on ordered rules for the system in which they chaotically compete with each other. This is also how open hardware and software standards work: interested parties cooperate to structure the standard (such as the USB connector geometry, or the HTML programming language) and then compete within that arena. Their cooperation creates a larger total market, saves them money, gives them more robust supply chains, and reduces waste from incompatible systems. The EU recently legislated all mobile phones and some other electronics use USB-C chargers, shown in Figure 24.5, in order to reduce the proliferation of charging hardware.

Organize fractally: A fractal is a form or pattern that is similar to itself at multiple levels. Fractals can plan for growth from the bottom up, or for detail from the top down. The logarithmic spiral of a snail's

Figure 24.5 USB-C plugs, while not designed by biomimicry, are an open standard, letting manufacturers cooperate and compete simultaneously

Source: Marcus Urbenz on Unsplash.

shell did not evolve because it is pretty, but because it allows for perpetual growth without changing shape. When the snail grows to twice its original size, it does not need to tear down walls to expand its house, as most architects would; instead, it just keeps adding more shell in the same shape. Likewise, the detailed complex shape of a tree can be approximated by a simple branching algorithm. Fractals often don't look alike at different scales, but they still have self-similarity. They can be applied to design in many ways, for example, structural columns branching for efficient material use in Figure 24.6; or Barcelona's urban planning "superblocks" for walkable streets that build community (Salat et al., 2014); or multi-level organizational interventions driving companies toward sustainability strategy (O'Brien et al., 2023).

Design for emergence (swarm): Many small, simple things can work together to act like one large sophisticated thing. This results from "emergent properties," described in Chapter 6, making the whole greater than the

Figure 24.6 Tree-like fractal beams in Stuttgart airport reduce material use for the same strength

Source: CatalpaSpirit, Wikimedia Commons.

sum of its parts. As Kelly (1995) describes it, an individual bee has a small brain and simple behavior, but combining many bees into a swarm creates a super-organism with more intelligence and quite sophisticated behavior. Emergent phenomena are difficult (perhaps impossible) to predict analytically, but they can be simulated in virtual models.

Designing for emergent phenomena can do more with less. For instance, Encycle control systems reduce peak electricity use by scheduling heating, cooling, and other equipment to turn on and off at scheduled times (Figure 24.7). Many systems do that, but Encycle uses an algorithm inspired by bee swarms, where there is no central schedule, there is a distributed system of simple rules resulting in collective behavior. This has made it more effective while costing less money.

Designing for emergence can also help avoid unintended consequences, such as the suburban sprawl and traffic that unexpectedly came from the invention of the automobile. Most complex systems have emergent properties whether they are designed or not, and trying to predict them during design can help find and prevent problematic ones.

TRIZ for biology: A method of classifying problem-solving principles, called TRIZ, claims there are just 40 methods ever used by any people to innovate new products, buildings, and systems (Russo & Spreafico, 2020). However, evolution works very differently from human minds. Researchers at the University of Bath extended TRIZ to biology, cataloging and analyzing many more than 40 methods that nature has used to "invent" new solutions in countless organisms around the world (Vincent et al., 2006). They

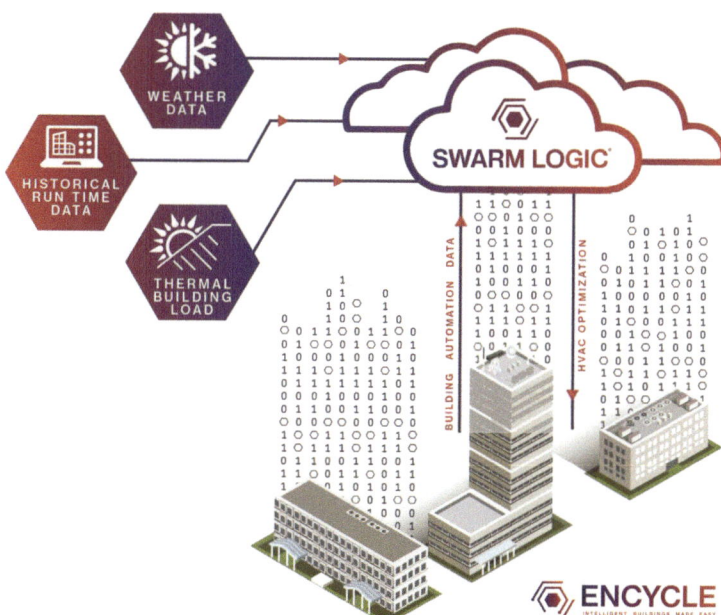

Figure 24.7 Encycle HVAC controllers use "swarm logic" algorithms to improve efficiency with decentralized control.

Source: encycle.com

found that "while technology solves problems largely by manipulating usage of energy, biology uses information and structure, two factors largely ignored by technology."

24.4 Mechanical Principles

Steven Vogel's (2000) book *Cats' Paws and Catapults* lists almost 20 nature-based design principles that can be extremely useful for mechanical engineers. Here are some highlights, quoted or paraphrased from the book:

"Nature uses fewer flat and more curved surfaces than we do." Natural forms round corners and taper shapes to avoid stress concentrations, thus using less material for the same strength and toughness, as with the tree in Figure 24.8.

Figure 24.8 A tree's curved surfaces and rounded corners versus bricks' flat planes and right angles

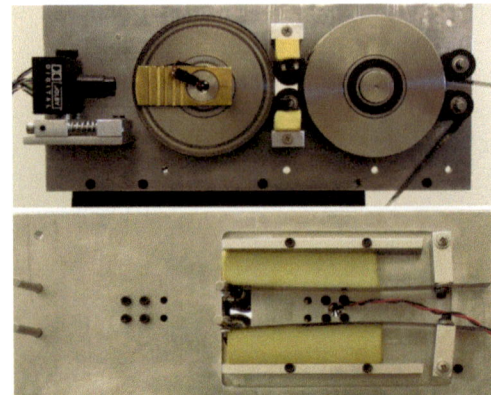

Figure 24.9 A biomimetic vibration damper for film soundtrack readers

"Nature's objects bend, twist, or stretch at predetermined places" while industrial mechanisms have rigid parts moving on sliding contacts. Figure 24.9 shows a vibration damper for film projectors that used this principle to replace $50 of machined arms, high-precision bearings, springs, and air piston with roughly $1.50 of spring steel and viscoelastic foam, which would also last longer and be more easily repairable.

We usually load materials in compression, nature very often loads in tension. Buckminster Fuller called this "tensegrity". Long thin objects usually fail by buckling long before reaching their ultimate strength limit, so tensegrity can enable radical reduction in material use, sometimes 10x or 100x in the case of the spokes on a bicycle wheel in Figure 24.10. Vogel also says, "Structures with tensile sheaths outside and pressurized fluid inside are both more common and more diverse in nature's designs than in ours." Figure 24.10's inflated bicycle tire is an example of pressurized fluid tensegrity.

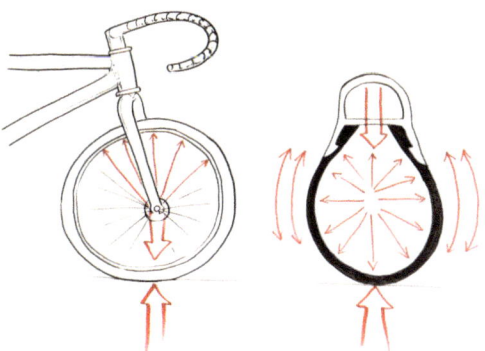

Figure 24.10 A bicycle wheel uses tensegrity both for the spokes (the axle hangs from the top of the wheel) and the tire (air inflates the rubber tube for strength with flexibility)

Source: Autodesk Sustainability Workshop, 2011.

Nature's factories are not size-limited.

Unlike industry, nature's factories often produce things larger, not smaller, than themselves. Industrial engineering takes for granted that factories are buildings within which materials and components are processed into products. However, many natural organisms produce things larger

than themselves. This happens through growth, such as seeds growing into trees or mothers bearing children that outgrow them. It also happens through mobile construction, such as single spiders spinning webs or swarms of bees building hives. Researchers are already investigating how small robots can build architecture (Tibbits, 2018), such as Figure 24.11's filament structure manufacturer.

Figure 24.11 Maria Yablonina's mobile robotic fabrication system for filament structures

Source: mariayablonina.com

Resources and References

Resources for further study

- *Cats' Paws and Catapults*, by Steven Vogel (an excellent list of biomimicry design principles for mechanical engineering)

- *On Growth and Form*, by D'arcy Thompson (listing several physical and geometric principles)

- *Out of Control*, by Kevin Kelly (listing biomimetic systems design principles)

- *The Way Nature Works*, edited by Robin Rees (a picture book for ideas)

- Biomimicry 3.8. (2013). Biomimicry DesignLens: A toolkit of best practices. Biomimicry 3.8. Available at: https://biomimicry.net/the-buzz/resources/biomimicry-designlens/

References

Biomimicry 3.8. (2013). Biomimicry DesignLens: A toolkit of best practices. Biomimicry 3.8. Available at: https://biomimicry.net/the-buzz/resources/biomimicry-designlens/

Kelly, K. (1995). *Out of control: The new biology of machines, social systems, and the economic world*. Basic Books.

O'Brien, K., Carmona, R., Gram-Hanssen, I., Hochachka, G., Sygna, L., & Rosenberg, M. (2023). Fractal approaches to scaling transformations to sustainability. *Ambio*, 52(9), 1448–1461. https://doi.org/10.1007/s13280-023-01873-w

Russo, D., & Spreafico, C. (2020). TRIZ-based guidelines for eco-improvement. *Sustainability*, 12(8), 3412.

Salat, S., Bourdic, L., & Labbe, F. (2014). Breaking symmetries and emerging scaling urban structures: A morphological tale of 3 cities: Paris, New York and Barcelona. ArchNet-IJAR: *International Journal of Architectural Research*, 8(2), 77.

Tibbits, S. (2018). *Autonomous assembly: Designing for a new era of collective construction*. John Wiley & Sons.

Vincent, J. F., Bogatyreva, O. A., Bogatyrev, N. R., Bowyer, A., & Pahl, A.-K. (2006). Biomimetics: Its practice and theory. *Journal of the Royal Society Interface*, 3(9), 471–482.

Vogel, S. (2000). *Cats' paws and catapults: Mechanical worlds of nature and people*. WW Norton and Company.

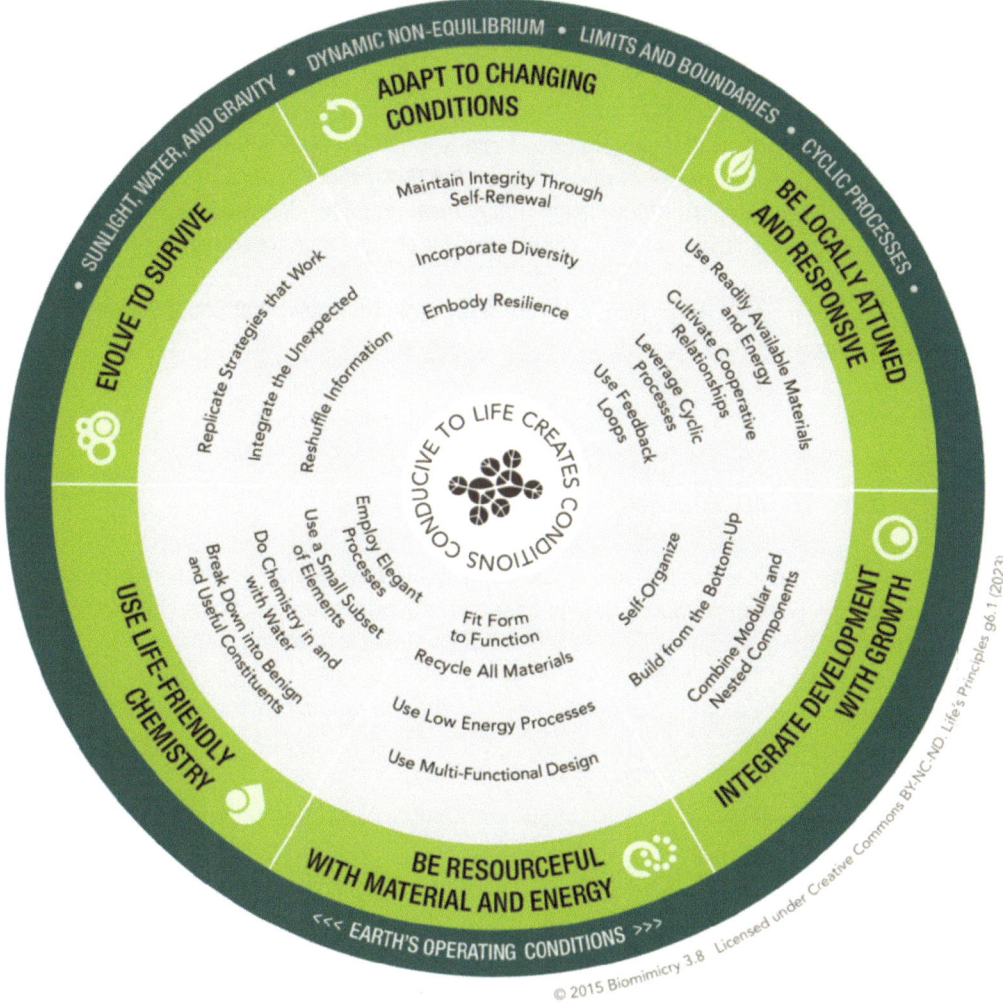

How to Apply #24: Biomimicry from Principles
Time Estimate: 30–120 Minutes

Goal: Generate new ideas to improve your product (or service) using a list of biomimicry principles.

This is the indirect method of biomimicry, brainstorming from general principles of nature that are listed by experts. Not as inspiring or personal as finding a specific design hero in nature, but faster and easier to apply.

STEP 1: Understand and Select Natural Principles
Time Estimate: 10–40 Minutes

Many scientists, designers, and engineers have already identified common ways in which nature designs differently from industry. Three good lists are the Biomimicry 3.8's "Life's Principles" list (see Figure 24.1), Steven Vogel's (2000) list of mechanical engineering principles in the book *Cats' Paws and Catapults*, and the list of other biological principles in this chapter.

Read through one or more of these lists, and pick a principle (or a few) that you think would be most fruitful for improving your product or system's sustainability.

STEP 2: Brainstorm from Natural Principles
Time Estimate: 10–40 Minutes

Now that you've chosen a natural principle, you can use them as design inspiration. Brainstorm how you could improve your product's design based on the principle you chose. Have at least 20+ ideas (50+ is better), and write or sketch them in whatever format you want.

If you chose more than one natural principle, have a separate brainstorm for each one.

STEP 3: Choose Solution and Illustrate
Time Estimate: 10–40 Minutes

After the brainstorm, choose a winning idea (or a couple) that you might want to further pursue. Illustrate it in a sketch, rendering, physical model, or other means to communicate both how it works and why it's a valuable idea.

Optional: to choose intelligently, we recommend using a Whole System Mapping decision matrix, LCA, or other sustainability measurements based on estimated impacts.

Checklist for Self-Assessment

To score your success on this exercise, see if you...

☐   *Clearly identified what principle(s) you brainstormed on.*

☐   *Brainstormed at least 20+ ideas.*

☐   *Chose and clearly described the winning design.*

☐   ***Illustrated the winning design.***

CHAPTER 25
Equity and Inclusion

Evren Uzer, Cynthia Lawson Jaramillo, and Michele Kahane

Goals

- Engage with diverse communities in a socially just manner

- Recognize positionality, power, and privilege in collaborations with clients or other stakeholders

- Recognize positionality, power, and privilege in collaborations within your design team

- Revise engagement methods toward equitable and inclusive design

DOI: 10.4324/9781003504672-27

Why It Matters

For whom do we design? Have we been equitable in our outcomes and inclusive in our processes? While few people are overtly racist, sexist, or otherwise discriminatory, almost everyone has unconscious biases. Calls to improve the diversity of design teams, to "decolonize" design, and to focus on matters of justice have become increasingly important and urgent in the 2020s. Designers, particularly those from historically privileged backgrounds, often overlook how their experiences shape their design decisions. Without realizing it, designers often create products, services, or spaces that cater to a specific demographic, inadvertently excluding or marginalizing others. By actively checking your privileges, you can become more inclusive in your work, ensuring your designs cater to (and benefit) a diverse range of people with different abilities, cultural backgrounds, and socio-economic statuses.

Summary

- Designing for inclusion means expanding who design is by and who it's for: including previously marginalized people both as users/stakeholders and as designers.

- Designing for equity means compensating for systemic injustices against marginalized people through design.

- Reflecting on "positionality" (social and political context such as gender, race, class, etc.) of yourself, your clients, your users, etc., helps collaboration conversations by uncovering privileges and biases.

- Six considerations are listed for community engagement: not "othering" communities.

25.1 Diversity and Inclusion

Improving equity and inclusion lets you improve the overall quality and integrity of your design outcomes, widen the appeal of your designs, and approach challenges with greater sensitivity, cultural competence, and social responsibility.

Design is mostly a commercial enterprise for companies, so it is usually part of modern global capitalism. While it is not run only by and for people with extreme wealth and power, it is mostly run by and for people with some wealth and power, excluding marginalized people who lack both. This biases design to not try to benefit or avoid harming marginalized people, even when they are the majority. To achieve the social justice in the UN Sustainable Development Goals, it is important and urgent to not always center design practice on global capitalism's dominant paradigm (affluent white western heteronormative users, clients, and design teams). If you want to design for a deep societal problem, the people closest to the problem (most harmed by it) often know the problem best; they have also usually had no power to change it. Designing for solutions requires engaging them in your process, both for input and to help make decisions. Some call this "decolonizing design."

Equity does not mean treating everyone equally, but treating people fairly. This includes compensating for unfairness in the circumstances, so everyone has the opportunity to reach an equal outcome (Figure 25.1). Designing for equity is also called "design justice." The different circumstances could be within the person or outside the person. For an example of internal circumstances, OXO GoodGrips kitchen implements have soft wide handles,

Figure 25.1 Equality (left) versus equity (right)

originally designed to make cooking accessible to elderly people with arthritis who struggled with traditional products. Incidentally, the designs also became popular with people of all ages without arthritis. Designing for "extreme" users with special needs often makes the design better for everyone.

For an example of external circumstances, Olympic oval racetracks do not start their runners on the same line, because even though that would be equal, it would be unfair. The width of the tracks means runners in outer lanes have to run further to reach the finish line—there is a bias against them in the system's structure. All professional racetracks stagger their starting lines to compensate for the structural bias of the track, because equity enables fair competition better than equality.

When circumstances are in your favor, internally or externally, you have some privilege. Recognizing your own privilege is hard; psychologists call it the "headwinds/ tailwinds" effect, where we are all very aware of barriers we struggle against, but do not notice advantages we benefit from (Davidai & Gilovich, 2016). By recognizing your own privileges (or your organization's),

as well as biases that might derive from those, you can use your position to include underrepresented voices and perspectives in design. This leads to more diverse designs, more creative designs, and brings justice to the design community.

Designers who work on complex social and environmental challenges often work with diverse communities with systemic biases, such as differences in wealth, power, race, etc. In most instances, designers are not from these communities and are invited in by others who are also, in various ways, "foreigners" with unequal power relationships between you, them, and the users or other community stakeholders. Such design engagements run the risk of failing from a sustainability and impact perspective by not accounting for such biases, and are often damage-centered and extractive (Tuck, 2009). To avoid and remediate such risks of extractive practices, you can use equity-centered mindsets and methods.

Diversity means having many different types of people involved in your work. Inclusion means bringing diverse people into your group and having them belong as they are, not making them conform to fit in. You can have diversity without inclusion.

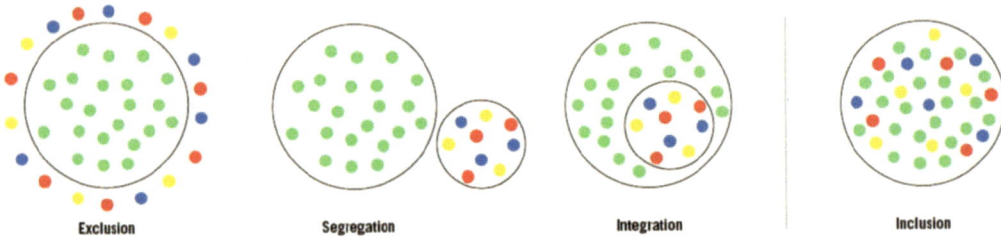

Exclusion   Segregation   Integration   Inclusion

Figure 25.2 Visual representations of exclusion, segregation, integration, and inclusion
Source: Moore (2017).

As Figure 25.2 shows, inclusion is different from creating a "separate but equal" group (segregation) or creating a subgroup within your group (integration). Inclusion requires different people mixing freely together in your group without losing their differences. The ultimate goal is a "pluriversality" where there is no central identity including other identities, but one group all sharing connection and responsibility. In your design team, who (what persons and with what characteristics) does or does not feel included in your process of designing and in the final design?

Designing for (and with) inclusion is a relatively new practice, growing especially after 2020, but methods and frameworks are emerging to support it. These transformations in design practice may be referred to as inclusive design, equity-centered design, or design justice (see Resources for Future Study at the end of this chapter for some notable approaches). These emerging methods that focus on reflection, relations, and trust building are beneficial both for inclusion between design teams and users or other stakeholders, as well as inclusion within design teams. This is especially as design studios and companies prioritize diversifying their staff and therefore have increasing differences of experiences, perspectives, biases, and privileges in their teams.

25.2 Equity and Inclusion in the Design Process

Most designers have been trained in the fairly linear "Double Diamond" design process (see Part I and Part II introduction pages), or in human-centered design or design thinking. These all start with a paying client or manager defining the design goal, moments of learning/ research about the topic and end users; then ideating, prototyping, testing, and iterating. Because they are all centered around clients and customers with some wealth and power, they generally don't try to benefit or avoid harming marginalized people. Even when they encourage you to redefine the problem, you usually stay within a narrowly-defined scope, not questioning your unconscious biases around who your design is for and why.

The Detroit-based studio Creative Reaction Lab (CRL) has been a disruptor to the status quo of design thinking. Founded "in support of the Uprising in Ferguson," their guiding principle is that if oppression, inequalities, and inequities are designed, they can be redesigned. Their design process is a non-linear framework with components nested and interconnected with one another. Juxtaposed with the more traditional Double Diamond, CRL's approach suggests a radical rethinking of the design process as a vehicle to arrive at more inclusive design.

CRL's design process includes all the elements of standard human-centered design, and is still centered around the design scenario at hand, but there are three important components that focus on equity and inclusion. First, CRL encourages you to acknowledge and unpack the powers and structures that keep underlying problems in place (problems that a design studio may be addressing in their project scope). Unless designers confront said powers, then what is designed will be useless in solving problems in the long term. Second, CRL encourages design teams to take stock of their identities (especially compared to those of the community), and consider how a more diverse group of co-creators may be better suited to lead the design. Here, designers become the facilitators of the process within which communities design. Third, CRL's process considers communities' histories of trauma. Listening to and understanding the complex impacts of traumatic history on a community can facilitate healing and enable transformative problem solving. To try CRL, see Resources for Further Study.

25.3 Positionality and Design

Positionality is an individual's social and political context, including gender, race, class, ethnicity, ability, geographic location, and other cultural background. In short, their position in a group or the world. Positionality influences people's perspectives, interactions, and design ideas. Teams and organizations also have positionality. Designers can benefit from reflecting on positionality: it allows you

to critically assess your social identity and its impact on your practice, allowing you to better understand your biases, privileges, and perspectives (Jacobson & Mustafa, 2019). This lets you question your assumptions, power dynamics, and ethical responsibilities, leading to more rigorous ethical design decisions. This in turn builds design integrity, generally leading to more inclusive and socially responsible design outcomes.

In addition, considering positionality fosters meaningful community engagement and collaboration. It makes you better equipped to engage with diverse people (both colleagues and clients/communities) in a respectful and culturally sensitive manner. This is called "cultural competence." It enhances co-creation by helping you navigate power dynamics and build trust with community stakeholders, leading to more effective design outcomes through more community support. Considering positionality also helps you integrate multiple perspectives in the design process (Soedirgo & Glas, 2020). This helps you actively seek out and amplify underrepresented voices, leading to more inclusive and representative design outcomes—solutions you likely would not have thought of, which are more responsive to the community's needs and experiences.

The Positionality Wheel (Figure 25.3) is a simple method for design teams to recognize their various social identities, so they might consider how they affect design decisions and collaboration. It can be used during the design process with potential users, communities, and other relevant collaborators or stakeholders; even within the design team itself. It is by Dr. Lesley-Ann Noel, author of *Design Social Change* (Noel, 2023) and co-editor of *The Black Experience in Design* (Berry et al., 2022).

Figure 25.3 The Positionality Wheel by Dr. Lesley-Ann Noel
Source: based on Noel (2023).

25.4 Collaboration and Community Engagement

Equity and inclusion in design require you to practice different forms of collaboration and engagement that help you achieve socially sustainable design propositions. Practicing mutually beneficial and ethical community engagement ensures that the voices,

needs, and priorities of the community are respected and placed at the forefront. It helps prevent external entities and actors from imposing their agendas, which might not align with the community's actual needs or desires. This kind of engagement leads to deeper understanding of the community, users, stakeholders, unique challenges, and preferences. The resulting solutions are culturally sensitive and highly adapted to specific user requirements, meaning they are not only more innovative but more likely to be used and supported.

The most effective collaborations begin with a long-term commitment and relationships built on trust. Trust is specific to each community and may look different in each collaboration—no two collaborations are the same. There is no cookie-cutter method to working with communities; however, there are some aspects you can consider, and there are tools and methods you can use for inclusionary and non-extractive ways of working in and with communities. Through active listening, and meaningfully engaging with community members, the design

process can lead to more meaningful and sustained outcomes.

25.5 "Community Engagement 101"

"Community Engagement 101" is an ethical framework for collaboration and community engagement, by this chapter's co-authors. It provides a starting point for designers to engage with community stakeholders and build a true community agreement. The framework's main tenets are below, and illustrated in Figure 25.4.

1. **Do not "other" communities**. "Othering" means representing or perceiving someone as alien to you and your group. Do not "parachute in" to places, certainly not for very brief periods of time, and disappear afterwards. Instead, co-create with the community: meaningfully involve them in deciding what the work will be and how it will be done; how they can lead a collaborative process.

2. **Mutual benefit begins with a self-reflection and critique on identity, positionality, and privilege**. Unpacking and directly self-critiquing your own identity and positionality enable you to better understand the identity and positionality of those you design for and design with. This, in turn, lets you better design for their needs and context better.

3. **Expertise exists in everyone and does not necessarily correlate with formal credentials**. It doesn't matter what degree or how many years of experience you might have—the community will always know more about themselves, who they are, their histories, their traumas, what they need, and what they don't need. Do away with the mentality of "we're here to help and tell you what you need," and instead celebrate all forms of knowledge and life experience as expertise. Especially in handcrafts, where some Indigenous communities have traditions going back hundreds or thousands of years.

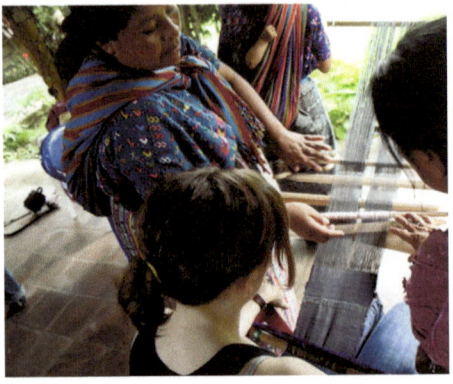

4. **Impact is always more important than intention**. The field of social design is replete with designers who intended to do good, but who cannot show their work's beneficial impact. It doesn't matter what the intention was if the impact is negative.

5. **All communities deserve to be meaningfully involved, not just as a destination**. Universities and companies often want to engage with adjacent communities for marketing purposes or to feel good about themselves. Especially problematic when such communities are not even welcome in the university or company. This is treating a community like a destination, not a partner. How might communities not be the destination, but be the people designing?

6. **Practicing equitable design takes time**. Designers need to create time and space to reflect, process, and engage in deep inquiry with the partner organization and/or community stakeholders, as well as amongst the design team. In client-based work, insist on longer timelines if equity and mutual benefit is one of the intended outcomes.

Figure 25.4 Community engagement 101.
Source: Cynthia Lawson Jaramillo.

25.6 Integrating Inclusion and Equity

The mindsets and activities in this chapter are the beginning steps of a long journey of integration into all aspects of design practice to help you create socially sustainable long-term impact through your design projects. The practice involves recognizing exclusion, recognizing our own privileges, and recognizing where we exclude others, even unintentionally. This not only helps marginalized groups, but helps everyone, just as curb cuts designed to make sidewalks accessible to wheelchair users also help people who can walk but have a baby stroller, bicycle, or other wheels. These mindsets can help any team collaborate better by discovering and accommodating diversity beyond obvious categories of race or gender to minute differences in working styles or personalities. And when actively working against exclusion to bring marginalized people into our design processes and teams, we can expand our creativity as well as our empathy and our benefits to humanity.

Resources and References

Resources for Further Study

- Noel, L.-A. (2023). *Design social change: Take action, work toward equity, and challenge the status quo*. Ten Speed Press.

- The New School Collaboratory. (2019). Community Engagement 101 Project Profile. Available at: https://thenewschoolcollaboratory.org/in-depth/community-engagement-101/

- Creative Reaction Lab. (2016). Field guide: Equity-centered community design. Available at: https://crxlab.org/shop/p/field-guide-equity-centered-community-design

- Agid, S., & Chin, E. (2019). Making and negotiating value: Design and collaboration with community led groups. *CoDesign*, 15(1), 75–89.

- Berdiel, F., & Lawson, C. (2010). Designing collaborative development: lessons from interdisciplinary teaching and learning. In *Proceedings of GLIDE'10 Conference*, New York, vol. 1, no. 2, pp. 45–55.

- Costanza-Chock, S. (2020). *Design justice: Community-led practices to build the worlds we need*. MIT Press.

- Light, A., & Akama, Y. (2012, August). The human touch: participatory practice and the role of facilitation in designing with communities. In *Proceedings of the 12th Participatory Design Conference: Research Papers*, vol. 1, pp. 61–70.

- Rojas, J., & Kamp, J. (2022). *Dream play build: Hands-on community engagement for enduring spaces and places*. Island Press.

References

Berry, A. H., Collie, K., Laker, P. A., Noel, L.-A., Rittner, J., & Walters, K. (2022). *The Black experience in design: Identity, expression & reflection*. Allworth Press, U.S.

Davidai, S., & Gilovich, T. (2016). The headwinds/tailwinds asymmetry: An availability bias in assessments of barriers and blessings. *Journal of Personality and Social Psychology*, 111(6), 835.

Jacobson, D., & Mustafa, N. (2019). Social identity map: A reflexivity tool for practicing explicit positionality in critical qualitative research. *International Journal of Qualitative Methods*, 18.

Moore, S. (2017). *One without the other: Stories of unity through diversity and inclusion*. Portage & Main Press.

Noel, L.-A. (2023). *Design social change: Take action, work toward equity, and challenge the status quo*. Ten Speed Press.

Soedirgo, J., & Glas, A. (2020). Toward active reflexivity: Positionality and practice in the production of knowledge. *Political Science and Politics*, 53(3), 527–531.

The New School Collaboratory. (2019). Community Engagement 101 Project Profile. Available at:. https://thenewschoolcollaboratory.org/in-depth/community-engagement-101/

Tuck, E. (2009). Suspending damage: A letter to communities. *Harvard Educational Review*, 79(3). https://pages.ucsd.edu/~rfrank/class_web/ES-114A/Week%204/TuckHEdR79-3.pdf

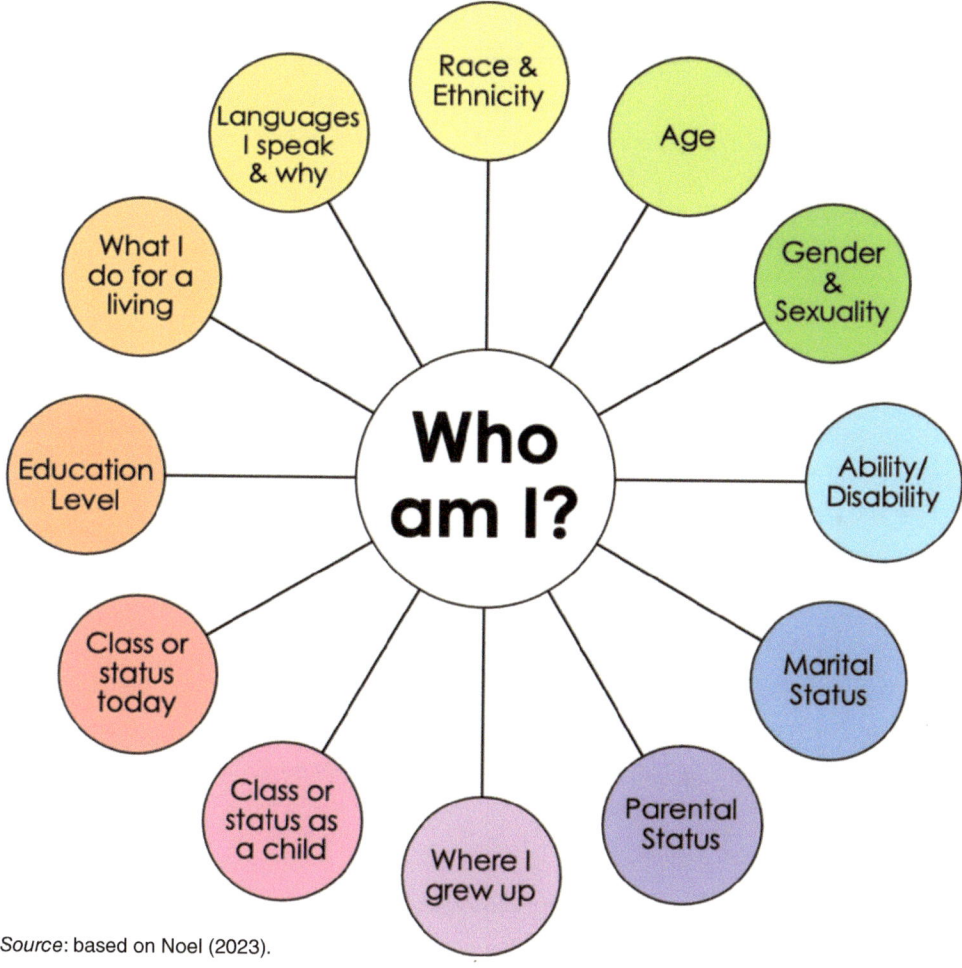

Source: based on Noel (2023).

How to Apply #25: Exercise: The Positionality Wheel
Time: 40–60 Minutes

The Positionality Wheel by Dr. Lesley-Ann Noel helps design teams reflect on potential biases, aspects of your identity you take for granted, and understand systemic inequities through the privileges you experience. If you're looking to design in ways that are equitable and mutually beneficial, it helps you understand yourself and your blind spots. It's useful to do this within your design team before doing a community-engaged project, and the community you work with might also want to do this with you. For more details and other exercises, see *Design Social Change* (Noel, 2023).

STEP 1 Read the Positionality Wheel
Time Estimate: 1–3 Minutes

Sit in small groups—ideally 3–5 people. While participants may not feel comfortable sharing out all aspects of their identity, it is critical to reflect on the exercise, as that is where the active learning happens. Each person should get a copy of the Positionality Wheel (Figure 25.3) to fill out. (Note that if people are not comfortable sharing all aspects, it could be useful to have a conversation about why, and what that reveals about the power dynamics.)

Everyone in the group individually familiarizes themself with the Positionality Wheel. Read the identity-related categories with your group, and discuss if necessary to clarify. While you can decide with your collaborators how to interpret each category, here are the original intentions:

- Race + Ethnicity: What is your racial and ethnic ancestry? Consider how you identify as well as how you present.
- Age: How old are you?
- Gender + Sexuality: What is your gender (as distinct from sex)? Was this your gender assigned at birth? How do you identify in terms of sexuality?
- Ability/Disability: Do you have different abilities of body or mind? Pay special attention to potentially invisible disabilities.
- Marital status: Are you single, married, divorced, partnered, or otherwise sharing your personal life with one or more people?
- Parental status: Are you a parent?
- Where I grew up: What was/were your childhood geographic location(s)? Were these places urban, suburban, or rural?
- Class or status as a child: How would you describe your socioeconomic status as a child?
- Class or status today: How would you describe your socioeconomic status today? How is it different from your parents or grandparents?
- Level of education: What is your highest level of education completed?
- What I do for a living: What is your profession or occupation?
- Languages I speak and why: List the languages you speak with your level of fluency (native, fluent, conversational, beginner). Also indicate why you speak that language (e.g., parental background, work, or personal interest).

STEP 2 Write Down Your Positionality
Time Estimate: 5–10 Minutes

Individually fill out each circle in the wheel with a word or brief phrase that describes you regarding that category.

STEP 3 Discuss the Group's Positionality
Time Estimate: 30–50 Minutes

Take turns going around the group and discussing the following four questions:

- In my identity, what am I aware of and what am I not aware of?
- Which is the part of my identity that gives me the most privilege?
- Where are we similar and where are we different?
- What biases or blind spots might we carry in our work as a team?

Designate a note-taker who can share highlights at the end (keeping specific identities as private as members of your group would like). If there are multiple groups, each group will report back to the larger group. Reflections may include: which question generated the most conversation? How does the conversation inform approaches we should take to a future project and/or in our design studio/practice? Did anyone feel uncomfortable? Feel unsafe? Create safety? Learn something they didn't know about themselves, others, or their culture? Listen to others? Listen to yourself?

STEP 4 Decide on Questions to Mitigate Blind Spots
Time Estimate: 5–15 Minutes

As a group, briefly discuss and decide how you might address the blind spots and biases in your design practice and your organization. Simple blind spots might have actions you can commit to, but more likely you'll have questions to carry forward, which will have better and better answers and actions over time, as you all deepen your understanding. Decide on one or more questions to carry forward in your collaboration.

Optional Reflection

Time Estimate: 5–15 Minutes

Reflect on how this exercise informs how you should approach your team's design process. Additionally, the question about biases and blind spots may inform the need for additional membership in the design team to address some of the biases the team will otherwise have.

Variations on the Positionality Wheel

The parts of identity discussed in the Positionality Wheel may vary according to project themes. For example, if the community engagement will center around digital applications, then you could consider adding a circle about "Digital Access" to the wheel. Furthermore, the community with whom you are working might also want to fill out a wheel (as a group or several individuals), if they think it will help them work with you or other stakeholders.

Positionality is a continuous process: as the societal structure and norms and personal understanding evolve, our positionality also shifts. Therefore, this exercise can be adapted to different phases of design work, as a continual part of your design practice.

Checklist for Self-Assessment

To score your success on this exercise, see if you…

☐   *Listed responses to all or most of the prompts on the Positionality Wheel.*

☐   *Mapped out/discussed your team's similarities and differences.*

☐   *Discussed which aspects of your identities grant you privileges.*

☐   *Discussed which biases or blind spots your privileges might bring to your design practice.*

☐   *Committed to ongoing questions addressing blind spots and biases in your practice.*

CHAPTER 26
Support and Collaboration

Jeremy Faludi

Goals

- Practice techniques for more effective collaboration with colleagues, bosses, employees, and partners

- Practice techniques for convincing colleagues to support your sustainability initiatives

- Practice supporting yourself and others in sustainability work

DOI: 10.4324/9781003504672-28

Why It Matters

Design and engineering are just a tiny piece of the puzzle—the greenest design in the world is nothing without the right material sourcing, manufacturing, business models, recovery systems, etc. Thus, to do sustainable design, you must collaborate with others inside and outside your organization, at all hierarchy levels. Often, you must also convince them of the value of sustainability. To do these well, and avoid burning yourself out, you can learn to communicate and collaborate more effectively. Collaboration skills can also help you improve inclusion by listening better, advocating for others, and building psychological safety to work through conflict.

Summary

- Acquiring support and collaborating effectively do not just happen, they are skills you can learn.

- Gathering support involves what to ask for, how to ask, and who to ask. You usually need to make the business case for sustainability.

- Support is not just money, time, and resources, but also psychological support for collaborators and yourself.

- Collaborating well requires psychological safety, dependability, structure / clarity, meaning, and impact.

- Collaboration can be improved by reflective listening, Nonviolent Communication, and Dialogue, among other methods.

26.1 Support and Collaboration

Bringing sustainable design from vision to action requires gathering and giving support, holding the vision, and collaborating effectively. These are intertwined, and together become leadership. While often disregarded as "soft skills." these social skills can be quite sophisticated and are critical to high-performing technical teams. Especially leading them.

Collaboration skills do not come automatically, but can be learned. They help teams work through conflict, and some kinds of conflict are essential to high-performing team dynamics. Good collaboration requires not just resource support, but also psychological support for collaborators and yourself. This is particularly important in sustainability, where working against destructive systems can be emotionally challenging.

26.2 Gathering Resource Support

Because designers cannot bring sustainable products to market alone, you need resources from your management, supply chain, marketing team, and so on. Asking for resource support, and providing it to others, are forms of leadership. You have to know what to ask for, how to ask, and who to ask.

26.2.1 What to Ask For

What to ask for varies by project, but common resources you'll need are time, money, materials, tools or training, etc. Also, to avoid greenwashing, you'll need to align on measurable evidence-based outcomes for the environment or society. Even when you're asking for the company to stop doing a destructive thing, you may want a positive alternative to replace it.

Measurable science-based targets are important because you can't manage what you don't measure. Examples could be life cycle assessment-based targets for carbon reduction, ethical labor certification scores, or anything else based on empirical evidence rather than marketing claims. You can start by proposing small deliverables for one short-term project, then work your way up to long-term large-scale deliverables and resources.

Determining the required resources is conceptually straightforward: simply estimate the time, people, money, and/or tools you need for your sustainable design initiative. These may be highly uncertain, but state your assumptions and talk with others to refine your estimates. You'll need to balance affordability with the viability of making meaningful improvements, and knowing the limits of available resources can help you choose deliverables for an achievable project scope. If you're asking to stop something (like making fewer new products to stop planned obsolescence), it may not take resources, but may reduce profits, effectively equivalent. However, this initial exercise of resource estimation is just the beginning: now you have to convince others to move forward and dedicate those resources.

26.2.2 How to Ask

Most often, people ask for sustainability support by connecting its value to financial profit and risk, which most institutions care more about than sustainability; especially companies. This is "making the business case" for sustainability, see Figure 26.1. Usually the core value is money; sustainability often saves money by reducing wasted materials and energy. Other common sustainability-linked values are legal risk reduction (both from lawsuits and from regulatory compliance), supply chain risk reduction (for critical raw materials or controlled substances), building brand image/customer loyalty (which builds profits and can open new markets), employee recruiting/retaining, and driving innovation (Whelan & Fink, 2016).

Instead of shaming or making people feel guilty for not valuing sustainability, clarify the values of the people you're pitching to with "active listening" or "reflective listening" (Rogers & Farson, 2015) to find what inspires them or motivates them. People usually want to do what's "right" but have different visions of what that is. Getting people to take a stand for sustainability in business often involves appealing to their personal values and aspirations.

When you listen to the other person, re-state their key points in your own words, and ask them to verify if you understood them. This looping for understanding not only can improve your factual accuracy of understanding, it greatly improves empathy on both sides. Empathy also helps you adjust your conversation to their values. You can initially pitch sustainability projects like any other project serving these values, until over time they accept the value of sustainability itself. One guide to this is ClimateVoice's Employee Climate Action Guide (2023).

Persistence is required, because asking people to value sustainability over money, time, aesthetics, and other design factors

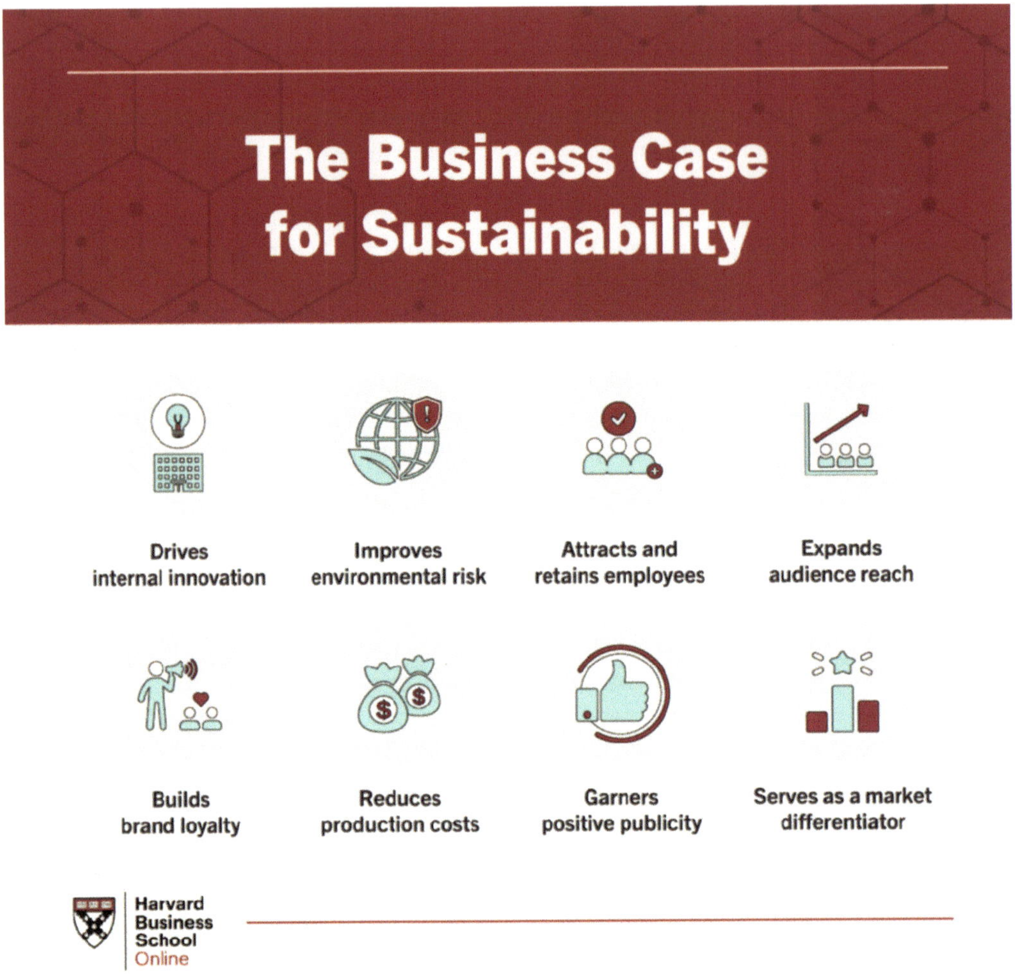

The Business Case for Sustainability

Drives internal innovation

Improves environmental risk

Attracts and retains employees

Expands audience reach

Builds brand loyalty

Reduces production costs

Garners positive publicity

Serves as a market differentiator

Harvard Business School Online

Figure 26.1 Potential business benefits of sustainability
Source: Cote (2021).

is asking them to shift core values, which does not happen overnight. Don't expect to have these conversations one time, have them dozens of times as part of healthy and respectful relationships. In these conversations, you'll learn what framings of sustainability resonate with both you and your colleagues. "Sustainability" has different interpretations, and is a long dry word without emotional resonance. However, it can contain values such as: care, stewardship, relationship, mindfulness, beauty, respect, reciprocity, and more.

Unfortunately, logic and facts alone are not powerful persuaders; while you do need the right facts in your business case, you can be more persuasive by tying them to emotions and personal values. A classic on psychological persuasion is

Influence: Science and Practice by Robert Cialdini (2009). It lists seven principles of persuasion: Reciprocity, Commitment and Consistency, Social Proof, Liking, Authority, Scarcity, and Unity.

Further details on some of Cialdini's persuasion tactics are found in Chapter 20, though you'll apply them differently in conversations than in design. It's important to apply them ethically and transparently, because tricking or bullying people into your values undermines your work's integrity.

26.2.3 Who to Ask

Part of knowing how to ask is knowing **who** to ask. Identify the key decision-maker(s) or influencer(s) for your design project or for your company overall. Who has the ability to grant your requests, or influence those who do, based on formal or informal relationships in the organization? To help with this, you can draw a "power map" of who these people are and how they connect to you and each other (ClimateVoice, 2023), see Figure 26.2. This can also help you understand their

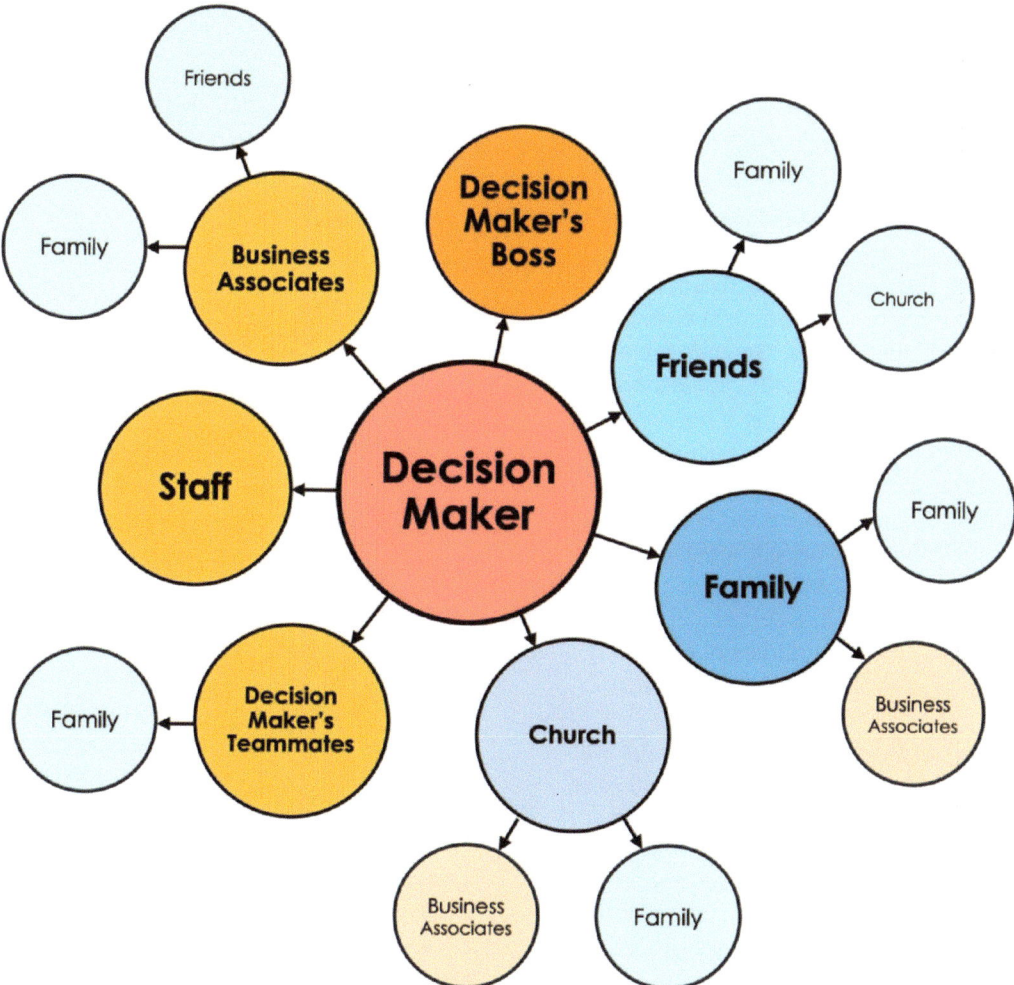

Figure 26.2 A simple power map for a single decision-maker

values, to make the business case or other arguments specifically for them. Most of Cialdini's persuasion strategies can apply equally to management or clients above you, employees below you, colleagues beside you, and external partners at other companies.

Combining what to ask for, how to ask well, and who to ask, is leadership. You don't need to be in a position of official power to lead, and it doesn't need to be a large part of your job to make a large difference in your organization.

26.3 Effective Collaboration

Once you get the resources you asked for, you need to execute the project well; and for that, you need effective collaboration. According to one of the largest studies on high-performing teams, Google's Project Aristotle (Friedman, 2019), high performance in creative technical teams is not primarily driven by the skill levels, experience, or creativity of the individuals on the team; rather, high performance depends more on their collaboration. Specifically, the factors that most determine a creative team's effectiveness are, in order: psychological safety, dependability, structure and clarity, meaning, and impact (see Figure 26.3).

Psychological safety is a shared belief that people can take risks (suggest ideas, voice concerns, ask questions) without being punished or shamed by others. Dependability means people do what they say they'll do, when they said they'd do it, at the quality level they promised. Structure and clarity are the team knowing who does what, and

Figure 26.3 Project Aristotle's factors most determining team effectiveness
Source: based on Friedman (2019).

how things get done (roles and procedures, including decision-making); both in the project and the overall company. Meaning is why the work matters to people personally. Impact is the work making a difference somewhere.

Sustainable design inherently has meaning and impact; however, it's important for team members to have a shared vision of how their work supports their values and how to measure it. Impact is clarified by discussing what measurable evidence-based metrics to use. Meaning is part of that conversation— you should choose metrics that objectively matter most for the context, which are not always CO_2, but could be land use, toxicity, or social metrics. Teammates can also say what different metrics mean to them personally, to help build shared values.

Structure and clarity come from explicitly discussed and agreed-upon lists of who does what, when, to what level of quality. They can be provided formally by team contracts and project management tools like Kanban boards, or can be provided informally in conversations. Most professional product development teams handle this well, and project management literature has many resources for this (PMI, 2021). When in doubt, discuss to clarify or alter agreements.

Dependability simply requires clear communication around tasks and deadlines, with realistic expectations of when and what quality level. If someone realizes they won't do a task on time or up to quality, they need to communicate as early as possible, to adjust the plan for what will happen. This also includes drawing boundaries for what someone is unwilling to do (e.g., too much overtime).

The factor least often trained for, despite being most important, is psychological safety. There are many techniques to build this. One is active listening, mentioned above; others are "nonviolent communication" and "dialogue". The Positionality Wheel in Chapter 25 also helps.

26.3.1 Nonviolent Communication

Nonviolent Communication (NVC) is a conversational procedure developed by Marshall Rosenberg (2005) for conflict resolution, from treaty negotiations to hostage release to interpersonal arguments. Environmental activists have also used it to build inclusivity through open dialogue (Wallace, 2003). NVC builds clarity and empathy between people by minimizing judgments about others, starting with objective facts that all can agree on, and emphasizing our own feelings and reactions to them (see Figure 26.4). This is especially helpful when working across cultures (be they ethnic or professional disciplines), where people have different assumptions and habits. Instead of avoiding conflict, it's a way to make conflict safer—it is not about being nice, it's about being clear and compassionate. If done consistently, it not only can help you resolve conflicts, but also it can make your team unafraid of conflict. Arguing without animosity enables greater risk-taking and higher standards, helping you create the best designs.

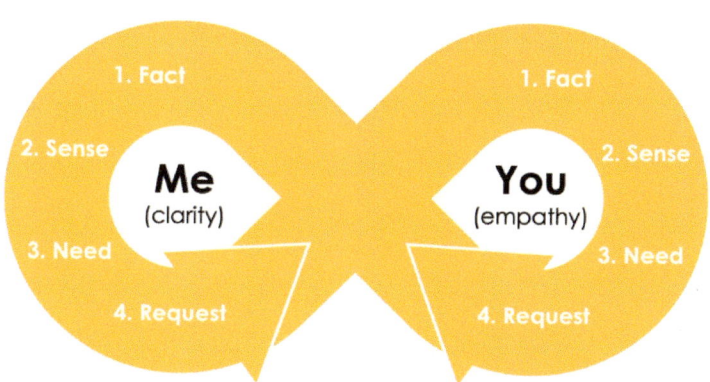

Figure 26.4 NVC helps works on multiple levels, for both clarity and empathy

NVC is a four-step process, as simple as one sentence each:

1. State what happened objectively (what a video camera would pick up).

2. State how that made you feel (a one-word emotion, not a story).

3. State your need in relation to that feeling (a one-word value, not a story of how to get it)

4. If you don't want things to continue that way, state your request for what you'd like to objectively happen in the future (again, that a video camera would pick up).

If you want help clarifying your feelings or needs, a list of common feelings and needs that most people have at different times is available on NonviolentCommunication.com (CNVC, 2015).

In this process, the other person can say no to your request. If they do, both sides should continue using NVC to continue working things through. Both people expressing their feelings and needs helps clarify the values of each side, what both sides want, and what is or is not available for negotiation. It also builds empathy, for more amicable and respectful communication even during conflict.

For example, if a teammate is often late to meetings, instead of saying "stop neglecting us, you don't care about this project, stop being so irresponsible," which is blaming and assumes things that may or may not be true, you could say, "you've been over ten minutes late to the last several meetings. I'm annoyed by that, because I need our meetings to be time-efficient and need consideration. Can you please arrive at meetings within one minute of the start time from now on?" The NVC phrasing minimizes judgments and clarifies why your request matters to you. The NVC phrasing also states positive values that you hold (e.g. time efficiency), and allows the other people to express their values and needs. Finally, the objective request doesn't ask the late person to change who they are, it merely asks for a behavior change.

Common mistakes in NVC are: In (1), leaking your judgments into your description of what happened, rather than being objective (just what a video camera could record). In (2), saying "I feel like..." or "I feel that..." which generally leads to a story, not an emotion. In (3), skipping the step or also telling a story there, rather than a need. In (4), commanding rather than requesting, or not asking for something a video camera could record.

You can also use NVC to appreciate teammates. Using steps (1)–(3) of NVC after a good experience can be a way to make your positive feedback more clear. For example, instead of saying "I love your UI design, you make the best buttons! What a graphics genius!," you could say, "the spacing and sizes of the buttons are clearly different, but still aligned across the top. That makes me feel confident that users can choose the right options without training, which we need for learnability." The NVC phrasing doesn't say this person is better than others, or that they couldn't do even better in the future. It concretely describes the positive actions or results, and how those connect to your feelings and values, sidestepping issues of ego and personal judgment. This helps shift people from mindsets of fixed talent to a growth mindset. Its added clarity might even guide the person's future work.

26.3.2 Dialogue

Teams interact differently at different times. The book *Dialogue* by William Isaacs (1999) classifies all conversations into four types: "politeness" (avoiding conflict, monologues), "debate" or "breakdown" (head-on conflict), "inquiry" (reflectively exploring the process of the conversation itself), and "flow" (generative dialogue where ideas build on each other). See Figure 26.5.

All four states can be useful and appropriate at different times, but teams often get stuck in politeness or debate, because underlying assumptions or expectations are not voiced. Surfacing these unspoken aspects can open up creativity, lead to constructive criticism, improve connection, and lead to positive resolution. Speaking about the unspoken aspects of a conversation is reflective dialogue. It enables a meta-level conversation exploring why the conversation is the way it is. This opens the door to generative dialogue ("flow"), both ideating the subject of the conversation and how the conversation itself might be improved. Innovation is best driven by generative dialogue, and designers are skilled idea generators.

For example, your team may be stuck arguing with each other about some design detail; then someone may mention the fact that only two people from your team of six are actually arguing, with everyone else silent, and may ask the other four teammates their opinions,

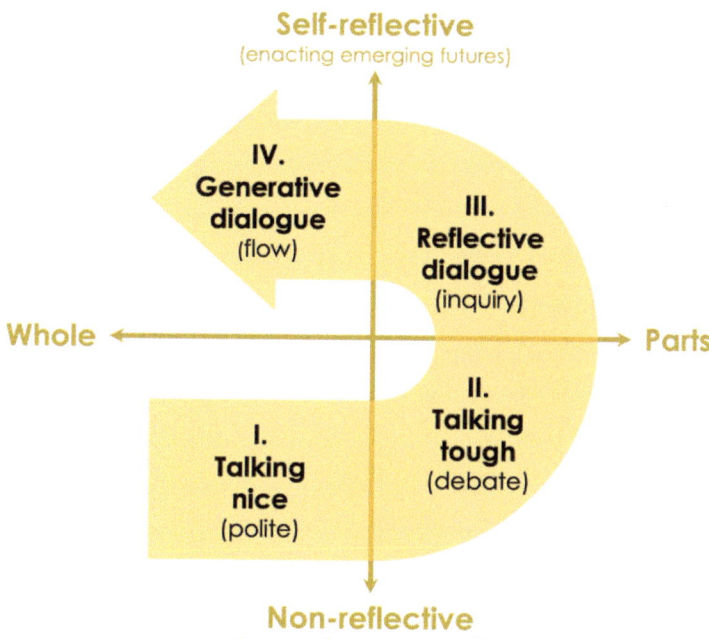

Figure 26.5 The four states of conversation in Isaacs's *Dialogue*. The arrow shows sequence of time and deepening connection.

Source: based on Isaacs (1999).

resulting in a new idea everyone agrees on. Or your team may be stuck with a mediocre design, and someone mentions that no one is arguing about anything, and asks why; the resulting discussion could open the door to constructive criticism in agreed-upon forms, and brainstorms of better solutions. In both cases, reflective dialogue pulls the team out of politeness or debate and lays the groundwork for flow.

When the sticking points are emotions or deeply-held beliefs, it helps to use active listening and/or nonviolent communication in the reflective dialogue of inquiry. To go even deeper into team dynamics, another framework that parallels the Double Diamond design process but in social connection is Theory U (Scharmer, 2009).

26.4 Supporting Yourself and Others

Sustainability work often requires emotional strength, support, and reasonable expectations. You will never fix all the world's problems, even with the largest company or national government behind you. Even if you could, the effects of success are often distant and long-term, and are often invisible because they are the absence of problems. Progress is often incremental, even when meaningful, and it can be exhausting because systems, structures, and people are so reluctant to change. Finally, there's also usually no objective threshold for when you are "good enough," because climate change and other impacts are not binary, but are degrees of damage or restoration.

Working in sustainability also means you cannot ignore the environmental and social problems of the world, you must face and feel them regularly. This can cause cognitive dissonance between having a "normal" lifestyle and knowing the damage it causes elsewhere, or can lead to lifestyle choices that are extremely austere or alienate those around you.

Because of these difficulties, people sometimes drop out of sustainability work for easier jobs. However, we need more people working in sustainability; so instead of dropping out, we should support people in the industry better, including ourselves. We can do this both by focusing more on progress and positive achievements, and by building skills to handle difficult emotions.

Focusing on progress and positivity is simple: spend more time thinking, talking about, and showing positive events. This could be small-scale and short-term, such as throwing an office party for cutting your product's material use by 5% from last year's model. It could also be large-scale and long-term; for example, the refrigerator and air conditioner manufacturers who replaced chlorofluorocarbons with non-ozone-depleting refrigerant gases in the 1990s have prevented perhaps a million people dying of skin cancer every year in Australia and elsewhere now, by letting the atmosphere's ozone layer recover. People today don't notice, because the crisis was averted, but we could declare a global annual holiday, named after the Montreal Protocol legislation that forced companies into action, to celebrate this huge achievement. Don't focus on positivity so much that you greenwash, pretending everything's solved, but enough to keep people motivated.

Building skills to handle difficult emotions is complex, with hundreds of psychology books on the subject. Two techniques that are simple and work well together, with demonstrated effectiveness in psychology practice, are mindfulness and compassion (Neff, 2023). They have been used for decades by some environmental activists (Macy et al., 2014).

Mindfulness is being aware of the present moment and your reaction to it—acknowledging and accepting your thoughts, feelings, and physical sensations. Rather than running away from negative feelings, or being overwhelmed by them, or having a knee-jerk reaction to them, you can acknowledge them consciously. This gives you a space between stimulus and response where you can choose how to react, which is usually empowering and calming. Practicing mindfulness yourself can strengthen your emotional resilience and well-being; initiating it with others can help strengthen theirs, too.

Compassion is concern for someone else's well-being: both the empathy of understanding and actively wishing them well. It is not indulgence or avoidance of action, it can actually unlock higher performance by acknowledging the difficulties of growth

(Neff, 2023), see Figure 26.6. Compassion helps build psychological safety among colleagues, not only reducing negative judgments, but helping reduce people's fears of being judged. Part of this is also self-compassion: often we judge ourselves more harshly than anyone else. While we should hold ourselves and our colleagues to high standards—not greenwashing—we must also have compassion for the limitations of what is possible under the circumstances, and not let the perfect be the enemy of the good.

Supporting your colleagues and yourself is a form of leadership, whether or not you are in a management role. Evidence shows compassionate managers improve team performance (Hougaard et al., 2020). They also bring humanity to our working lives. No work will ever fix all the world's problems, but celebrating our accomplishments, acknowledging difficulties or flaws with compassion, and being mindful will let us support and inspire each other for further action. As Joanna Macy said: "You don't need to be extraordinary. If the world is to be healed through human efforts, I am convinced it will be by ordinary people, people whose love for this life is even greater than their fear" (Macy et al., 2014).

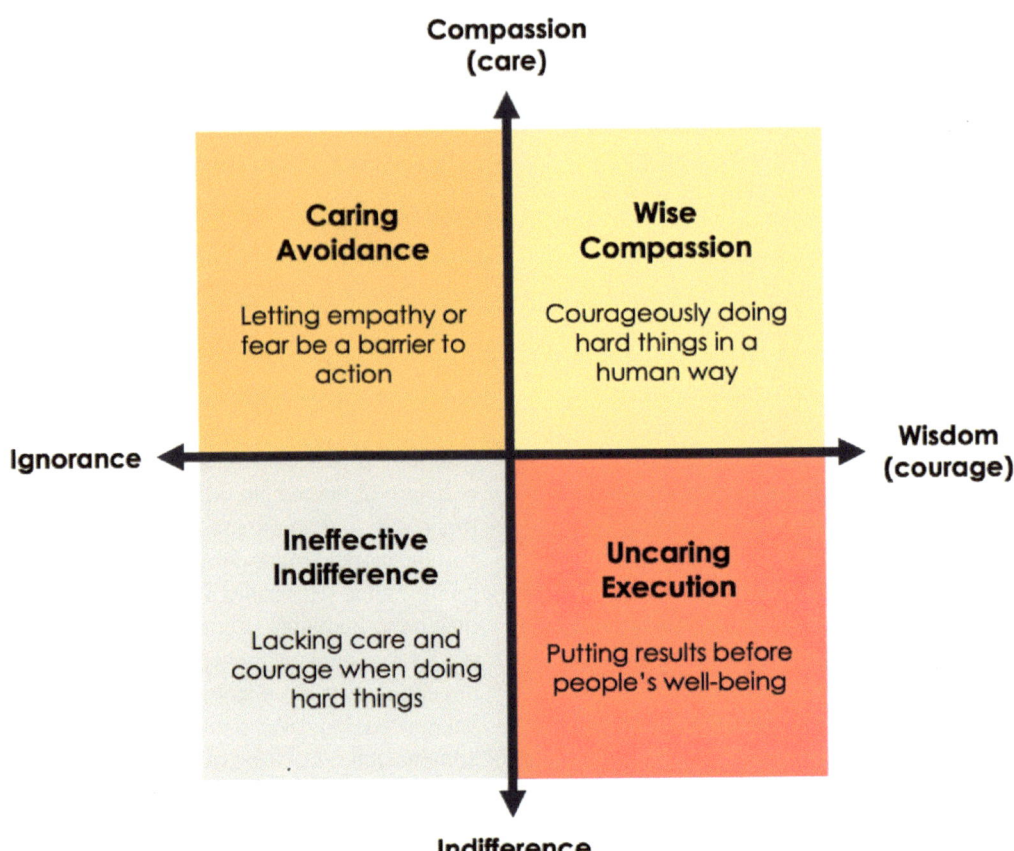

Figure 26.6 Distinguishing between compassion and indulgence ("caring avoidance") or indifferent action ("uncaring execution")

Source: based on Hougaard et al. (2020).

Resources and References

Resources for Further Study

- Climate Voice. (2023). Employee Climate Action Guide. Deloitte Development LLC. https://climatevoice.org/employee-climate-action-guide/

- Rosenberg, M. B. (2002). *Nonviolent communication: A language of compassion*. Puddledancer press. See also list of feelings at https://www.nonviolentcommunication.com/learn-nonviolent-communication/feelings/

- Isaacs, W. (1999). *Dialogue: The art of thinking together*. Currency.

- Scharmer, O. (2009). *Theory U: Leading from the future as it emerges*. Berrett-Koehler Publishers.

References

Cialdini, R. B. (2009). *Influence: Science and practice* (Vol. 4). Pearson Education.

Climate Voice. (2023). Employee Climate Action Guide. Deloitte Development LLC.

CNVC. (2015). Nonviolent Communication Training—Free Resources—Feelings and Needs We All Have. Puddledancer Press.

Cote, C. (2021, April 13). Making the business case for sustainability. *Business Insights Blog*, Harvard Business School. Available at: https://online.hbs.edu/blog/post/business-case-for-sustainability

Friedman, Z. (2019, January 28). Google says the best teams have these 5 things. *Forbes*.

Hougaard, R., Carter, J., & Hobson, N. (2020, December 4). Compassionate leadership is necessary—but not sufficient. *Harvard Business Review*.

Isaacs, W. (1999). *Dialogue: The art of thinking together*. Currency.

Macy, J., Brown, M., & Fox, M. (2014). *Coming back to life: The updated guide to the work that reconnects* (Revised ed.). New Society Publishers.

Neff, K. D. (2023). Self-Compassion: Theory, method, research, and intervention. *Annual Review of Psychology*, 74, 193–218.

PMI (Project Management Institute). (2021). *A guide to the project management body of knowledge* (7th ed,). Project Management Institute.

Rogers, C. R., & Farson, R. E. (2015). *Active listening*. Martino Publishing.

Rosenberg, M. B. (2002). *Nonviolent communication: A language of compassion*. Puddledancer Press.

Scharmer, O. (2009). *Theory U: Leading from the future as it emerges*. Berrett-Koehler Publishers.

Wallace, R. L. (2003). Social influences on conservation: Lessons from U.S. recovery programs for marine mammals. *Conservation Biology*, 17(1), 104–115.

Whelan, T., & Fink, C. (2016). The comprehensive business case for sustainability. *Harvard Business Review*, 21(2016).

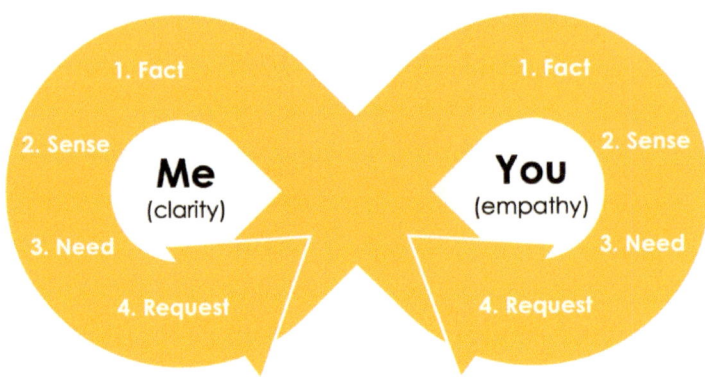

How to Apply #26.1: Nonviolent Communication Request and Appreciation
Time Estimate: 15–45 Minutes

Practice NVC by having a conversation with a colleague (or several).

STEP 1: Plan an NVC Request Conversation
Time Estimate: 5–15 Minutes

Plan out a conversation with a colleague using an NVC request. Writing it out before the conversation allows you to take time to prepare, check, and rewrite to practice the NVC process.

Your plan should contain, in this order, a few words or one sentence each:

1. Your objective concrete observation of their action(s), which a video camera could record.

2. Your feeling about it (see the list of feelings and needs cited in the chapter if necessary).

3. Your need related to it (see the list of feelings and needs cited in the chapter if necessary).

4. Your request for an objective concrete action by them (which a video camera could record).

Remember to avoid the common NVC mistakes listed in the chapter. And again, while you're expected to be respectful and polite, NVC is not about being extra nice, it's about being extra clear.

STEP 2: Have the NVC Request Conversation
Time Estimate: 1–15 Minutes

Have the actual conversation you planned, either as part of a larger conversation or separately, or as an email if needed. If your NVC request is a particularly fraught subject, start by asking them if they're okay with having a short conversation about a request you have, so you don't blindside them. Your conversation may not go as planned, but try to stick to the process of NVC throughout.

If time allows, repeat this for all colleagues you work with frequently and/or have unresolved issues with.

STEP 3: Plan an NVC Appreciation Conversation
Time Estimate: 5–15 Minutes

Plan out a conversation with a colleague using NVC to appreciate something they did or often do. Writing it out before the conversation allows you to take time to prepare, check, and rewrite to practice the NVC process.

Your plan should contain, in this order, a few words or one sentence each:

1. Your objective concrete observation of their action(s), which a video camera could record.

2. Your feeling about it (see the list of feelings and needs cited in the chapter if necessary).

3. Your need related to it (see the list of feelings and needs cited in the chapter if necessary).

Note that appreciation is different from positive judgment. You're not telling them how good they are (certainly not using superlatives like "best" or "most amazing"), you're telling them your emotion related to their action, and how it meets your needs or helps you meet your needs. As with Step 1, remember to avoid the common NVC mistakes listed in the chapter.

STEP 4: Have the NVC Appreciation Conversation
Time Estimate: 1–15 Minutes

Have the actual conversation you planned, either as part of a larger conversation or separately, or as an email if needed. Your conversation may not go as planned, but try to stick to the process of NVC throughout.

If time allows, repeat this for all colleagues you work with frequently, but at least do it for the person you made the NVC request of above.

STEP 5: Reflect on How It Went
Time Estimate: 5–10 Minutes

After the conversation(s), think back on how they went, and write brief notes for yourself on it. How did you feel afterwards? How do you think the other person felt? Do you think they will honor your request? Remember, large changes rarely happen at once, many conversations are usually required. Are there any other implications?

Checklist for Self-Assessment

To score your success on this exercise, see if you…

☐   *First wrote an NVC request, step by step, then made the request in conversation with the person.*

☐   *First wrote an NVC appreciation, step by step, then made the appreciation in conversation with the person.*

☐   *Wrote and said NVC steps 1 and 4 as things a video camera could record.*

☐   *Wrote and said feelings as one or two words, not a story.*

☐   *Wrote and said needs as a few words, not a story.*

☐   *Wrote a final reflection on the conversations.*

How to Apply #26.2: Project Aristotle Reflection
Time Estimate: 30–90 Minutes

Discuss with collaborators how well your team meets everyone's needs for the five factors in Project Aristotle, and reflect on how you might improve together.

STEP 1: Everyone in the team rate how well your team scores on Project Aristotle's five factors of team effectiveness
Time Estimate: 5–15 Minutes

Ask everyone on your team (or whatever set of collaborators you want to focus on) to score how well the team meets their individual needs for all five factors of team effectiveness from Google's Project Aristotle: psychological safety, dependability, structure and clarity, meaning, and impact.

Everyone should individually score each factor on a scale of 1–5, where 1 is very bad (seriously hampering your productivity or motivation), 3 is neutral, and 5 is very good (greatly helping your productivity or motivation). Small casual teams might do this scoring verbally in a meeting with one note-taker; for larger groups or privacy protection, build a survey everyone can fill out anonymously.

STEP 2: Discuss the team's scores, and what they mean
Time Estimate: 10–40 Minutes

Once you have the scores, graph them or otherwise display them for everyone to see. Discuss them to reflect as a group which are good, which need work, and why. Remember to not only discuss the problems, but also the good scores. Make sure everyone has a chance to voice their scores and their reflections.

As a team, write a one-sentence summary of which one(s) are best and why, and a one-sentence summary of which one(s) need the most improvement and why. Do **not** yet brainstorm ideas to solve problems, just diagnose the situation for now. Discuss as a group until the whole group agrees with the summaries. If they absolutely cannot agree on best or worst or reasons why, write that down.

STEP 3: Discuss how to act on your findings
Time Estimate: 20–40 Minutes

As a group, discuss how you might improve the factor(s) needing most improvement, how you might celebrate the factor(s) scoring best, or a combination. You might decide to "fix" the lowest-scoring factor, do nothing and celebrate high scores, make a high- or medium-scoring factor even better, or some combination. What agreements need to be made or re-set? What tacit understandings or ways of working need to be made explicit?

To improve scores, brainstorm one at a time on each factor you target. You could brainstorm both on new actions to start taking or old actions to stop, also noting good actions to continue doing. There may be different actions for different people in the team. These "actions" could be asking questions of each other, too.

After brainstorming many possibilities, narrow down to decide on one or more actions or questions to carry forward in your collaboration. Simple problems might have single actions you can commit to, but more likely you'll have questions to carry forward, with continued ideation bringing better and better answers and actions over time, as you all deepen your understanding.

Write down what ongoing question(s) or action(s) you commit to. Different people may commit to different things.

To help ensure this exercise has value to the team, decide together on a time to check-in about it.

Checklist for Self-Assessment

To score your success on this exercise, see if you...

☐ *All individually quantitatively scored your team on the five factors of effective collaboration.*

☐ *Wrote a one-sentence summary of which one(s) are best and why.*

☐ *Wrote a one-sentence summary of which one(s) need the most improvement and why.*

☐ *Wrote what ongoing question(s) or action(s) you commit to for the future. If different people commit to different things, say who will do what.*

Index

Note: Page numbers in *italics* refer to figures and those in **bold** to tables.